Earth Treasures

VOLUME 1: THE NORTHEASTERN QUADRANT

Connecticut, Delaware, Illinois, Indiana, Maine, Maryland, Massachusetts, Michigan, New Hampshire, New Jersey, New York, Ohio, Pennsylvania, Rhode Island, Vermont, and Wisconsin

VOLUME 2: THE SOUTHEASTERN QUADRANT

Alabama, Florida, Georgia, Kentucky, Mississippi, North Carolina, South Carolina, Tennessee, Virginia, and West Virginia

VOLUME 3: THE NORTHWESTERN QUADRANT

Idaho, Iowa, Kansas, Minnesota, Missouri, Montana, Nebraska, North Dakota, Oregon, South Dakota, Washington, and Wyoming

VOLUME 4: THE SOUTHWESTERN QUADRANT

Arizona, Arkansas, California, Colorado, Louisiana, Nevada, New Mexico, Oklahoma, Texas, and Utah

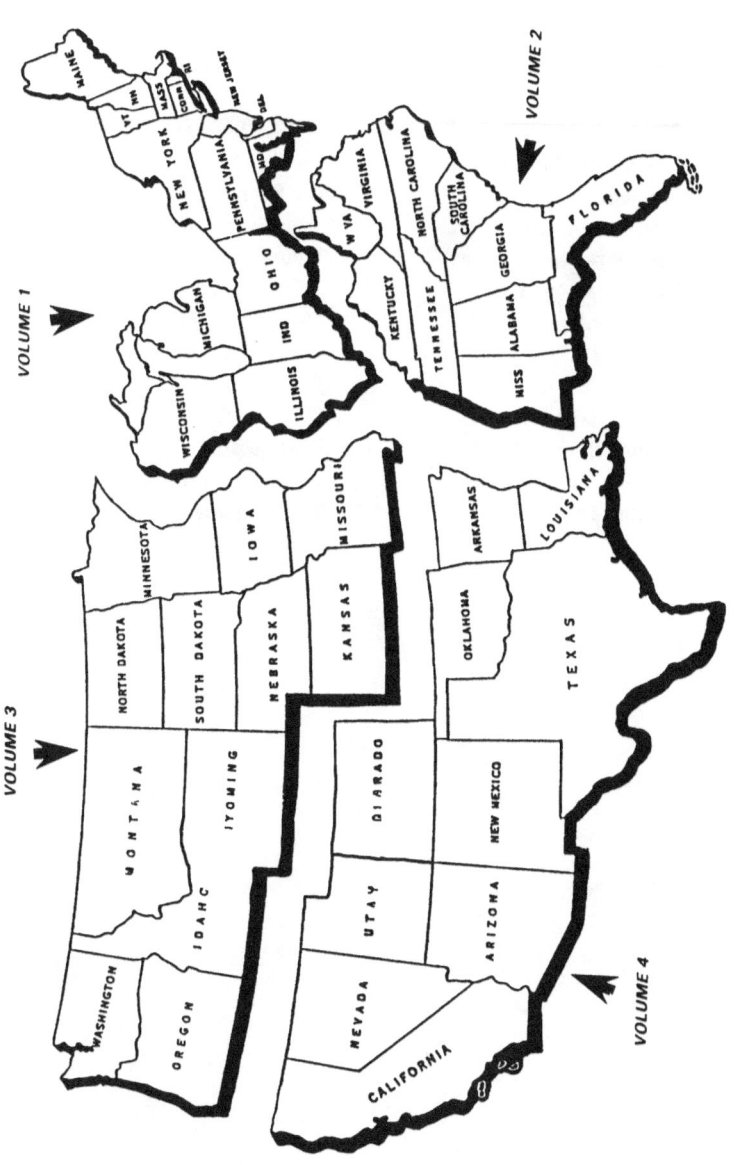

Also by Allan W. Eckert

EARTH TREASURES

VOLUME **2**

The Southeastern Quadrant

Alabama, Florida, Georgia, Kentucky, Mississippi, North Carolina, South Carolina, Tennessee, Virginia, and West Virginia

by Allan W. Eckert

AN AUTHORS GUILD BACKINPRINT.COM EDITION

Earth Treasures:
Volume 2: The Southeastern Quadrant
All Rights Reserved © 1987, 2000 by Allan W. Eckert

AN AUTHORS GUILD BACKINPRINT.COM EDITION

Published by iUniverse.com, Inc.

For information address:
iUniverse.com, Inc.
620 North 48th Street, Suite 201
Lincoln, NE 68504-3467
www.iuniverse.com

Originally published by Harper & Row, Publishers

Designer: C. Linda Dingler

ISBN: 0-595-08959-3

Printed in the United States of America

To a budding rockhound,
Lee Wayne White,
of Fort Myers, Florida

CONTENTS

HOW TO USE THIS BOOK

Earth Treasures is geared toward everyone interested in collecting rocks, minerals, and fossils—not only people who are devoted collectors and willing to spend hours, days, sometimes even whole vacations in the pursuit but also those who have only a mild interest and want results quickly without a great deal of wasted effort. Whether these specimens are for personal collections, for trade or sale, for scientific study, for display as cabinet specimens, or for tumbling or cutting into faceted gemstones or cabochons as jewelry makes no difference. It is the purpose of these volumes, through very precise directions and accurate county maps, to get you to the best and most accessible sites as quickly and easily as possible. Each of the four volumes covers one quadrant of the 48 contiguous states, and each volume is an entity in itself, complementing the others but not dependent on any of them.

This particular volume contains directions to nearly 1150 collecting sites and some 380 accurately scaled maps. Each collecting site included is indicated by number on the map for the county in which it is located. This site number also appears under the given county in the text, followed by specific written directions for getting to the site in question and a description of the type (and often the quality) of the rock, mineral, or fossil material to be found there, plus exactly where to search for the material *at the site* (in soil, matrix, road cut, pegmatite, shoreline, quarry, gravel pit, prospect, mine, etc.). Let's take, as an example, under the heading of Georgia, on the map of Spalding

County, the site indicated as number 2. Turning to the proper text page, the corresponding entry is as follows:

2. Vaughn area: 2 miles north on unmarked paved road to dirt road right; 0.15 mile east on dirt road to the T. J. Allen Prospect.

   Beryl: fine; clear; light blue (one crystal was cut to excellent 2-ct gem). Quartz, rose: massive. Quartz, smoky: good crystals to 1.5″ diameter. Tourmaline: blue; crystals to 1″ long. (6-9-granite -16-21-24)

The initial number, 2, corresponds to the number on the correct county map; in the directions, the first portion—*Vaughn area*—initially directs the reader to the closest municipality of any consequence within that county; the second portion gives distance and direction—2 miles north from the center of Vaughn on a paved but unmarked road; the third portion narrows the location area even more by directing the collector, upon reaching that 2-mile point, to turn right on a dirt road and follow it 0.15 mile eastward from the paved road to the collecting site, which in this case is the T. J. Allen Prospect. In the next section, devoted to *what* you will find there and *where* to look specifically, we see that four minerals can be found at this site—light blue beryl of gem quality, massive rose quartz, good smoky quartz crystals of substantial size, and blue tourmaline crystals to an inch in length; finally, the numerals in parentheses following the entry correspond to the mineral location key on the rear endpapers of this book, indicating that the specimens are found (6) in the soil and (9-granite) in vugs, pockets, or cavities in the matrix rock, which is a (16) pegmatite dike that has been worked as (21) a prospect, and that the prospecting has (24) exposed the dike to the collector. More expansively, this means that the four minerals in question, originating in a pegmatite dike, are found not only in pockets in the existing granite pegmatite, from which they can be removed, but also loose in the soil adjacent to where the pegmatite dike has been exposed.

Unless you begin finding considerable worthwhile material in an area where you are searching, it is usually somewhat of a waste of time to dig haphazardly. Wherever possible, put someone else's efforts to your own advantage; it is wise to benefit from the diggings done by others, such as in quarries, gravel pits, sand and clay pits, mine diggings and dumps, abandoned prospects, and excavations being done for highway, canal, railroad, and building construction.

By far the greater majority of the sites indicated here are reachable by auto, and many require very little strenuous hiking after you've parked. Some are reachable by Jeep trail (some for which four-wheel drive is recommended, others for which it is a prerequisite), and a few may demand more extensive walking. Every effort has been made to provide accurate directions, but when following them it should be borne in mind that conditions change constantly: route numbers may be revised or new roads may have been constructed that provide easier access than indicated here; similarly, old roads may have become closed or abandoned or, if unmaintained, may have fallen into impassable condition; some cannot be negotiated during inclement weather conditions; some collecting areas become closed, while others are opened; many areas require permission for entry, and some may charge a small fee or have restrictions concerning the type or amount of material that may be taken. It is your responsibility always to check for local regulations and possible restrictions and to acquire the proper permission if this is necessary.

For those who will be using this guide to stop at numerous locations for collecting (such as during vacation travels), it is wise to carry a supply of plastic bags (the heavy-gauge zip-lock type are especially useful) and a packet of self-adhesive labels. It is, unfortunately, not uncommon for a collector to pick up a selection of specimens from various sites visited and then later (perhaps many months later), when looking them over, discover that he has some especially fine material and no good recollection of exactly where he found those particular specimens. An acquaintance of the author turned out, to his dismay, to be a case in point. On a summer vacation drive through the western states he collected a great many specimens from various stops. Not until the following winter, when he was going through these collected materials and cutting some of them, did he realize that one chunk of quartz he had picked up had a thin vein of native gold running through it. He has no idea where he picked up that particular rock, not even the state. Variations on this theme have occurred to many collectors in the past. Therefore, when *you* are collecting specimens, don't mix them up in a sack or pouch or box containing specimens from several locations. Instead, place all specimens from one location in one bag and then stick a label on the bag and mark it with this guide's page number, location, and date. Thus a bag of specimens labeled 5/21/87—164/6 would later refresh your memory that the specimens in that particular bag were found on May 21, 1987, in the location marked as number 6 on the map on page 164 of this guide.

Bear in mind that many collectors, with justification, carefully guard their secret collecting locations and, if they speak of them at all, are reluctant to reveal their whereabouts except in very nebulous terms. As a result, occasionally some (fortunately, not too many) of the entries in this book will be found to be annoyingly vague. This is in no way meant to be a deliberate effort to mislead; it indicates only that the specimens seen or reported from the location are so good that it was deemed worthwhile to include the information, however limited, the directions provided herein to the site being the best available at the time this volume was prepared. In this same vein, there may also be an occasional site listed that is in error, since in some cases the information has been gathered by word of mouth from collectors and the author may not have been to the site himself. If there are such errors, letters of outrage may be sent to the author in care of the publisher of this book, and corrections will be made in subsequent printings. It is, after all, one of the aims of *Earth Treasures* to make it not only the most comprehensive guide available but also the most *accurate*. Furthermore, the author would appreciate hearing from any users of this back who have collecting sites of their own that they would like to have included in future printings. Such letters should contain all the pertinent information to the best of your ability, with *accurate* directions to the site. Address letters to the author, care of Harper & Row, Publishers, Inc., 10 East 53d Street, New York, NY 10022.

The purpose of this book and its three companion volumes is solely as a guide to collecting sites. A relatively comprehensive glossary of terms involved in rock, mineral, and fossil collecting will be found at the end of each volume. However, no information is provided as to identifying specific rocks, minerals, or fossils; there are texts available that do this remarkably well, providing keys to identification, hardness scales, specific-gravity charts, photo identification of specimens, and crystal formation information. It would behoove the collector to acquire such books as are needed for identification in the field or at home. Highly recommended as two of the best for use on collecting trips are *The Audubon Society Field Guide to North American Rocks and Minerals* by Charles W. Chesterman (Knopf, 1978) and *Guide to Rocks & Minerals* by Martin Prinz, George Harlow, and Joseph Peters of the American Museum of Natural History (Simon & Schuster, 1978). Two excellent hard-bound at-home books are *Gemstones of North America* and *Prospecting for Gemstones and Minerals*, both by John Sinkankas (Van Nostrand Reinhold, 1959 and 1970). The latter book is undoubtedly one of the better works available on the most successful techniques of collecting.

The county maps in this volume do not include every county of the state in question; only those counties with significant mineral, rock, and fossil collecting sites are included. Additional counties and their maps may be included in future editions. Each of the maps herein, unless otherwise indicated, is oriented with north at the top of the page. In some cases it has been necessary to divide a single county map into two or more portions; San Bernardino County in California, for example, is more than four times larger than the entire *state* of Connecticut.

Do not make the mistake of checking out only the precisely indicated site when visiting any of these locations. Quite often the guide will lead you to collecting areas where, with some diligent searching in adjacent or nearby areas, you may make some noteworthy mineral, rock, or fossil discoveries on your own that have previously lain undetected. In the event that you make such a discovery, there is ample room then to mark in your own site on the map covering that area in this volume. Consistent marking of this guide in that way will, over the years, provide you with a decidedly valuable record.

It should be stressed that because these sites are listed here does not mean that you automatically have the right to collect there. Unless the sites are on public lands (and even some of these require getting a collecting permit), all these sites are on land that is privately owned, and it is your responsibility to ask permission to collect. Failure to do so could result in a confrontation with an angry landowner or possibly even your arrest for trespassing. Remember, too, that collecting rocks, minerals, or fossils is not permitted in national parks and in many state parks. There may also be stringent regulations with respect to collecting on Indian reservations. It is up to you to learn and obey whatever local regulations exist. No matter where you collect, respect the land and anything on it and follow a few simple rules: ask permission, do not litter, be sure to refill any holes you dig, close any gate you open, do not cut or break down fences, do not tamper with or move any equipment, do not molest livestock.

The author, when on collecting expeditions, always carries a stock of release forms to offer to the landowner, quarry operator, mine manager or other person of authority at the collecting site. Signed by you in the presence of this person and given to him, this form will often allay whatever fear may exist on his part that a lawsuit will be brought against him if you are injured while on his property. It also assures him that you are a responsible person who will respect his property, equipment, or livestock. An example of the form will be found immediately following. Having a quantity of such forms on hand will open many potentially good collecting areas for you, as they have for me, to which you might otherwise be denied access. This is especially true in the case of commercial operations, such as mines, quarries, and gravel pits. (Bear in mind as well that most such commercial operations, if they will let you in at all, require that you wear a hardhat—sometimes safety shoes and safety glasses as well—and may refuse entry to you if you do not have such protection.)

RELEASE AND AGREEMENT FORM

[Sample]

_____, 19__

By signing this paper, I hereby release the owner, proprietor, supervisor, or other official of:

_____

_____

_____

of any responsibility, liability, or claim for damages while I am on the above property. Any personal injury or damage to my property that I may suffer, from any cause whatever, while on this property is deemed by me, the undersigned, as my own responsibility.

My sole purpose in entering this property is to look for fossils and/or rock and mineral specimens. I agree to respect the property at all times, to stay away from machinery and out of danger zones or places where I might interfere with any operations. I also agree not to park my vehicle where it will block machinery or traffic. Any gate opened by me will be immediately closed again. I will not, in any way, molest any equipment, material, or livestock. In all respects, I agree to treat this property with care and thoughtfulness.

I understand that my access to this facility is limited to the date on which I sign this release and agreement.

(signed) _____

(address) _____

_____

COMPASS DIRECTIONS

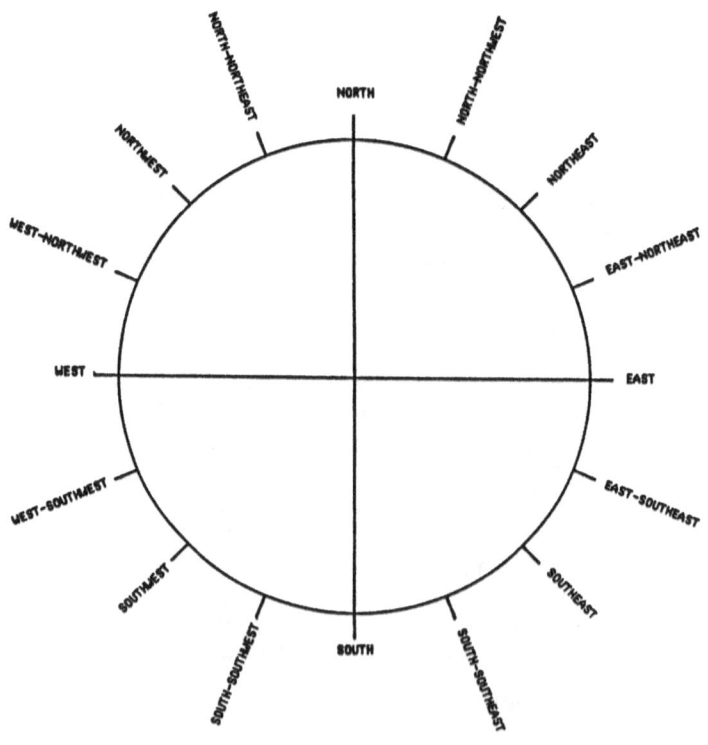

KEY TO PRINCIPAL ABBREVIATIONS AND MINERAL LOCATION NUMBERS

Principal Abbreviations

aka	also known as
BLM	Bureau of Land Management
Byp	bypass route
CF	collecting fee
cm	centimeters
CR	county road
ct	carat
I	interstate highway
lb	pounds
MM	mile marker
mm	millimeters
oz	ounces
RR	railroad
r/w	right-of-way
SR	state highway
US	U.S. highway
var.	variety
"	inches
'	feet

Mineral Location Numbers

1. On the surface and/or in the soil
2. In mine and/or in mine dump
3. In gravel deposit
4. In natural or man-made exposure and/or cut
5. In matrix or *in situ*
6. In soil (but usually *not* on surface)
7. In stream bed, lake bed, coastal shallows
8. In talus and/or natural debris
9. In vugs and/or pockets and/or cavities
10. In gravel pit
11. In sand pit
12. In quarry
13. In washes, ravines, gullies, draws, ditches
14. Weathered from matrix
15. In volcanic tuff or pillow basalt
16. In pegmatite dike
17. In gold-bearing sands and/or gravels
18. In/on gravel bar
19. In vein or seam
20. Loose on shoreline
21. In prospect or placer
22. In sediment deposits or sedimentary strata
23. In alluvial deposits
24. In exposed pegmatite dike
25. In clay
26. In sand
27. In schist
28. In shale and/or slate
29. In outcropping (natural or exposed by man)
30. In spoil deposits
31. In ash deposits
32. In cave and/or cave vicinity
33. In marl
34. In gneiss
35. In clay pit

36. In/on sand bar
37. In/on fill deposit
38. Occurring as drusy or crust
39. In natural bed
40. In conglomerates
41. In concretions
42. In/on ledges

ALABAMA

Statewide total of 125 locations in the following counties:

Bibb (3)	Coosa (7)	Marengo (1)
Blount (3)	De Kalb (2)	Marshall (2)
Calhoun (5)	Elmore (2)	Randolph (12)
Chambers (2)	Franklin (3)	St. Clair (2)
Cherokee (6)	Greene (2)	Shelby (2)
Chilton (1)	Jackson (4)	Sumter (4)
Clarke (5)	Jefferson (2)	Talladega (3)
Clay (27)	Lee (3)	Tallapoosa (10)
Cleburne (3)	Limestone (2)	Tuscaloosa (3)
Colbert (1)	Madison (3)	

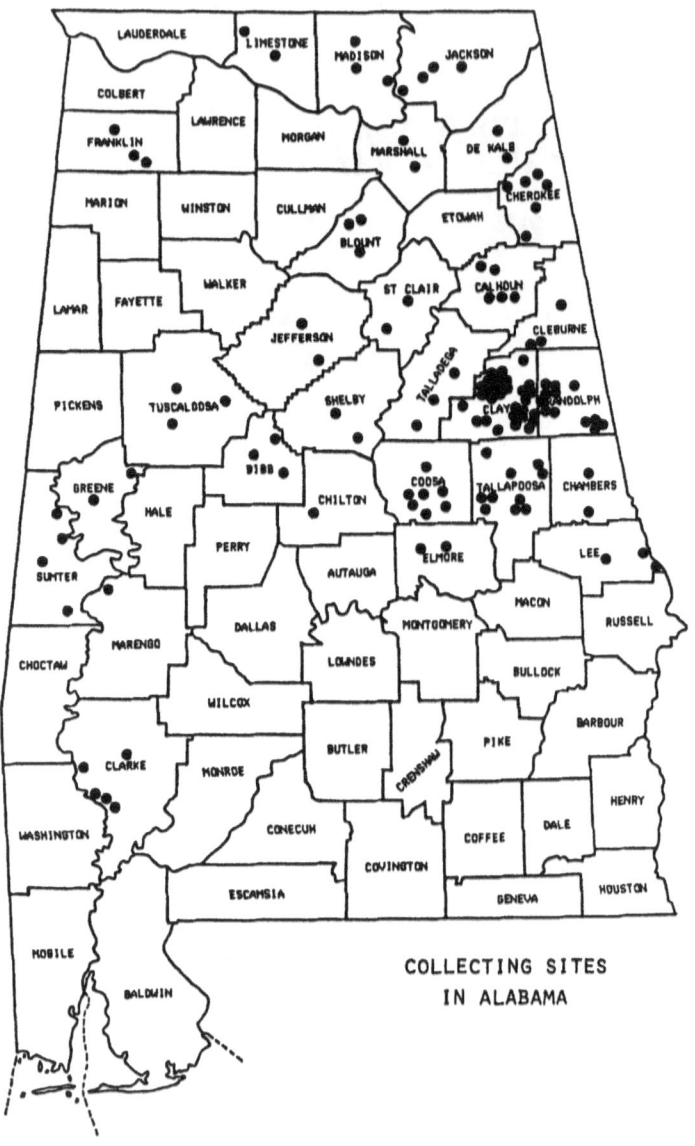

COLLECTING SITES
IN ALABAMA

(FOR SPECIFIC DETAILS ON THESE LOCATIONS,
SEE THE FOLLOWING COUNTY MAPS AND
RELATED TEXT.)

BIBB COUNTY

1. Centreville area: 5 miles north on SR-5.
   Chert. Oolite (Rice agate): silicified; rice patterns; shell-like designs. (12)

2. Little Cahaba River and its tributaries.
   Pearl. (7)

3. Sixmile area: mines along Sixmile Creek.
   Barite: massive, nodular, crystals. (2)

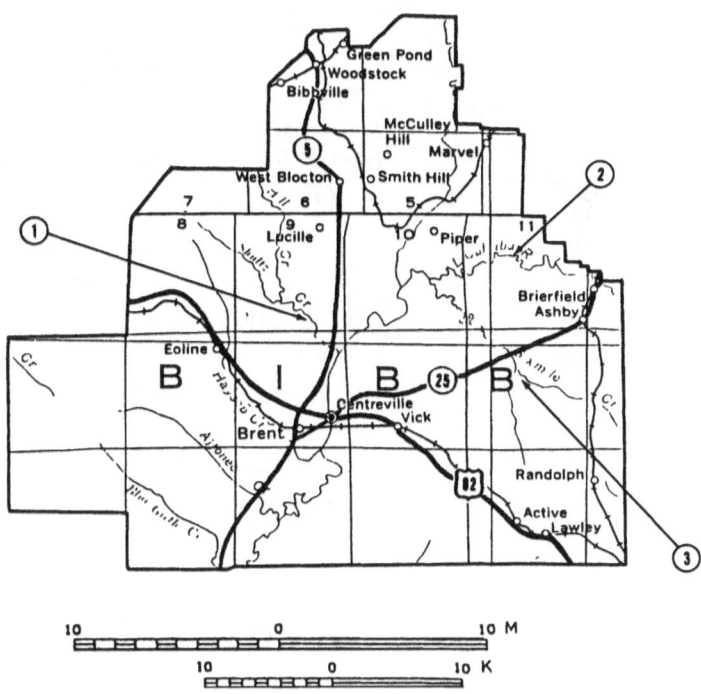

BLOUNT COUNTY

1. Blountsville area: 2 miles southwest on SR-26; toward Holly Pond.
   Chalcedony: nodules to 4" diameter. (1)

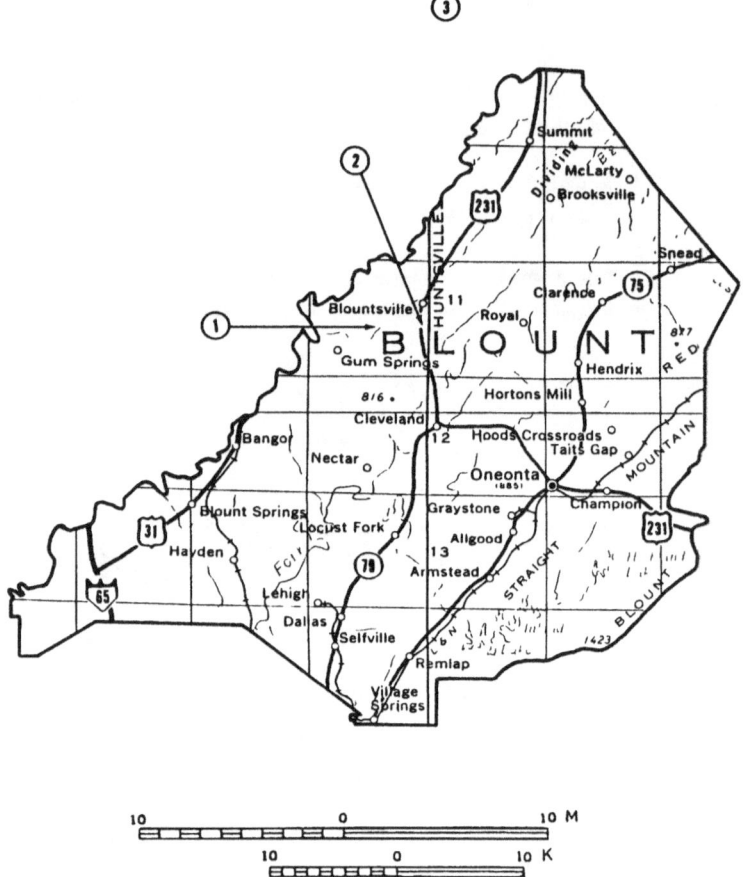

2. Blountsville area: 1 mile south on SR-26; in a series of pits and prospects.

> Agate. Carnelian: red, yellow. Chalcedony: blue. Chert: black, white. Sardonyx: yellow, brown. (12-21)

3. County-wide in lakes and streams, especially Locust Fork and its tributaries.

> Pearl. (7)

CALHOUN COUNTY

1. County-wide in dolomite and limestone quarries.
   Barite: good crystals. Calcite: well-terminated crystals. (9-12)

2. County-wide in iron mines.
   Hematite. Marcasite. Magnetite. Pyrite. (2)

3. County-wide in lead mines.
   Galena. (2)

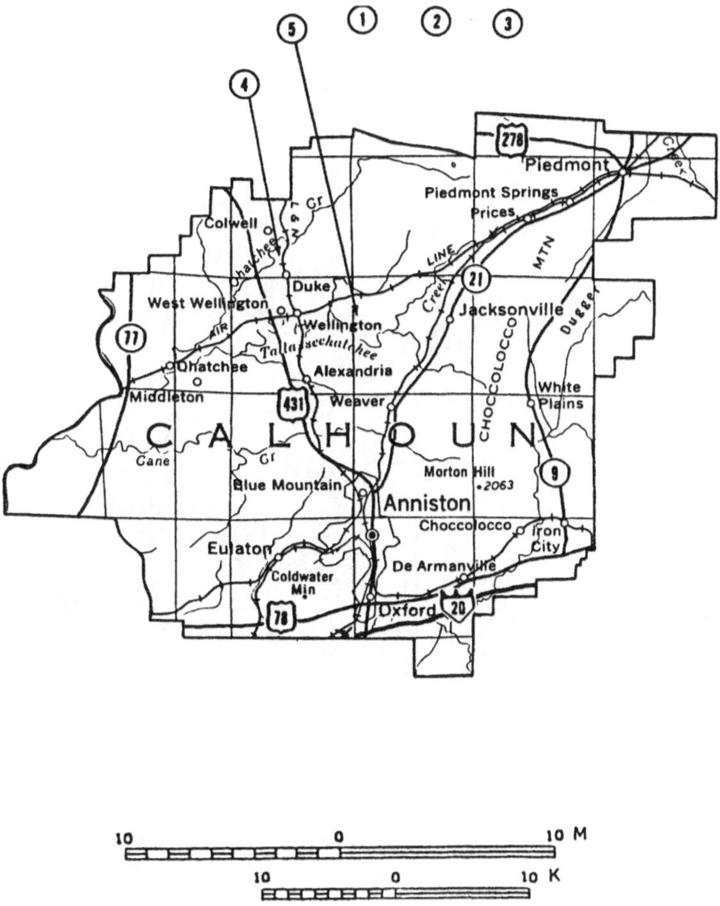

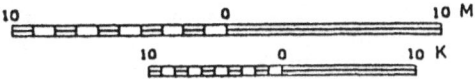

4. Duke area: northwest to the area of Angel Station.

    Barite: good crystals. (1-4-13)

5. Jacksonville area: 5 miles west; in a series of abandoned quarries.

    Galena. (5-limestone 9-12)

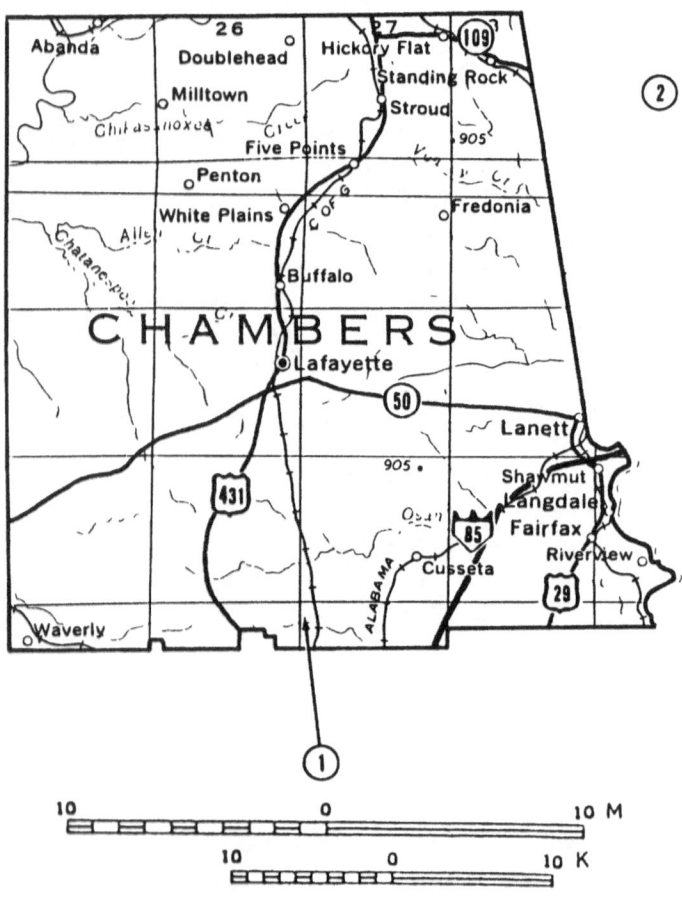

CHAMBERS COUNTY

1. County-wide in lakes and streams, especially in the Tallapoosa River and its tributaries.

    Pearl. (7)

2. Lafayette area: 10 miles south on US-431 to SR-147; 2.5 miles east on SR-147 to dirt crossroad (last before Lee County border); turn right; first mile of dirt road is literally paved with tourmaline (to such an extent that road is nicknamed "Tourmaline Lane"); also in areas adjacent to the dirt road.

    Tourmaline: black: crystals to 0.5", but most are smaller needles. (1-4-13-37)

CHEROKEE COUNTY

1. Cedar Bluff area: 3.6 miles east-northeast on SR-9; in field west of road.

    Quartz, rock: fine, small, clear crystals; some doubly terminated; some with manganese inclusions. (1-13)

2. Cedar Bluff area: north on SR-68 a short distance to CR-75 (at Lighthouse Restaurant); turn west and go 1 mile to creek; cross creek and turn left; search in road cuts on side of road away from Weiss Lake.

    Chert: agatized; varied colors. Fossils: marine. (4)

3. County-wide in lakes and streams, especially in Weiss Reservoir and in the Coosa River and its tributaries.

    Pearl. (7)

4. Gaylesville area: 0.5 mile west on SR-9 to side road on left; 0.5 mile south on the side road.

    Quartz, rock. (1)

5. Leesburg area: just north of town on SR-68; on the Lowe farm.

    Amethyst. (1-3)

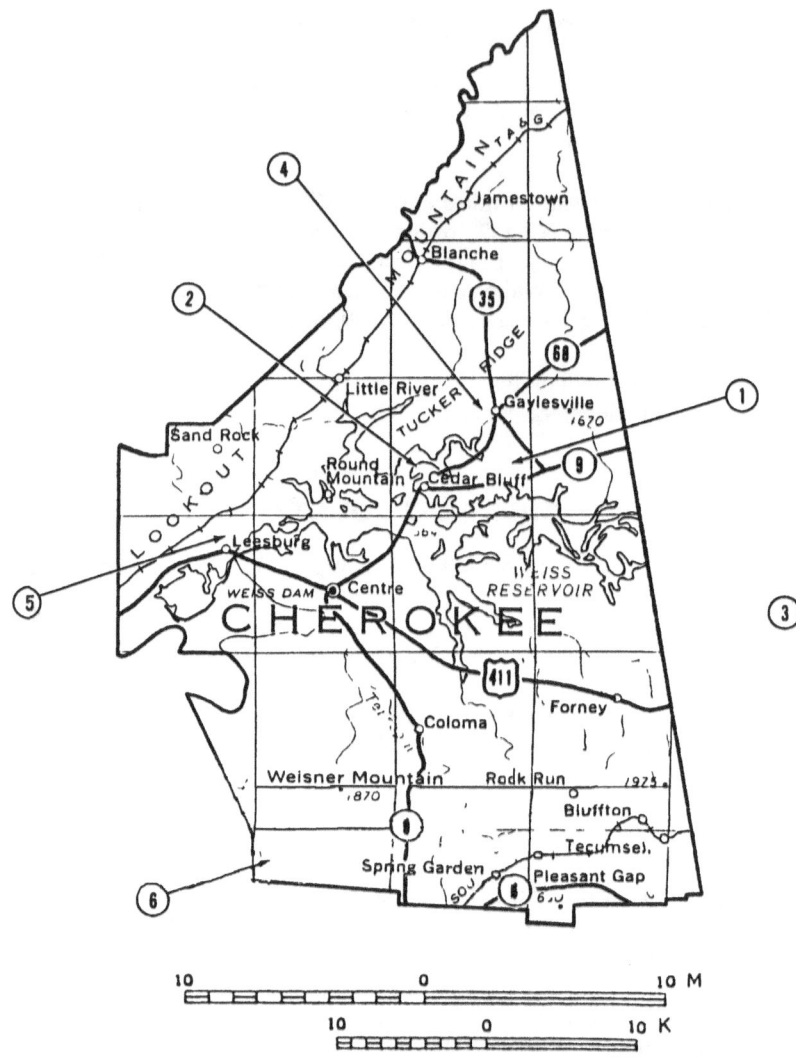

6. Piedmont area (Calhoun County): 7 miles west on US-278
   to Knightins Road; turn right; 2 miles north to CR-6; 0.8
   mile on CR-6 toward Grotville; 0.25 mile south of minnow
   ponds; take farm road through material that has been
   blasted out.

   Calcite. Fluorite. (21)

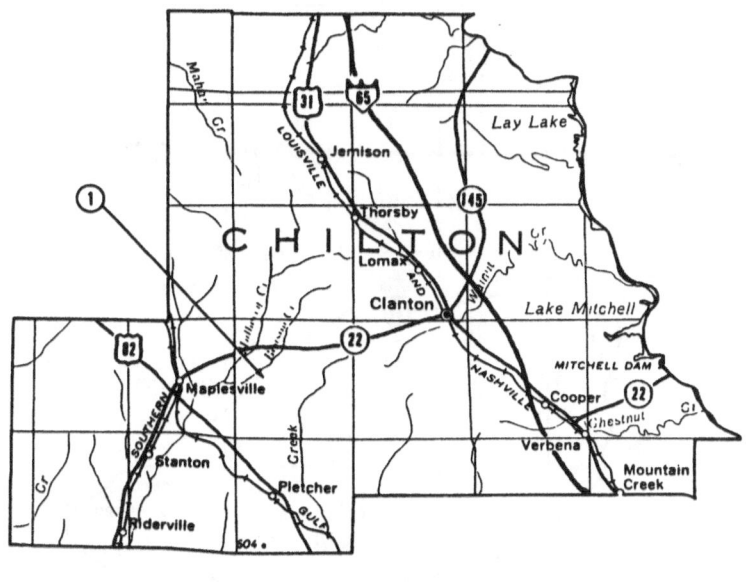

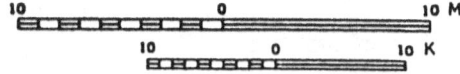

CHILTON COUNTY

1. Maplesville area: 4 miles east on SR-22; go just east of the SR-191 junction to bridge; collect upstream (especially) and downstream in Mulberry Creek and tributaries.

   Gold. (17)

CLARKE COUNTY

1. County-wide in lakes and streams.

   Pearl. (7)

2. Salitpa area: 3.8 miles south; turn left on dirt road; collect on both sides of road.

   Agate: various grades; some unusually fine; some to 5 lb. (1-13-25)

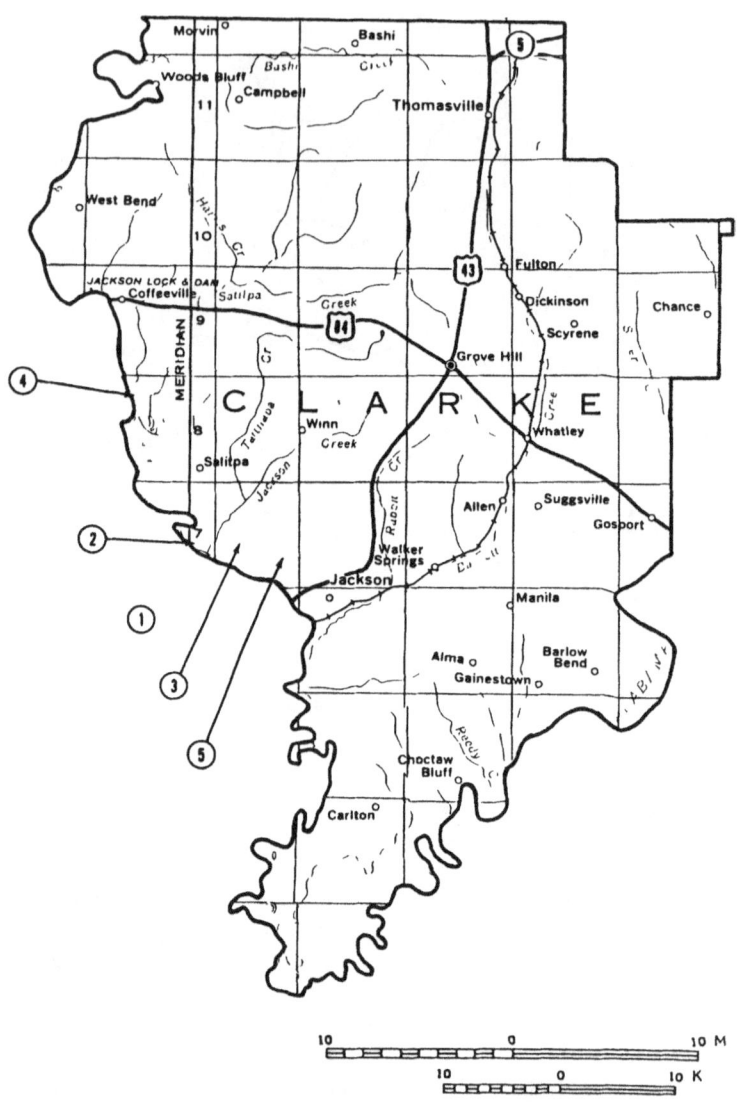

3. Salitpa area: 4.4 miles southeast on SR-69; in road cut.
   Agate. (14-19)

4. Salitpa area: 5.1 miles northwest on SR-69; in road cut.
   Agate. (14-19)

5. Salitpa area: 6.6 miles southeast on SR-69; beds off road on right.

     Septarian: nodules to 16″ diameter. (1-13-39)

CLAY COUNTY

1. Ashland area: along Pleasant Grove Road.

     Quartz, rock: large crystals. (1-13)

2. Ashland area: at the Gibson Prospect.

     Garnet. (21)

3. Ashland area: at the M & G Mine.

     Apatite: chartreuse. Garnet. Quartz, smoky. (2)

4. Ashland area: at the Shirley Prospects.

     Garnet. Kyanite. Tourmaline. (21)

5. Ashland area: south of Pleasant Grove; east of Buzzard Creek; at abandoned gold mine.

     Turquoise: sky blue; some mottled with network of dark limonite. (2)

6. Ashland area: west on SR-77; road cuts in the Talladega Mountains.

     Azurite. Malachite. (4-road)

7. Bluff Spring area: southeast; east and southeast of Harlan; in mile-wide strip of altered Talladega stratum containing thin quartzite layers.

     Chiastolite: raised crystals on rock surfaces. Garnet. Quartz, rock. (1-13-14)

8. County-wide in lakes and streams.

     Pearl. (7)

9. County-wide; scarce.

     Graphite: fine crystals. (27)

10. Cragford area: go east and follow road as it turns north;

collect in the first road cut on left and in the deep RR cut visible from here.

Garnet: crystals to 0.25″ diameter. Pyrolusite. (4)

11. Delta area: 2.5 miles northwest; on slope of Cheaha Mountain; at the Smith Mica Mine.

Beryl. Kyanite. Quartz, rock. (2)

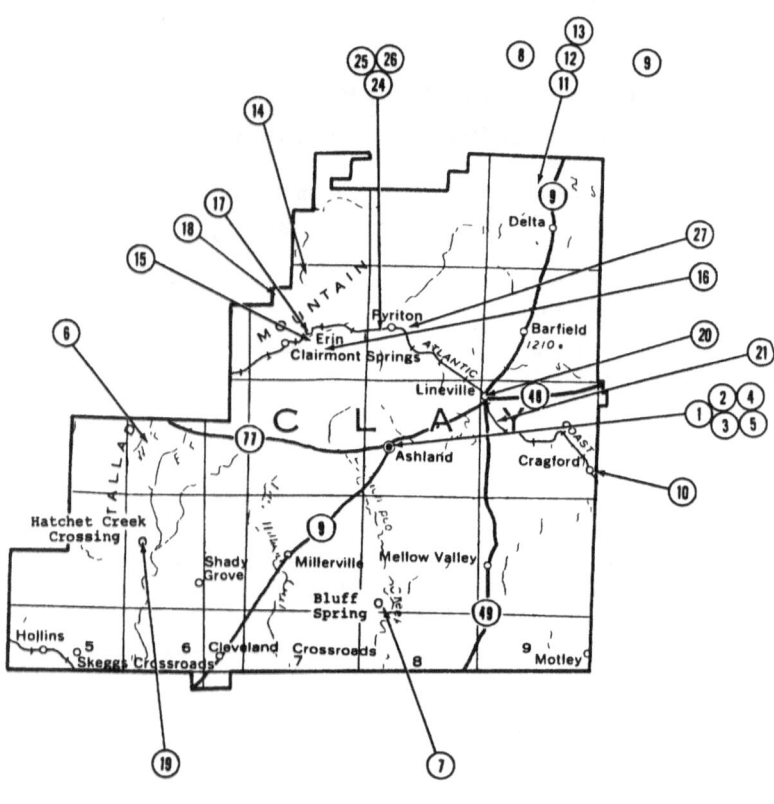

12. Delta area: 2.5 miles northwest on SR-9 to 0.5 mile south of Cleburn County border; on slope of Cheaha Mountain; at the old Delta Mine.

   Beryl. Cassiterite. Kyanite. Quartz, rock. (2)

13. Delta area: 2.6 miles northwest; on slope of Cheaha Mountain; at the Smith No. 1 Mine.

   Beryl: green, blue-green. Garnet: red; to 1" diameter. Kyanite: bright blue crystals in pink feldspar. Tourmaline: black. (2)

14. Erin area: 2.5 miles north of Pleasant Grove Church.

   Turquoise: chartreuse to pale green. (4-RR- 19)

15. Erin area: just south; along edges of and east of Gold Mine Creek.

   Apatite. Beryl. (3-29)

16. Erin area: southeast at 11 mica mines along the road to Grabun.

   Albite. Apatite: green; gem quality. Beryl: golden, green. Garnet, almandine. Garnet, rhodolite. Microcline. Tourmaline. (2)

17. Erin area: in RR cut of the Atlantic Coast RR.

   Turquoise. (4)

18. Erin area: to Talladega County border, then upstream on Gold Creek to headwaters.

   Actinolite. Chlorite: bright crystals. Garnet, pyrope. Gold. Olivine. Quartz, green (Aventurine). Sillimanite. Soapstone. Wavellite. (17-20)

19. Hatchet Creek Crossing area: 0.25 mile south on CR-7; east of road in small brook.

   Azurite. Malachite. (1-7-13-20)

20. Lineville area: north; just south; on the Hobbs farm.

   Turquoise: blue; gem quality. (1)

21. Lineville area: 1.5 miles southeast on SR-48; turn to the Barfield Mine.

   Feldspar. Garnet. Quartz, rock. Tourmaline. (2)

22. Lineville area: 1.5 miles southeast on SR-48; turn to the Gibbs Mine.

> Turquoise. (2)

23. Lineville area: 1.5 miles southeast on SR-48; turn to Griffin Mines.

> Rhodonite. (1-16)

24. Pyriton area: just west; at the L & M Mining Association Mine.

> Azurite. Barite. Chalcocite. Chalcopyrite. Copper. Garnet. Malachite. Pyrite. Sulphur: crystals. (2)

25. Pyriton area: just west; at the National Pyrite & Copper Company Mine.

> Azurite. Barite. Chalcocite. Chalcopyrite. Copper. Garnet. Malachite. Pyrite: excellent. Sulphur: good crystals. (2)

26. Pyriton area: just west; at the Southern Sulphur Company.

> Garnet: very fine crystals; gem quality; unusually abundant. Gold. Sulphur: excellent crystals. (5-quartz-21-28)

27. Pyriton area: east to church; turn north to first graded road east; follow it east to logging road where there is a sign for Lake Simon; turn north here and watch for pegmatites shortly.

> Beryl. Feldspar. (16-24)

CLEBURNE COUNTY

1. County-wide in lakes and streams.

> Pearl. (7)

2. Hollis Crossroads area: 1.5 miles southeast on US-431; then 0.5 mile east on CR-4199 to dirt road left; 0.5 mile north on dirt road; pegmatite is in ditch on left side of road.

> Beryl. (16)

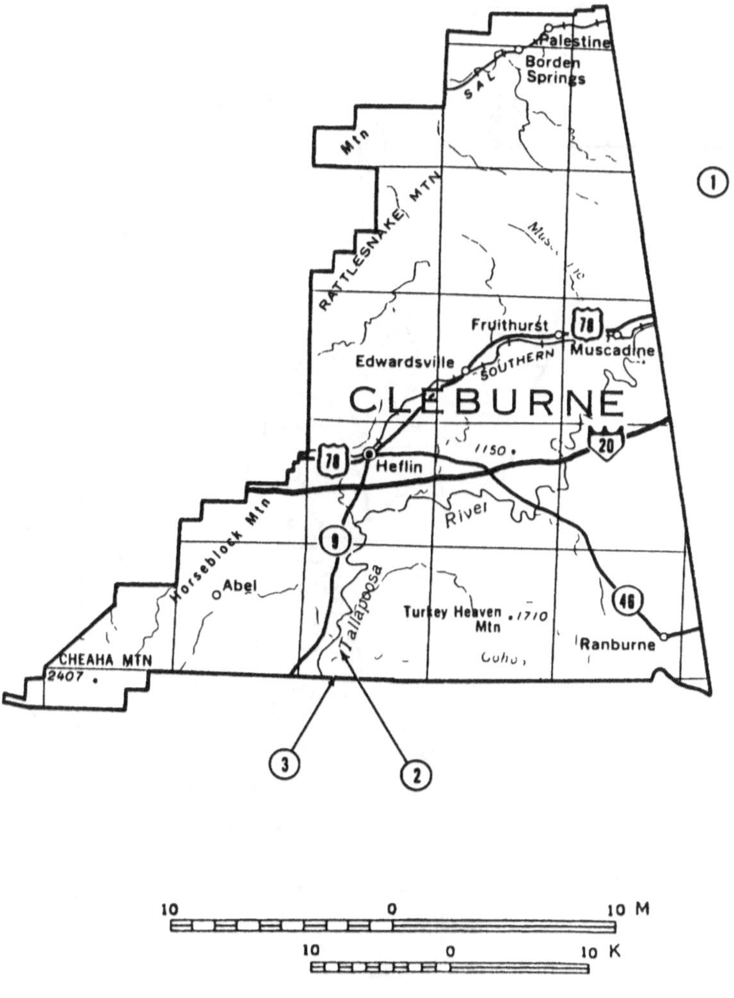

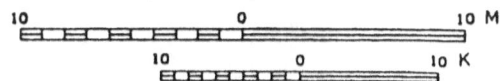

3. Hollis Crossroads area: just southeast; straddling Randolph County border; at the Stone Hill Copper Mine (aka Woods Copper Mine).

Albite. Apatite. Biotite. Copper. Garnet: crystals to 0.5″ diameter; also as fine-grained masses. Sphene. (2)

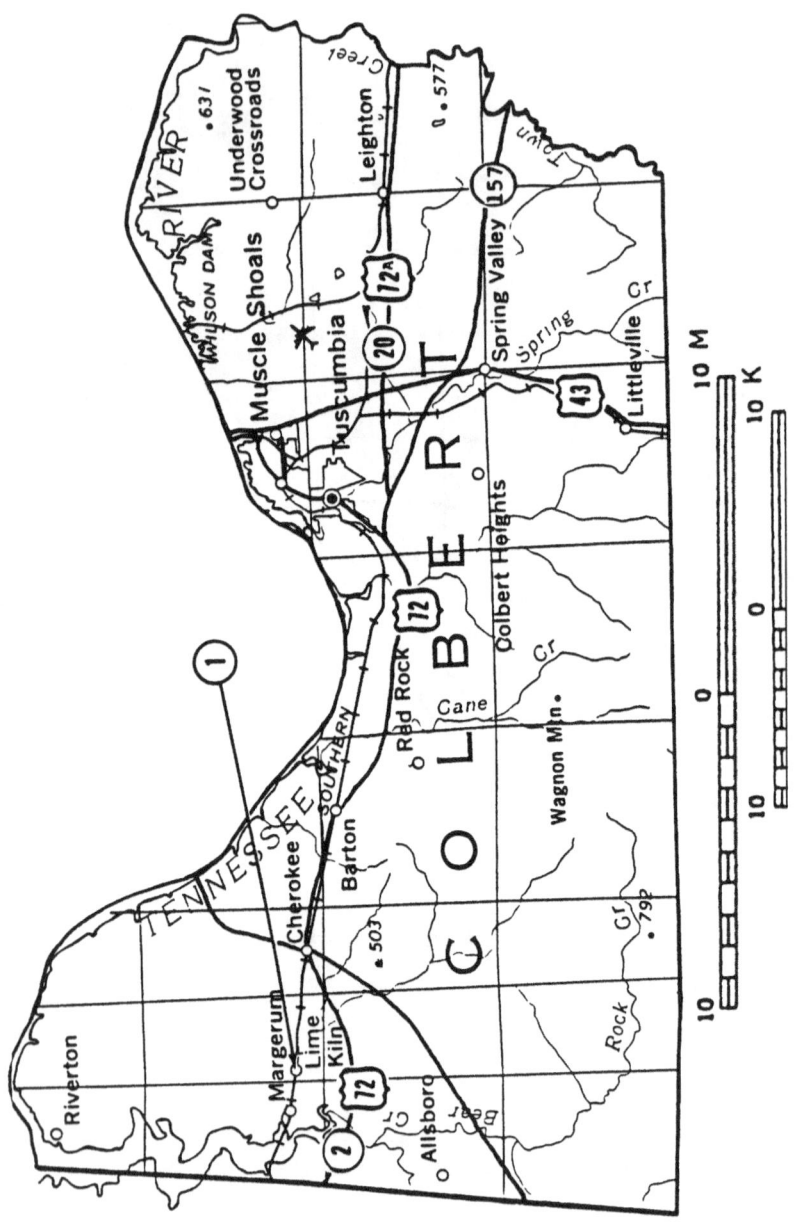

COLBERT COUNTY

1. Margerum area.
   Bauxite: crystals. (19-29)

COOSA COUNTY

1. County-wide in lakes and streams.
   Pearl. (2)

2. Hissop area: 0.75 mile southwest on the Eliza Coggins property.
   Beryl: golden. (16)

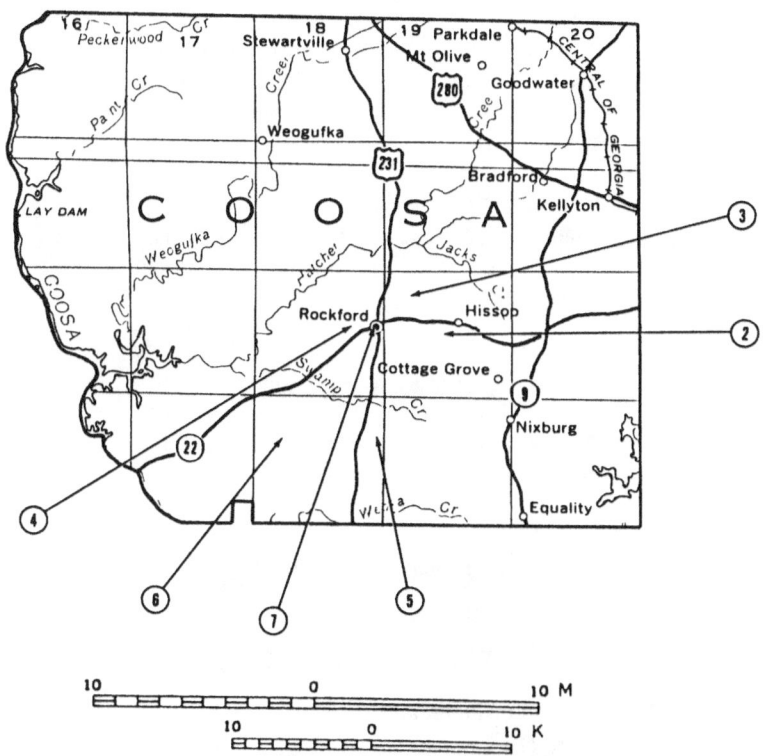

3. Hissop area: 2 miles northeast on SR-22; then 1 mile north-east on the Crewsville road; then left 0.75 mile to fork; bear left to end of road at the abandoned J. H. Thomas Prospect.

   Beryl: golden. (2-16-21)

4. Rockford area: 1 mile west on SR-22; then right on mine road to the Pond Mine.

   Corundum: vivid brown crystals; asterism; chatoyant. Feldspar. Garnet. Moonstone. Quartz, rock. Tourmaline. (2)

5. Rockford area: 5 miles south on US-231 to CR-14; take it 0.6 mile west to fork; bear right and continue another 1.4 miles; at open pit mine.

   Beryl: bright yellow; pale green; transparent and opaque. (2)

6. Rockford area: 6.5 miles southwest; south on US-231 to Pentonville; 0.5 mile west on SR-14 to fork; bear right and go 1.5 miles to the Williams Prospect and Williams Mine.

   Beryl. (2-21)

7. Rockford area tin mines.

   Albite. Apatite: clear green. Epidote. Garnet. Lepidolite. Quartz, milky. Quartz, rock. Sillimanite. Topaz. Tourmaline. (2)

DE KALB COUNTY

1. County-wide in lakes and streams.
   Pearl. (7)

2. Fort Payne area.
   Chert: fine-grained. Hematite. (1-3-7-13-20)

ELMORE COUNTY

1. County-wide in lakes and streams.
   Pearl. (7)

10   0   10 M

10   0   10 K

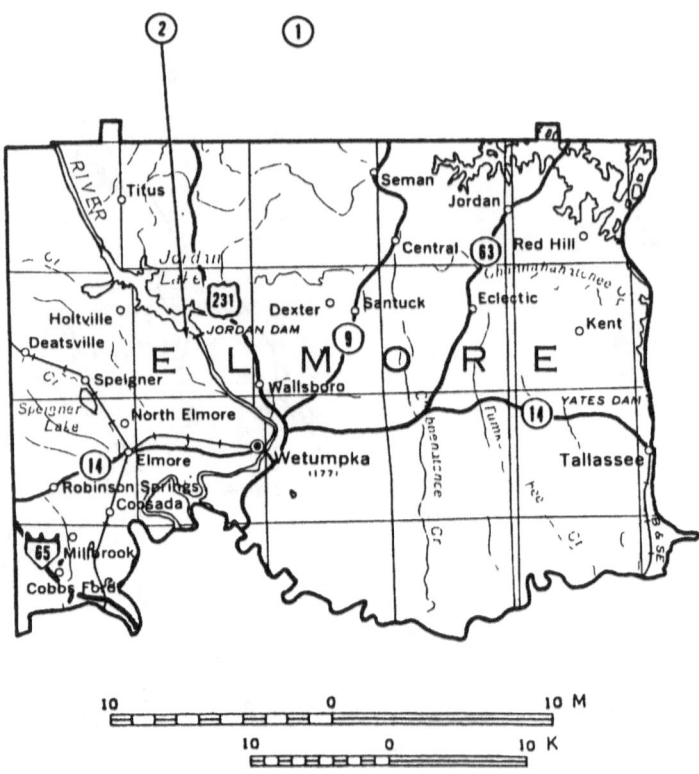

2. Holtville area: 3.25 miles southeast on SR-111 to gravel road left; follow it 0.25 mile to dirt road right within 100 yards of lighthouse at Jordan Dam; follow dirt road 100 yards downstream; foot trail to base of bluff; collect in bluff wall on west bank, in talus, and on water's edge.

   Garnet: good crystals to 0.25″ diameter. (1-3-8-13-20)

FRANKLIN COUNTY

1. County-wide in lakes and streams.
   Pearl. (7)

2. Phil Campbell area gravel pits.
   Carnelian. (10)

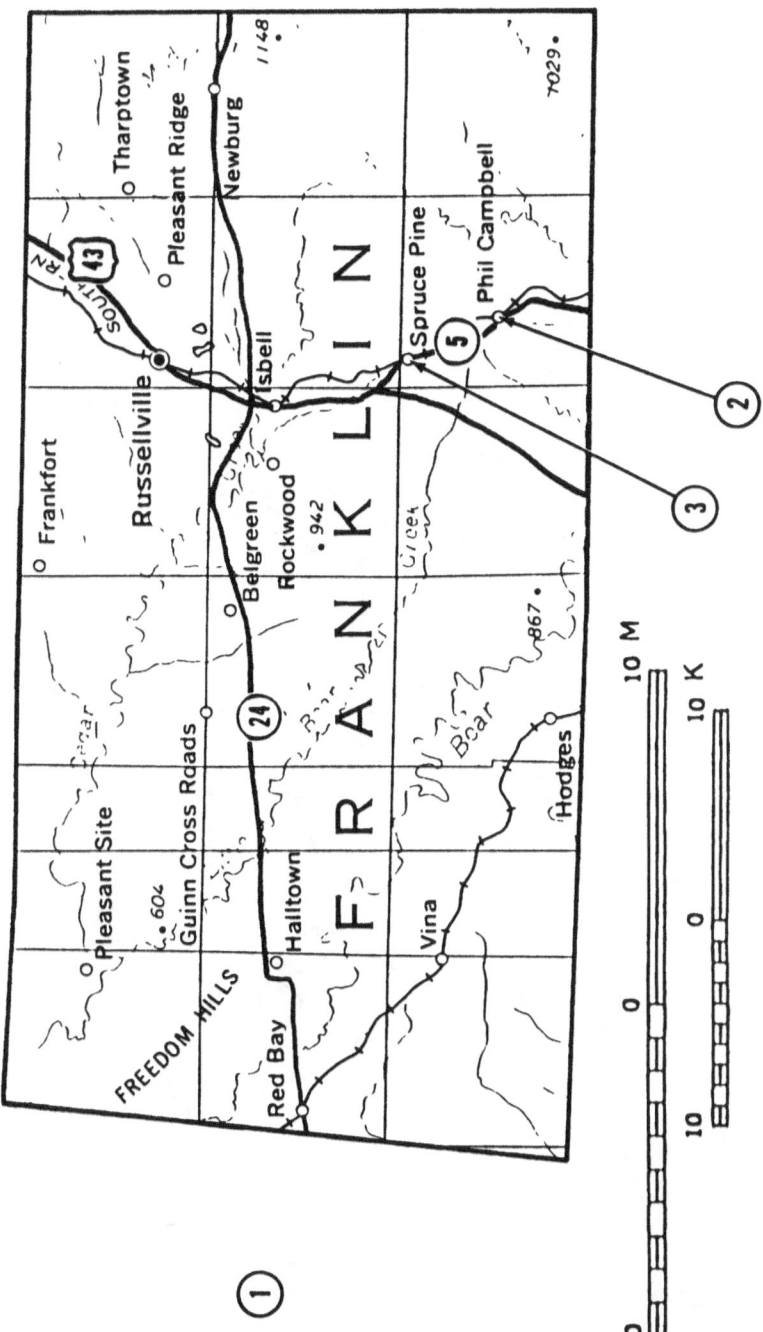

3. Spruce Pine area gravel pits.
   Carnelian. (10)

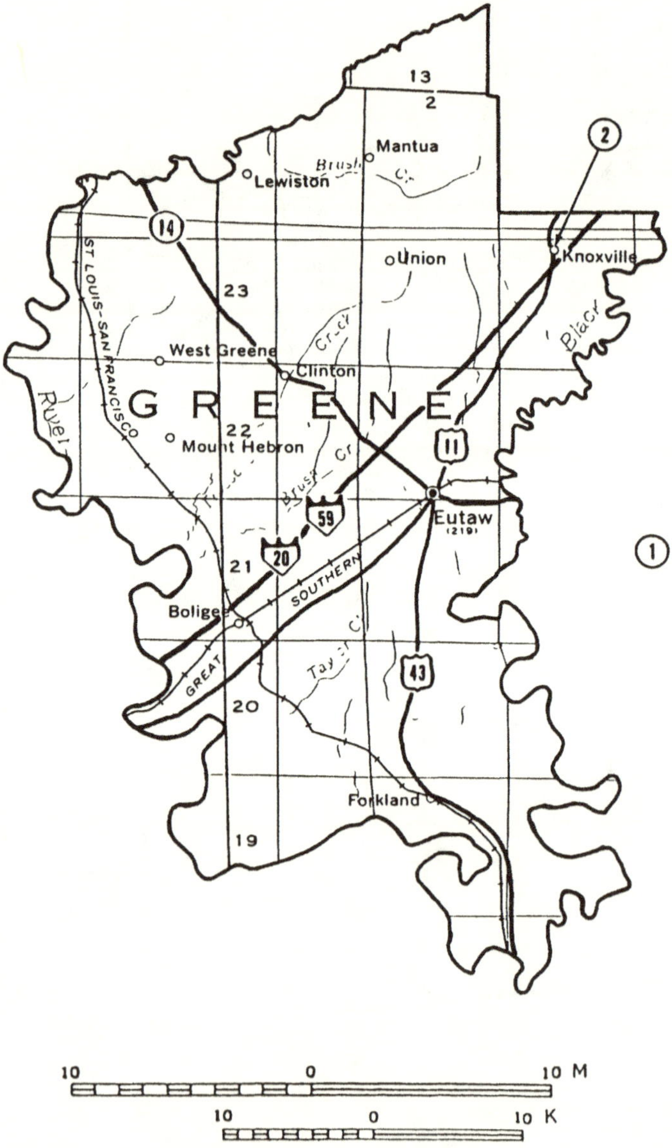

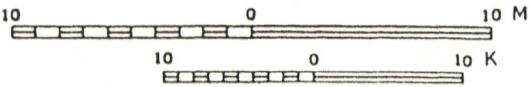

GREENE COUNTY

1. County-wide in lakes and streams.
    Pearl. (7)

2. Knoxville area: south along US-43 in road cuts and wherever there are outcroppings of Demopolis chalk, all the way south to Marengo County border.
    Marcasite: rosettes. Pyrite: rosettes. (4-29)

JACKSON COUNTY

1. County-wide in lakes and streams.
    Pearl. (7)

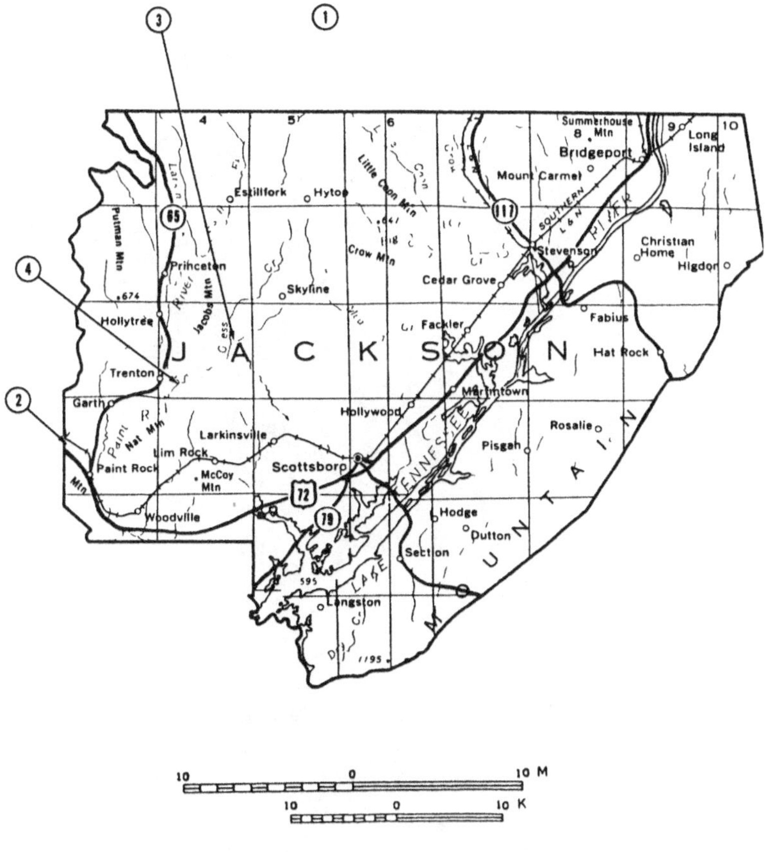

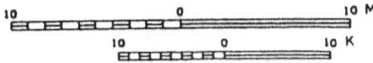

2. Paint Rock area: 2.2 miles north-northwest on US-72 to
   small mountain near the Madison County border.

   Chert: colorful; compact grain. (1)

3. Skyline area: 4.1 miles southwest; on slopes and in drain-
   ages of Jacobs Mountain, as well as downstream in Guess
   Creek.

   Agate (aka "Paint Rock Agate"). Jasper. (1-3-7-13-20-
   39)

4. Trenton area: 1.2 miles east-southeast; along shorelines at
   confluence of Paint Rock Creek and Guess Creek.

   Agate (aka "Paint Rock Agate"). Jasper. (1-3-7-13-20-
   39)

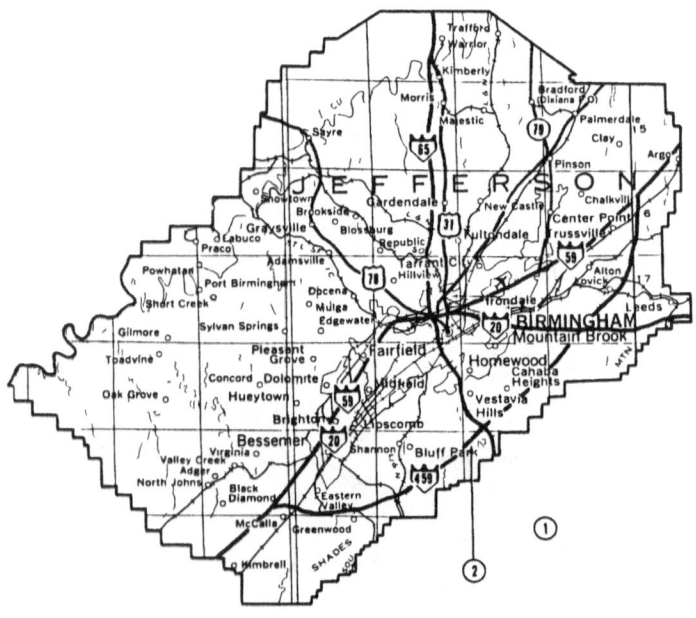

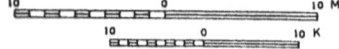

JEFFERSON COUNTY

1. County-wide in lakes and streams.
   Pearl. (1)

2. Vestavia Hills area: 4.2 miles south; on west bank of Cahaba Creek; 300 yards north of the mouth of Little Shades Creek.
   Diamond: 1 stone; unverified as 0.875 ct; date unknown. (20)

LEE COUNTY

1. County-wide in lakes and streams.
   Pearl. (7)

2. Goat Rock Dam area: 1 mile south on west bank of the Chattahoochee River.
   Diamond: 1 stone; 4.5 ct; distorted white octahedron; found in 1978. (18)

3. Smiths area: 5 miles southeast; along west bank of Chattahoochee River.
   Diamond: 1 stone; 4.6 ct; fine; pale green; found in 1901. (17)

LIMESTONE COUNTY

1. County-wide in lakes and streams.
   Pearl. (7)

2. Good Springs area: 1 mile south to Dobbins Branch; downstream on both sides.
   Quartz: nodules to 7". (1)

MADISON COUNTY

1. County-wide in barite quarries.
   Barite: good crystals. (5-massive barite -12)

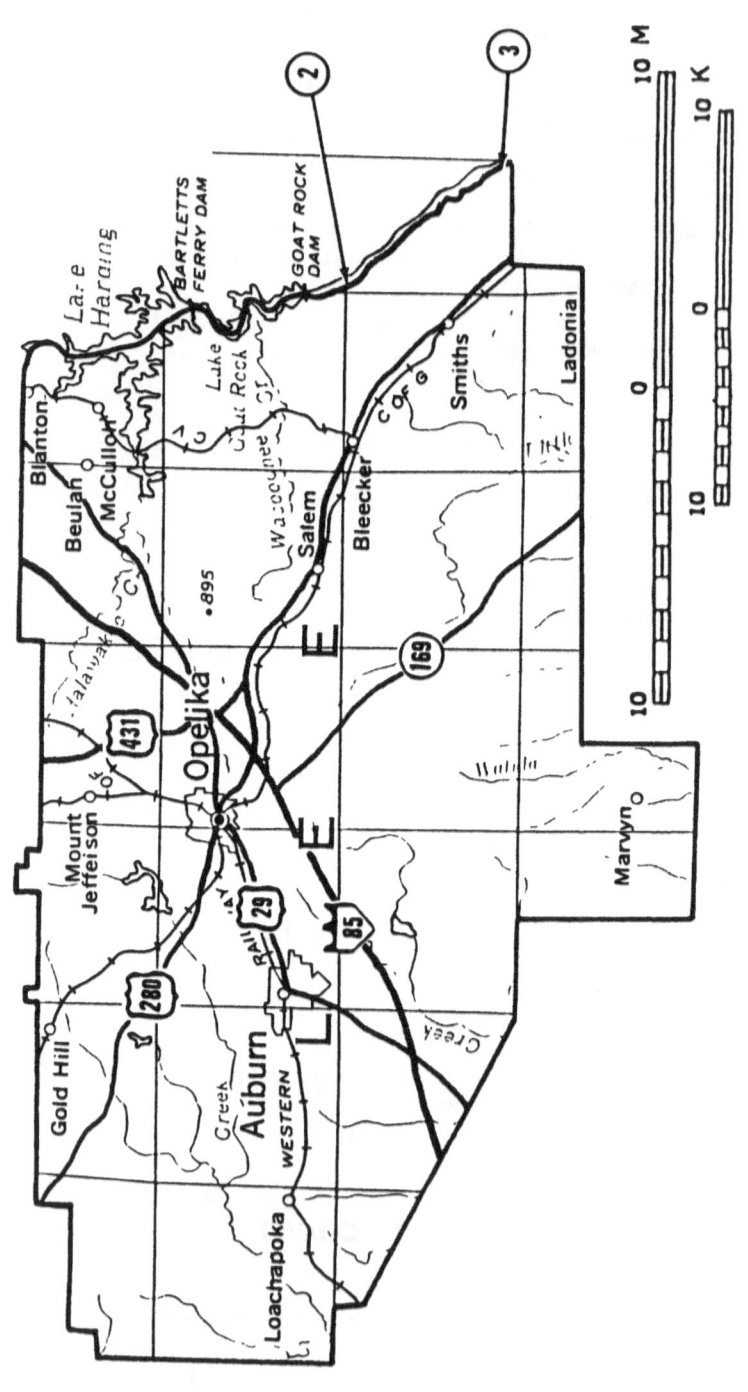

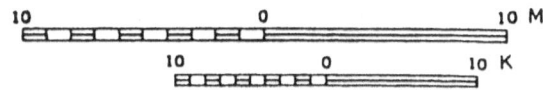

2. County-wide in streambeds.

   Agate (aka "Paint Rock Agate"). Chert: gem quality.
   (3-7-18-20)

3. Gurley area: southeast on SR-72 to large hill; collect on
   slopes.

   Chert: gem quality. (1-13)

MARENGO COUNTY

1. Demopolis area: 6.7 miles southwest along unimproved road following Tombigbee River; in the area locally known as Bartons Bluff.

    Calcite: crystals. (1-13-14-29)

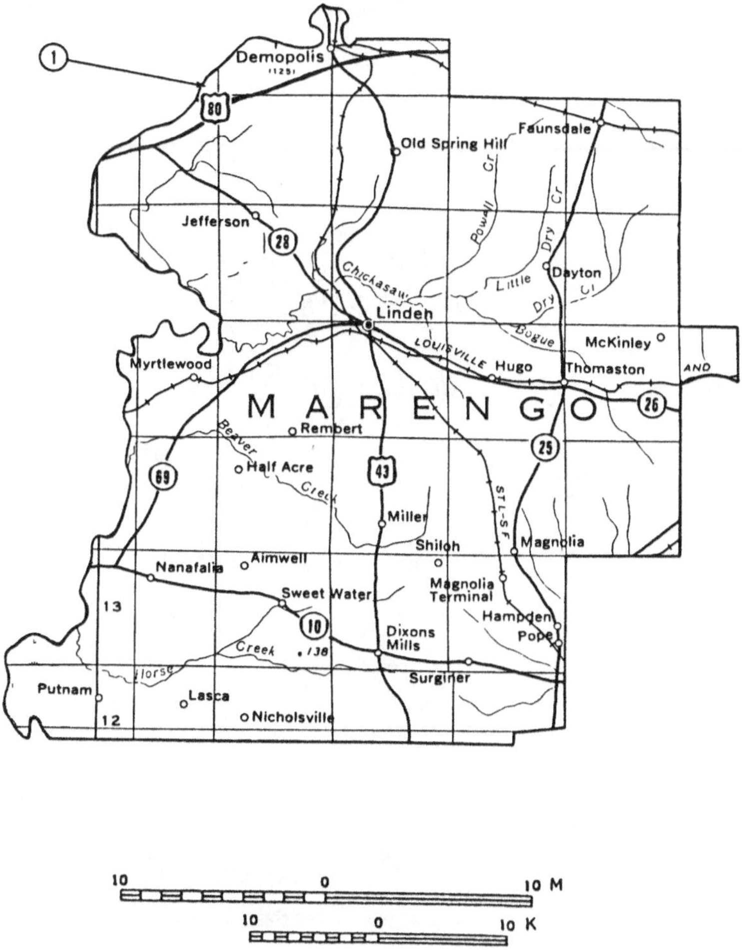

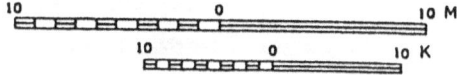

MARSHALL COUNTY

1. County-wide in lakes and streams.
     Pearl. (7)

2. Guntersville area.
       Quartz, rock: fine clear crystals; small. (1-13)

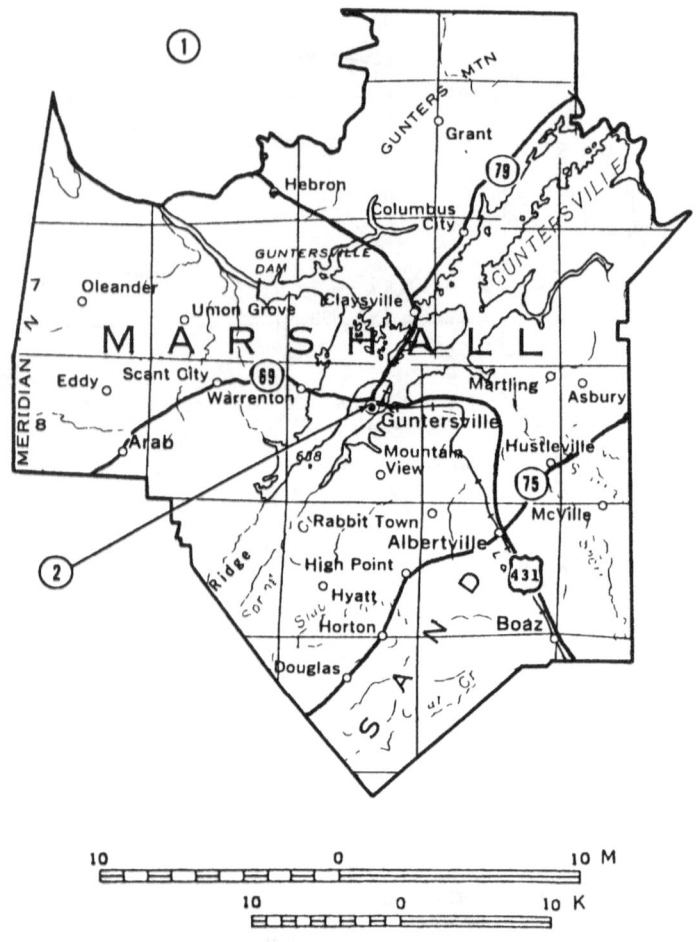

RANDOLPH COUNTY

1. County-wide in lakes and streams.
    Pearl. (7)

2. Milner area: 2 miles east on CR-82; 1.5 miles west of SR-1;
    just north of Rice Mill.
        Apatite: sky blue. Garnet, rhodolite: pink. (1-13)

3. Milner area: at the old Stone Hill Copper Mine.
   Azurite: fine blue crystals. Calcanthite. Cuprite. Malachite. (20)

4. Milner area: just west; on dumps of the Vickers Prospect.
   Tourmaline. (21)

5. Milner area: on the dumps of the J. J. New No. 2 Mine.
   Garnet: deep red. (2)

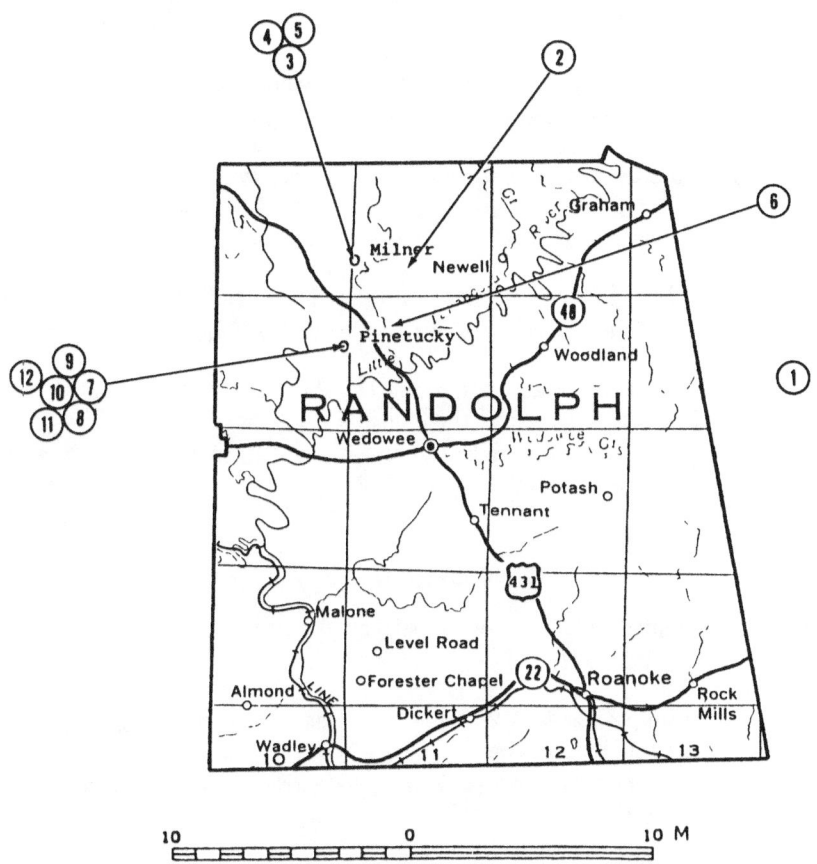

6. Pinetucky area: 2.5 miles northeast; at the Pat Ayers Prospect No. 2 Mica Mine.

    Aquamarine; fine gem quality. (2)

7. Pinetucky area: go toward Milner; at the Jones Mica Mine west of the road to the old mines; on pegamatite running northeast to southwest.

    Apatite. Beryl. Garnet, rhodolite. (2-16)

8. Pinetucky area: just off the Pinetucky-Micaville Road; at the Great Southern Mine.

    Apatite: chartreuse; glassy. Aquamarine: blue. Beryl: pale blue. Garnet: bright red. Tourmaline: black. (2)

9. Pinetucky area: northwest; at the Arnott No. 1 Prospect.

    Garnet: massive. Tourmaline. (21)

10. Pinetucky area: northwest; at the Liberty Mine.

    Apatite: light green. Garnet: bright red. Pyrite: bright brassy crystals. Tourmaline: black. (2)

11. Pinetucky area: northwest; at the Haynes Mine.

    Tourmaline: excellent gem-quality crystals. (2)

12. Pinetucky area: south along the Pinetucky-Milner Road; just west of road at the Jones No. 1 Mine.

    Apatite: superb bright blue masses. Garnet, rhodolite: pink. Tourmaline: black. (2)

ST. CLAIR COUNTY

1. County-wide in lakes and streams.
    Pearl. (7)

2. Brompton area: 1.25 miles east; near Prescott Siding; on the Isbell property.

    Diamond: 1 stone; 2.41 ct; pale green; found in 1905. (1)

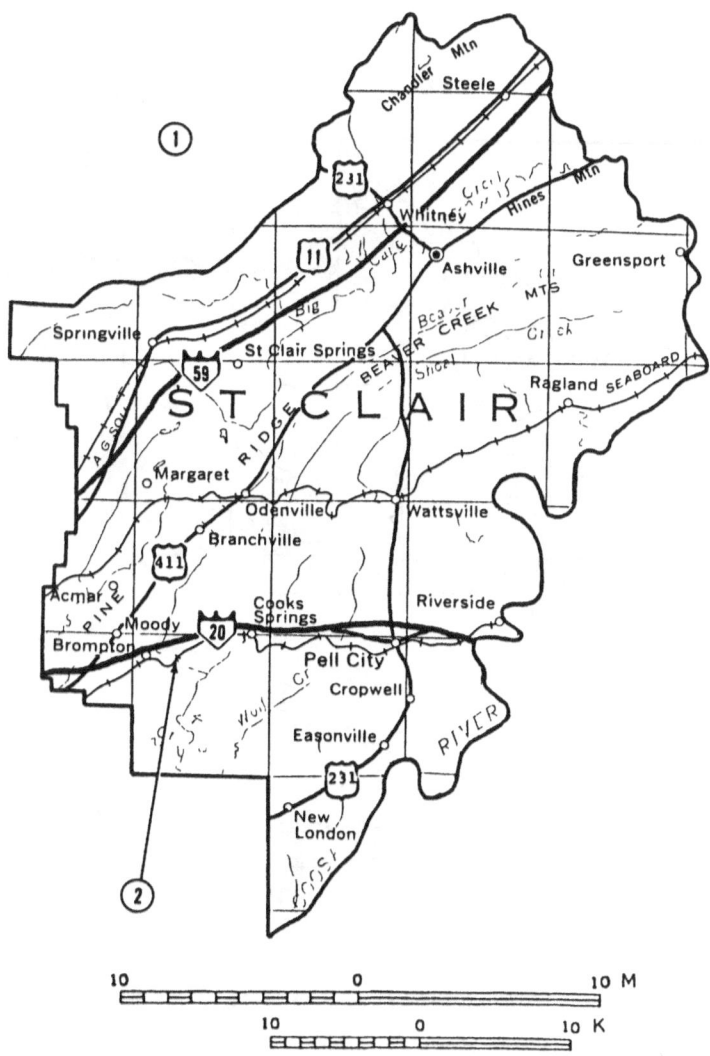

10   0   10 M

10   0   10 K

SHELBY COUNTY

1. County-wide in lakes and streams.
   Pearl. (7)

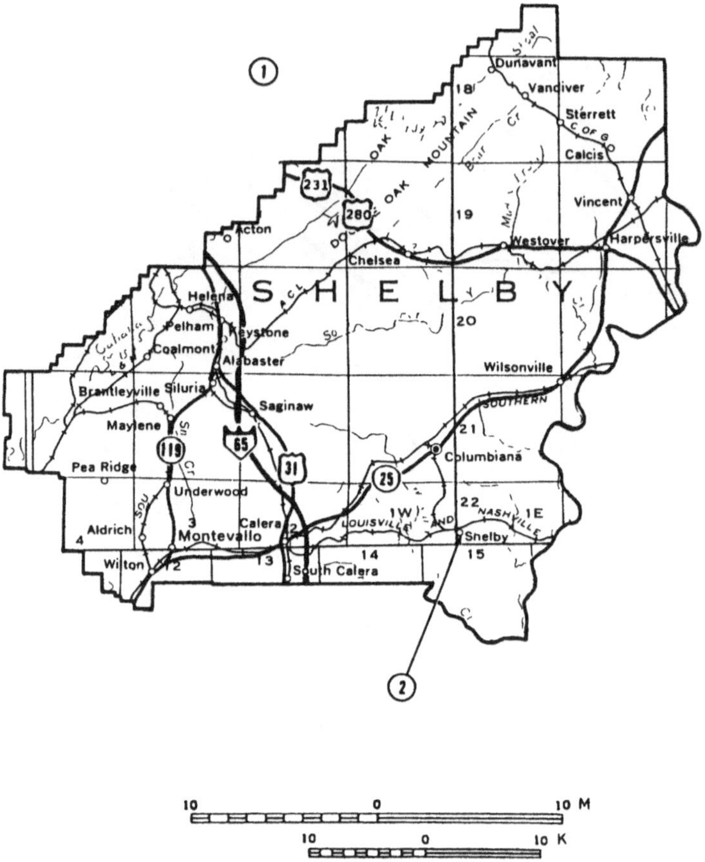

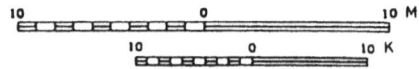

2. Shelby area.

  Diamond: 1 stone; 4.27 ct; fine; found in 1900. (17)

SUMTER COUNTY

1. County-wide in lakes and streams.
  Pearl. (7)

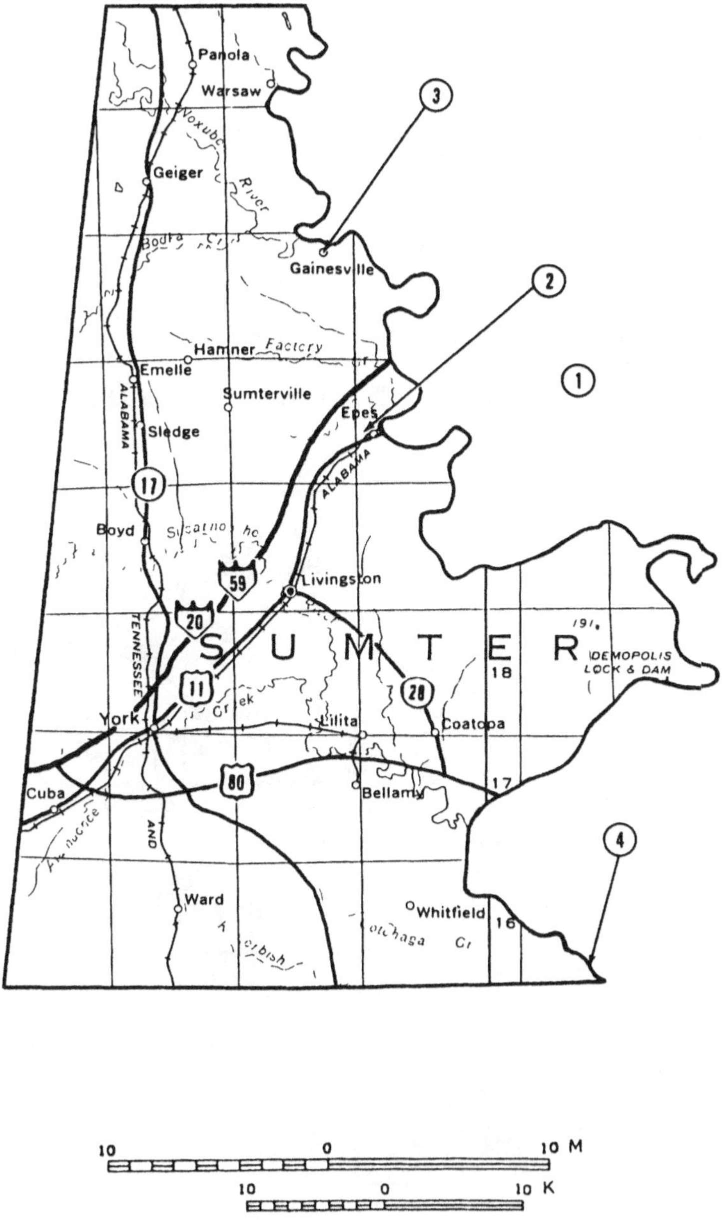

Panola
Warsaw
③
Geiger
Woxub... River
Boyd...
Gainesville
②
Hamner Factory
Emelle
Sumterville
①
Sledge
Epes
ALABAMA
ALABAMA
⑰
Boyd Sucarno... ho
⑤⑨
Livingston
⑳
S U M T E R
191
DEMOPOLIS
LOCK & DAM
TENNESSEE
⑪
York Creek 18
Lilita
㉘
Coatopa
⑧⓪
Bellamy
17
Cuba
AND
⓪
Ward
Whitfield 16
otchaga Cr
④

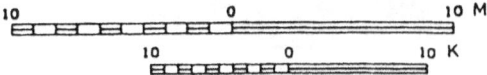

10   0   10 M

10   0   10 K

2. Epes area: southwest along US-11; in road cuts and out-croppings of Demopolis chalk all the way to York.

    Marcasite: rosettes. Pyrite: rosettes. (4-29)

3. Gainesville area: west on SR-116; in road cuts and outcrop-pings of Demopolis chalk all the way to SR-17.

    Marcasite: rosettes. Pyrite: rosettes. (4-29)

4. Whitfield area: upstream in the Tombigbee River from the southeast corner of the county to the northeast corner of the county; in stream gravels, quarries, and gravel pits along the course of the river bordering the entire county on the east.

    Pyrite: very fine; bright brassy crystals. (3-10-12-20)

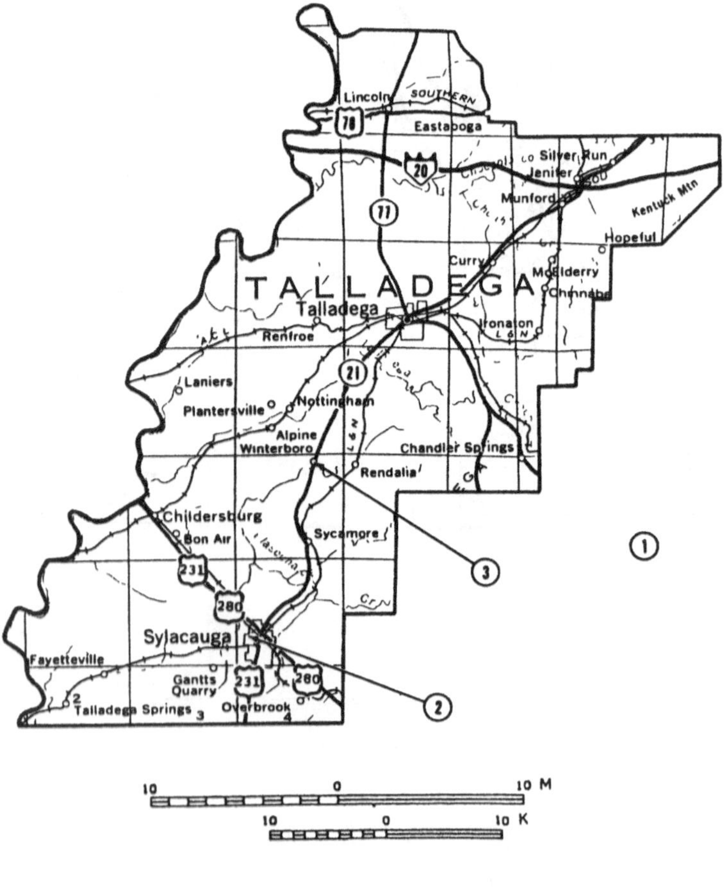

TALLADEGA COUNTY

1. County-wide in lakes and streams.
   Pearl. (7)

2. Sylacauga area: 1 mile northwest on CR-8; then south about 0.5 mile to quarry.
   Marble: quality. (12)

3. Winterboro area quarries.
   Soapstone. (12)

TALLAPOOSA COUNTY

1. Carrville area: 8.8 miles north to Martin Dam; collect along the east shoreline of Lake Martin.
   Actinolite. Bronzite. Cleavelandite. Epidote: fine crystals. Feldspar. Hematite, specular. Quartz, rock: fine. (20)

2. County-wide in lakes and streams.
   Pearl. (7)

3. Dadeville area: at the Kidd Mine.
   Garnet: purplish. Pyrite. Quartz, rock. Sericite: green. (2)

4. Dadeville area: at the Mica Hill Mine.
   Granite, graphic. Quartz, smoky. (2)

5. Dadeville area: near Camp Hill; at the Doc Heard Prospect.
   Garnet. Tourmaline. (21)

6. Dudleyville area mine dumps.
   Margarite. Tourmaline. (21)

7. Dudleyville area: on the Garfield Heard farm.
   Soapstone. (1)

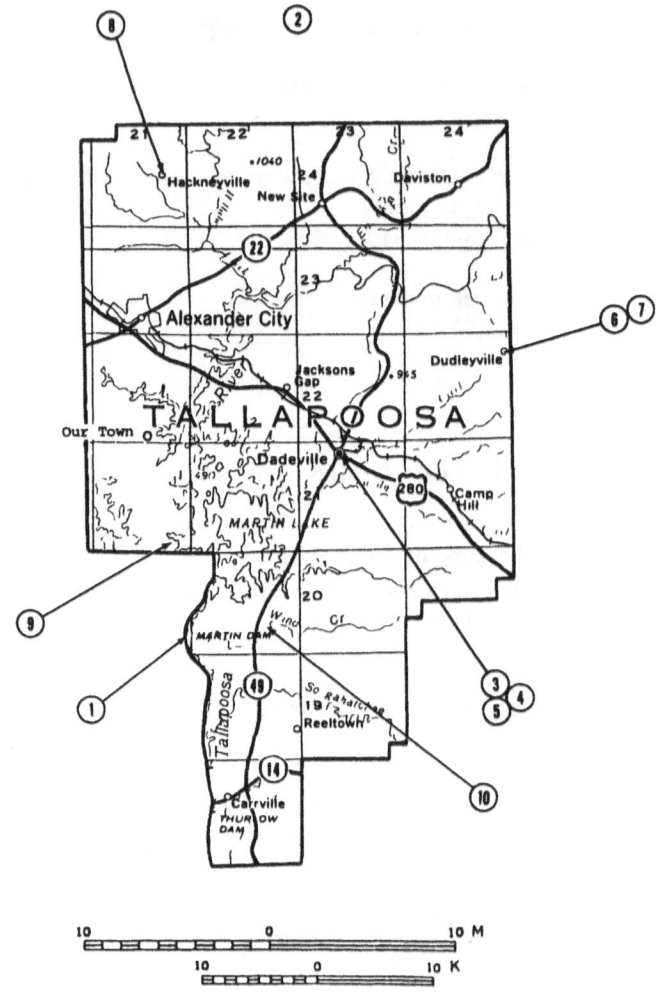

8. Hackneyville area: 1 mile north of Wades Mill.
    Chiastolite. (1-13)

9. Our Town area: 5.2 miles south on SR-63 to paved road at
    left by Catherine's Grocery; go east 1.25 miles to narrow
    dirt road at left; 0.7 mile on dirt road past bulldozer cuts
    to end, which is in loop at Garnet Cove on Lake Martin;

search bulldozer cuts (walls of cuts and spoil piles) and on shoreline when lake level is low.

Garnet: to 1″ diameter; those to 0.5″ diameter are quite common; many crystals cracked and/or deformed by twinning. (1-4-20-29)

10. Reeltown area: 6 miles north-northwest to Wind Creek; on shore and in surrounding fields, especially north and west of the creek.

Epidote. Quartz, rock. (1-3)

TUSCALOOSA COUNTY

1. Brookwood area: 4 miles west on SR-216 to side road north; follow it to iron strip mines.

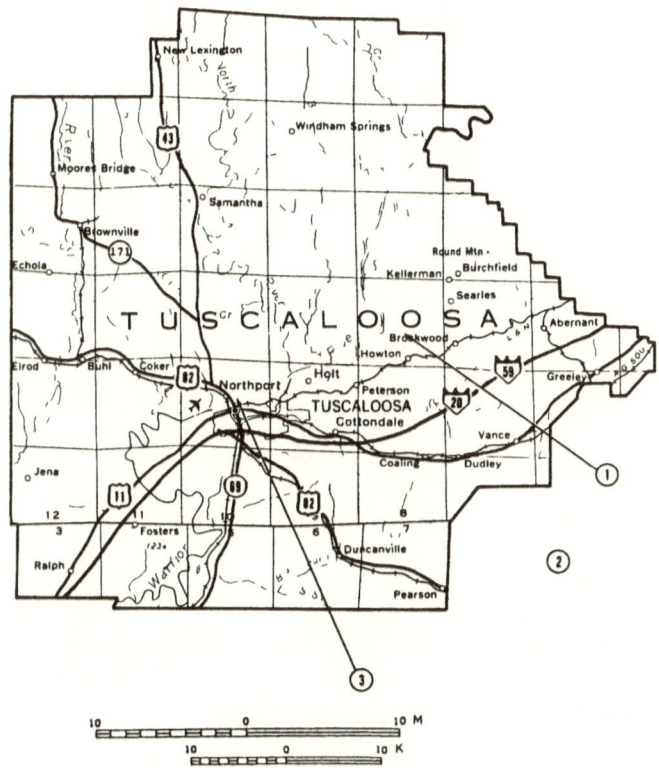

Agate: sagenitic. Chalcedony. Jasper. Petrified wood. (2)

2. County-wide in lakes and streams.
     Pearl. (7)

3. Tuscaloosa area: north on SR-216 to 0.25 mile past sign for 4-H Club; just beyond to strip mines.
     Petrified wood. Quartz, rock. (2)

FLORIDA

Statewide total of 83 locations in the following counties:

Alachua (2)	Hernando (4)	Polk (15)
Bradford (2)	Hillsborough (5)	Sarasota (2)
Charlotte (2)	Indian River (1)	St. Johns (2)
Citrus (5)	Jackson (3)	Sumter (2)
Dade (1)	Lafayette (1)	Suwanee (4)
Duval (1)	Levy (2)	Taylor (2)
Gadsden (4)	Marion (2)	Volusia (1)
Gilchrist (1)	Pasco (3)	Washington (5)
Hamilton (6)	Pinellas (5)	

COLLECTING SITES
IN FLORIDA

(FOR SPECIFIC DETAILS ON THESE LOCATIONS,
SEE THE FOLLOWING COUNTY MAPS
AND RELATED TEXT.)

ALACHUA COUNTY

1. County-wide in excavations and in road and RR cuts.
    Chert: good quality. Fossils, vertebrate: ivory (usually black). (4)

2. County-wide in limestone quarries.
    Chert: concretions; boulders. (12)

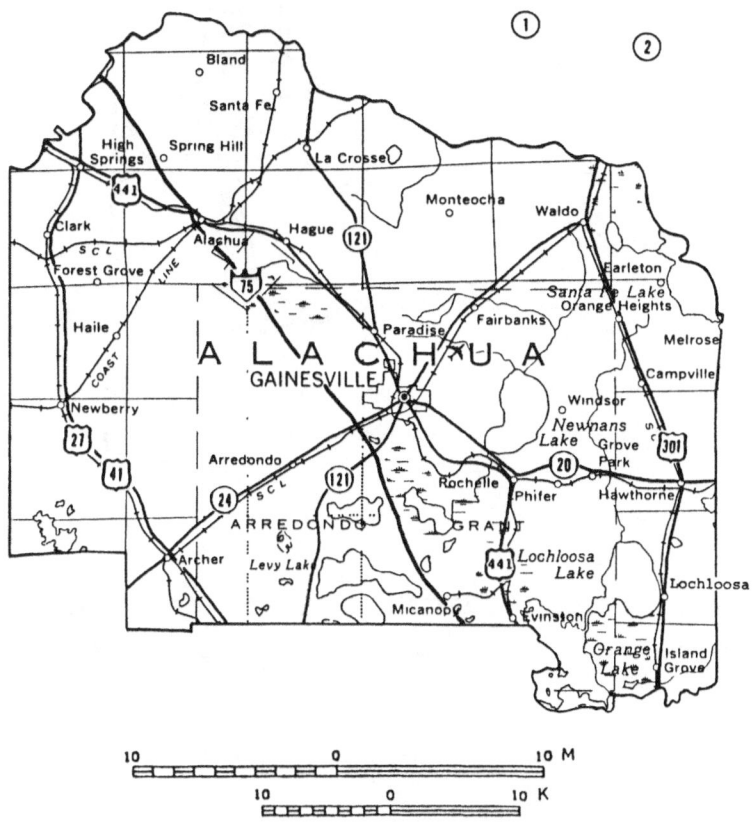

BRADFORD COUNTY

1. Lawtey area: at the Highland Mine.
    Ilmenite. Zircon. (2)

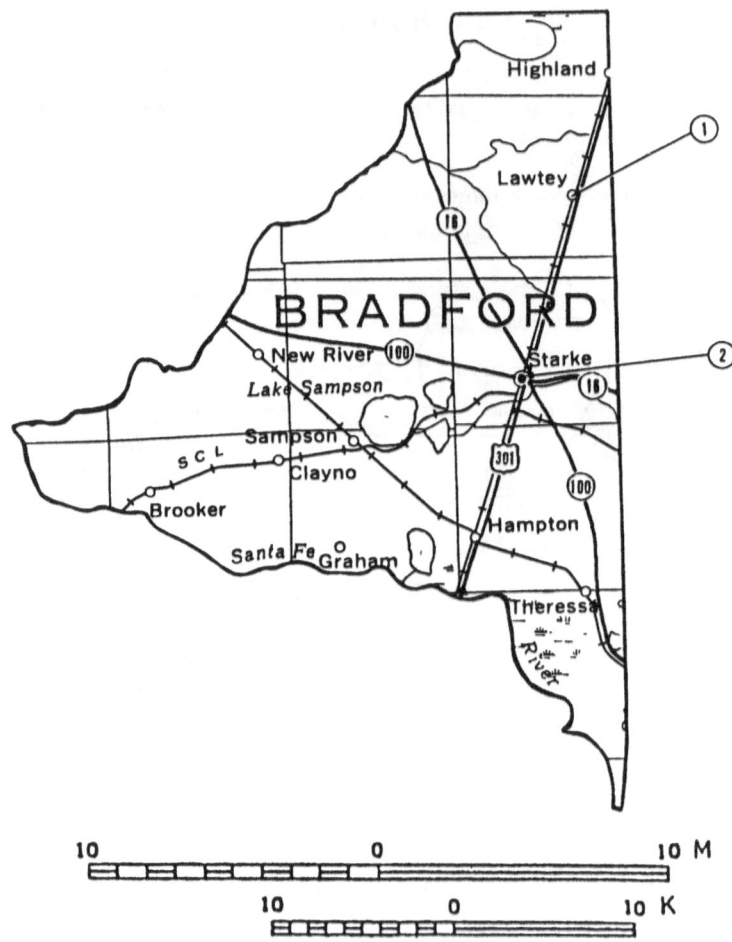

2. Starke area: at the Trail Ridge Plant of E. I. du Pont de Nemours & Co.

    Staurolite. (1-4-13)

CHARLOTTE COUNTY

1. County-wide in lakes and streams.

    Pearl. (7)

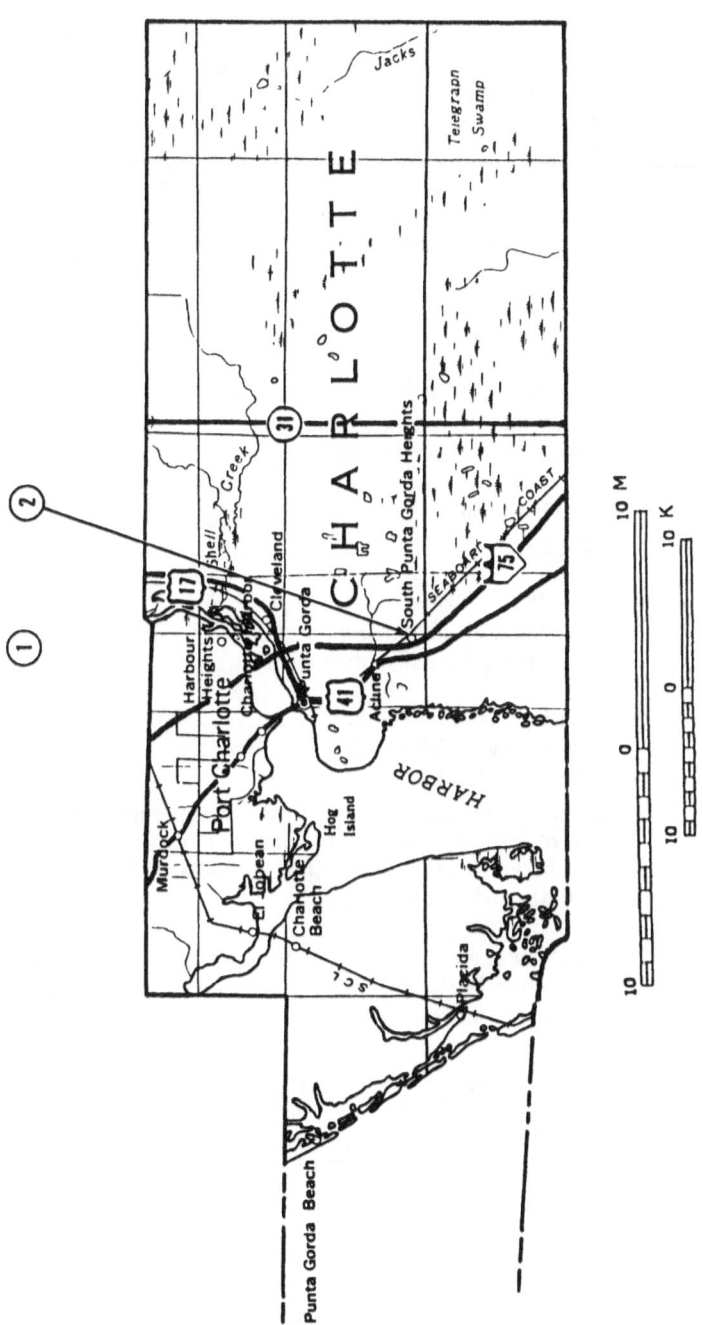

2. Punta Gorda area: 5 miles southeast; along east side of I-75; in fill/spoil mounds of Charlotte County Department of Transportation fill pits and those just north on the adjacent Wayne ranch.

> Calcite: fine honey-brown crystals; transparent; to 0.75″ long; some in stellate clusters; some prismatic. Fossils: marine invertebrates; some with transparent brown calcite replacement or fillings. (30-37)

CITRUS COUNTY

1. County-wide in excavations and in road and RR cuts.

> Chert: nodules; some gemmy material. Fossils, vertebrate: ivory (usually black). (4)

2. County-wide in lakes and streams.

> Pearl. (7)

3. County-wide in limonite quarries.

> Chert: nodules; some gemmy material. Fossils, vertebrate: ivory (usually black). (4)

4. Crystal River area: at the huge inactive limestone quarry now owned by GTE Satellite Systems Corporation, on Rock Crusher Road.

> Calcite: excellent crystals; white opaque; also transparent in clear, yellow, and honey. Chert. Fossils: marine invertebrates; often filled or replaced by calcite crystals. (5-9-limestone-12)

5. Crystal River area: 8 miles north on US-19 to just south of the Cross-Florida Barge Canal; 1 mile south of the Levy County border; 0.3 mile east of US-19; in quarry and spoil piles.

> Calcite: opaque white, transparent rich honey-colored and bright yellow; in crystals and masses. Fossils, marine: excellent quality; many filled or replaced by calcite crystals. (5-9-limestone-12)

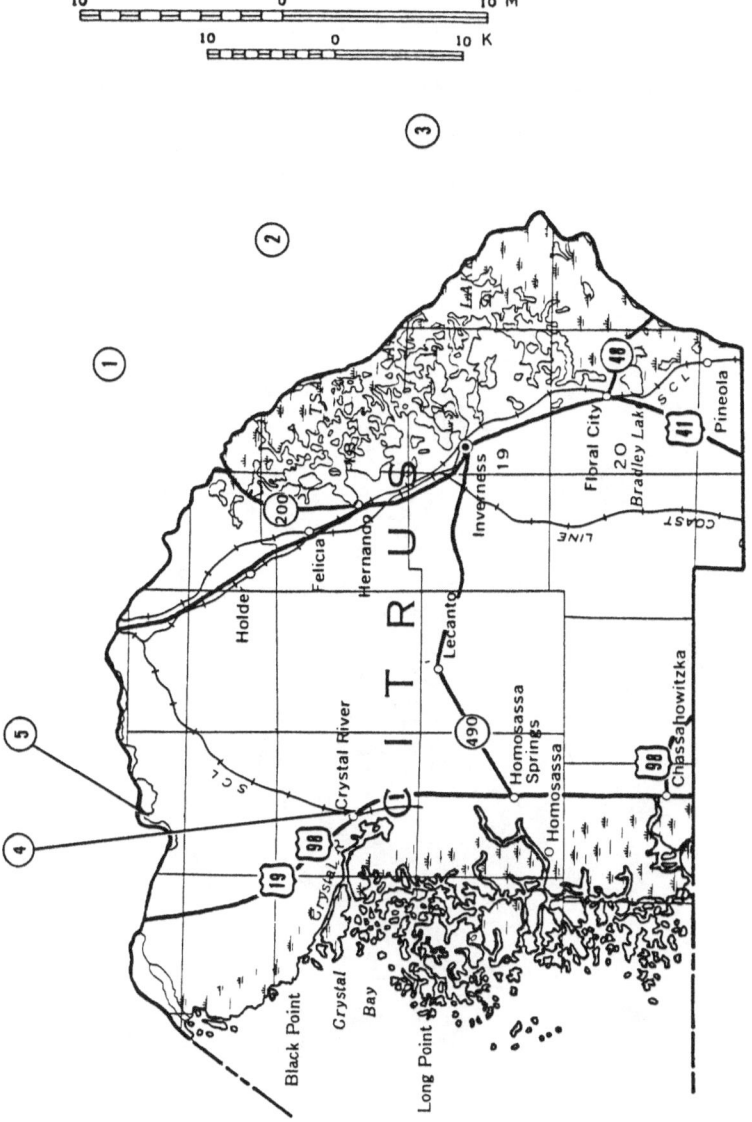

DADE COUNTY

1. Miami area: east at Municipal Auditorium and over bridge
   toward Port of Miami; immediate right turn at east foot of
   bridge and follow to water's edge; collect south from
   below bridge along shore and in spoil deposits.

   Chalcedony (aka "Miami Agate"): replacing coral.
   Fossils: coral; silicified. (20-30)

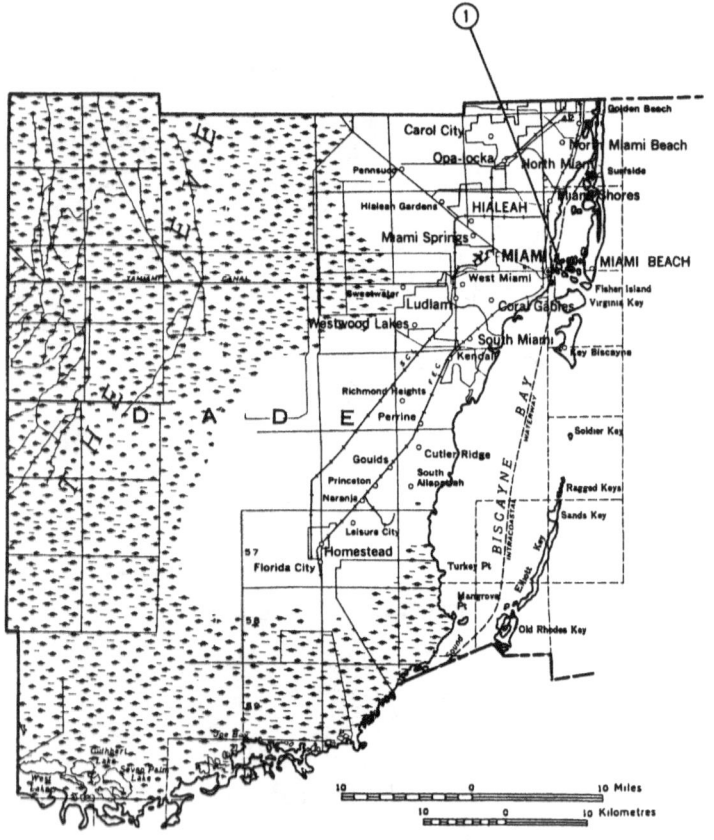

DUVAL COUNTY

1. South Jacksonville area; at the Skinner Mine.
   Ilmenite. Rutile. Zircon. (2)

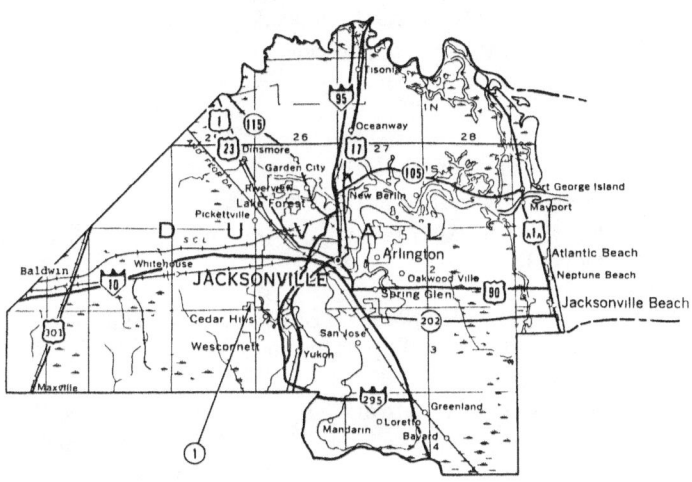

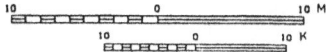

GADSDEN COUNTY

1. Havana area: north on US-27; at fuller's earth mines.

   Fossils: invertebrate and vertebrate; varied shells; bones; some ivory (usually black). Petrified wood: silicified. (2)

2. Hinson area: north toward the Georgia border; in a series of abandoned fuller's earth mines.

   Fossils: invertebrate and vertebrate; shells of various kinds; bones; some ivory (usually black). Petrified wood: silicified. (2)

3. Jamieson area fuller's earth mines.

   Fossils: invertebrate and vertebrate; numerous shells; bones; some ivory (usually black). Petrified wood: silicified. (2)

4. Quincy area: 8 miles north on SR-65; east of the highway and northeast of the Withalacoochee River; at a series of fuller's earth mines.

   Fossils: invertebrate and vertebrate; numerous shells; bones; some ivory (usually black). Petrified wood: silicified. (2)

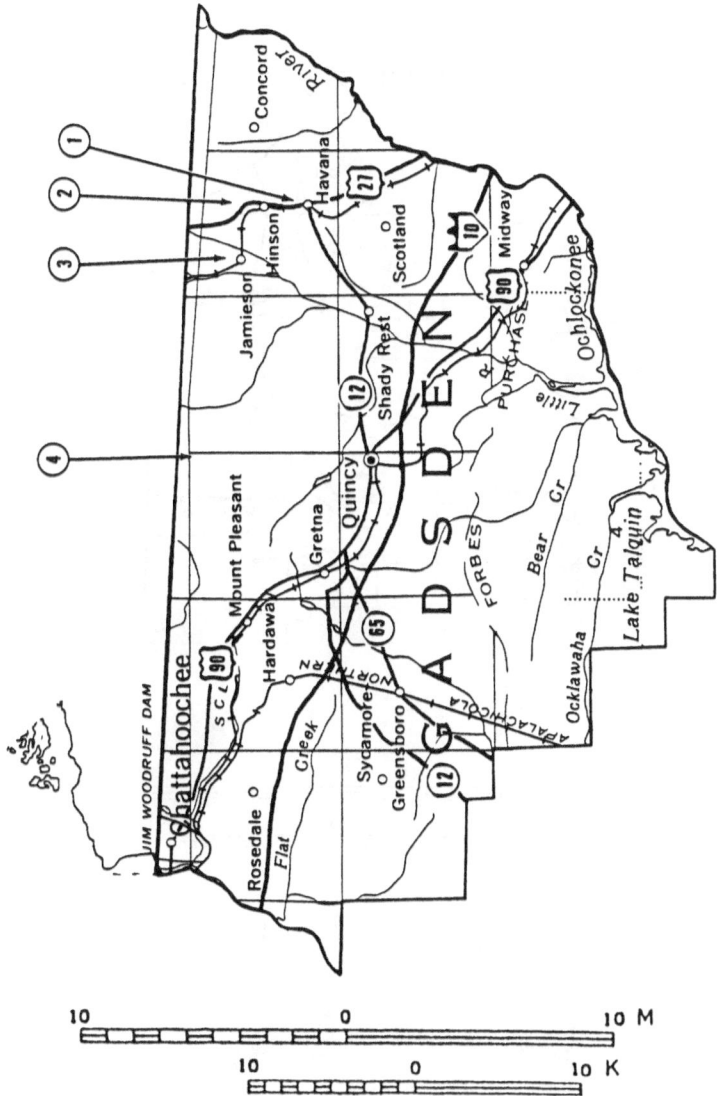

GILCHRIST COUNTY

1. County-wide in quarries.

> Chert. Fossils, marine: agatized corals; some geodes.
> Petrified wood: silicified. (12)

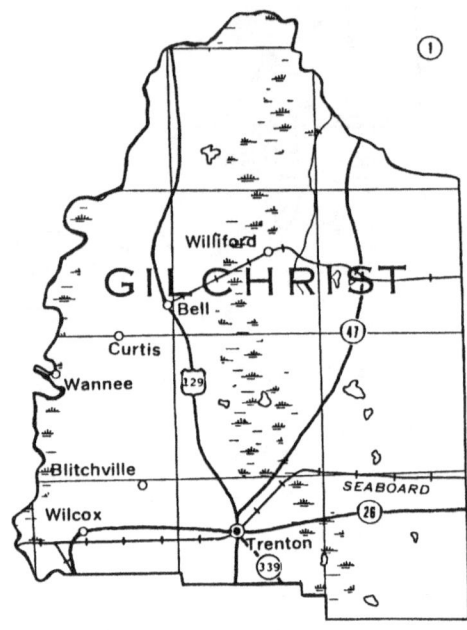

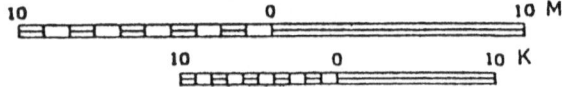

HAMILTON COUNTY

1. County-wide in lakes and streams.
   Pearl. (7)

2. Jasper area phosphate mines.
   Fossils, marine: coral geodes (Tampa Bay Agate var.).
   (2)

3. White Springs area phosphate mines.
   Fossils, marine: coral geodes (Tampa Bay Agate var.).
   (2)

4. White Springs area: from north gate of Stephen Foster
   Memorial on US-41 go 1.75 miles east on sand road to
   crossroad just after second wooden bridge; turn right; collect in creekbed.
   Agate. (7)

5. White Springs area: northeast and northwest on SR-135 to
   SR-6; collect downstream to Suwannee River on both sides
   of all tributaries crossed by SR-135.
   Fossils, marine: coral heads; silicified; some to 4' diameter. (14-20)

6. White Springs area: upstream along both banks of the
   Suwannee River for 15 miles.
   Fossils, marine: coral heads; silicified; some to 4' diam-
   eter. (14-20)

HERNANDO COUNTY

1. Brooksville area excavations and in road and RR cuts.
   Calcite: crystals; stalactitic; also as crystal fillings
   within fossil shells; honey-colored. Fossils, marine:
   echinoid geodes with crystal interiors.

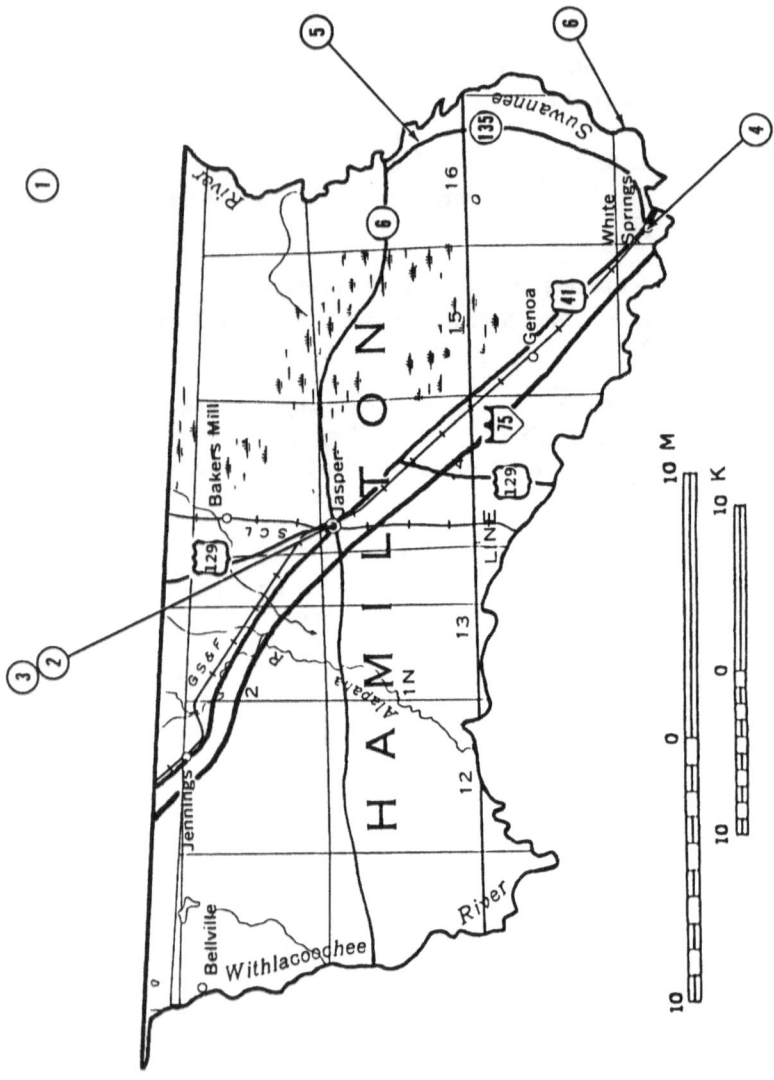

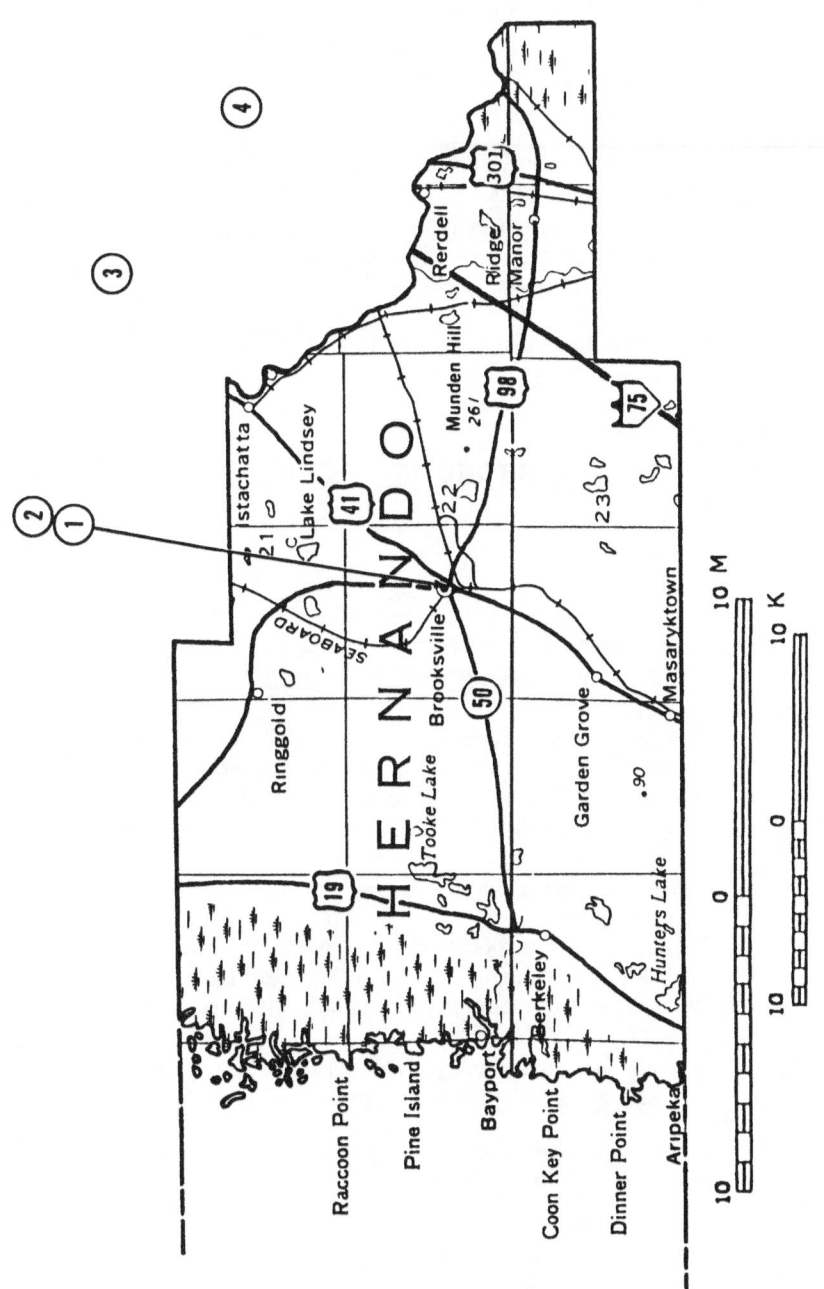

2. Brooksville area quarries.

> Calcite: crystals; honey-colored; stalactitic; also as fillings in fossil shells. Fossils, marine: echinoid geodes with crystal interiors.

3. County-wide in any sort of dredging or deep-digging construction operations.

> Calcite: transparent brown crystals. Fossils: marine invertebrates; many filled or replaced by transparent calcite crystals. (1-13-30)

4. County-wide in lakes and streams.

> Pearl. (7)

HILLSBOROUGH COUNTY

1. County-wide in lakes and streams.

> Pearl. (7)

2. Plant City area: 1 mile south on SR-39 to left turn on mine road; 2 miles east on the mine road.

> Gypsum: crystals. Vivianite: crystals, crusts. (2-38)

3. Tampa area: along Bayshore Boulevard and eastern section of Interbay Boulevard; collect on exposed tidal flats at low tide; probing and digging are most effective.

> Fossils, marine: coral geodes (Tampa Bay Agate var.); also shells filled with or replaced by chalcedony (agate, carnelian, sardonyx). (39-tidal)

4. Tampa area: Hillsborough River banks upstream along both shores from mouth; collecting best from shallow-draft boat at low tide; probing and digging are often productive.

> Fossils, marine: coral geodes (Tampa Bay Agate var.); also some silicified brain corals and finger corals.

5. Tampa area: in Hillsborough Bay; along shores of Davis Island; especially at Ballast Point.

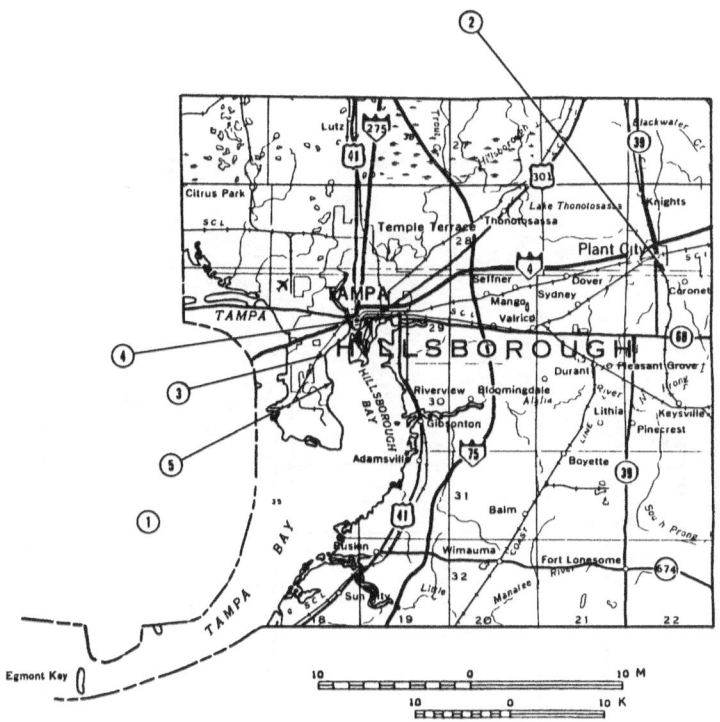

Chalcedony (aka Tampa Bay Agate): blue, pink, white, orange, brown; translucent; often as hollow coral linings and/or replacements. Fossils, marine: corals; silicified. (22-30-37)

INDIAN RIVER COUNTY

1. Vero Beach area: at the Florida Minerals Company Winter Beach Mine.

    Ilmenite. Rutile. Zircon. (2)

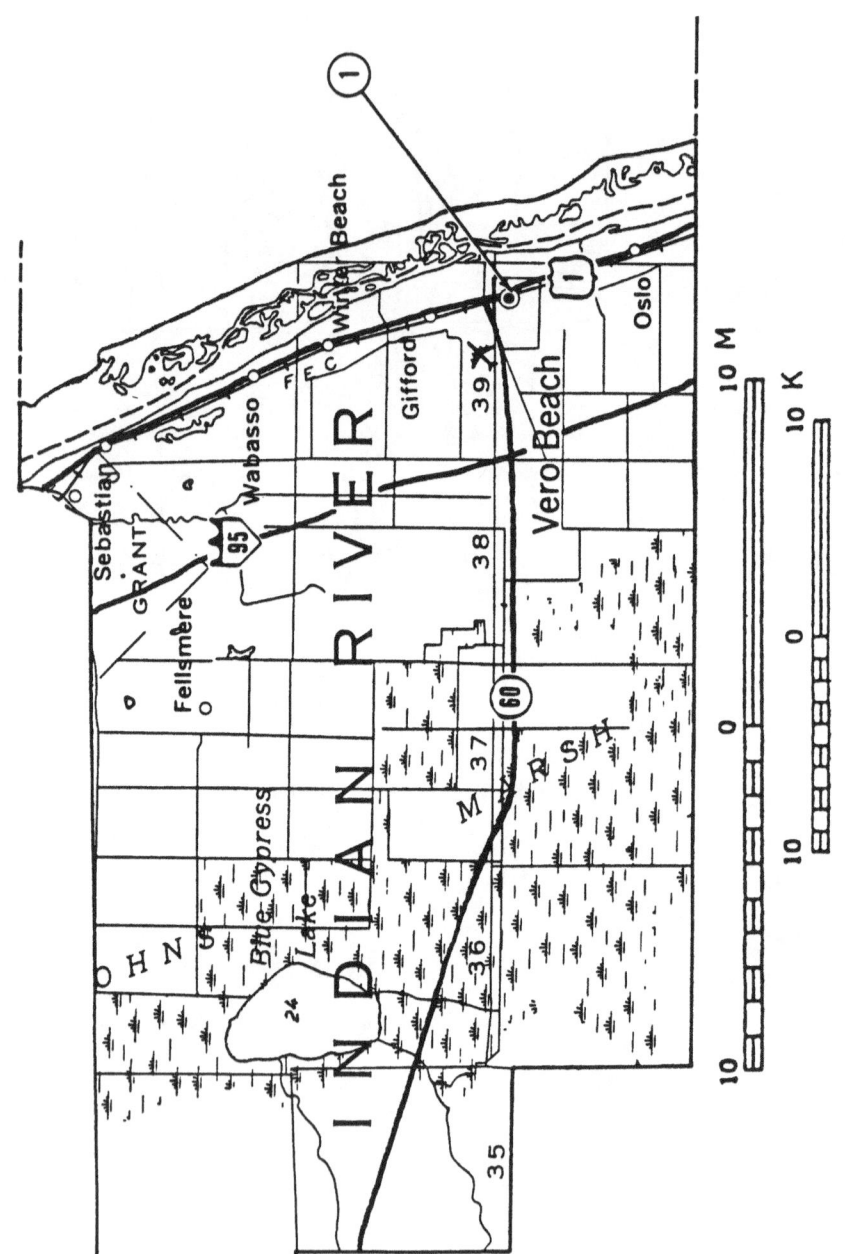

JACKSON COUNTY

1. Cottondale area streambanks.

     Chert, red: fossiliferous. (20)

2. County-wide in limestone quarries.

     Calcite: crystals; honey-colored; often as filling in or
     replacement of fossil shells. Fossils, marine: numer-
     ous varieties of shells, often filled with calcite. Lime-
     stone, fossiliferous: crystal-lined vugs. (9-limestone
     -12)

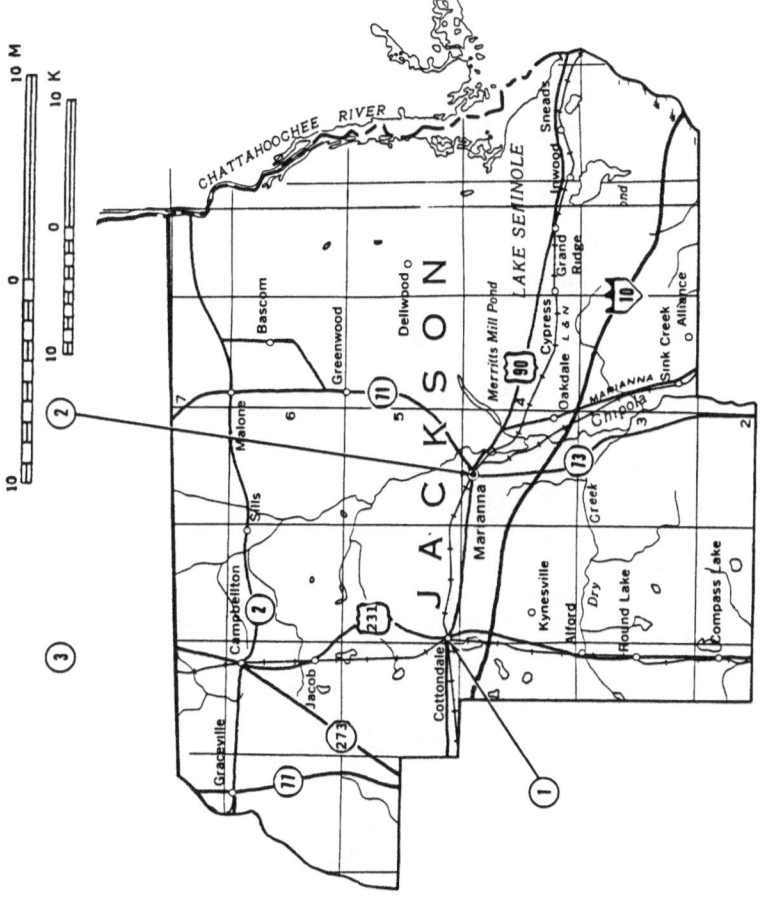

3. Marianna area limestone quarry.

    Limestone, fossiliferous: good polishing quality. (12)

LAFAYETTE COUNTY

1. County-wide in quarries.

    Chert. Fossils, marine: coral geodes (Tampa Bay Agate var.); other agatized shells. Petrified wood: silicified. (12)

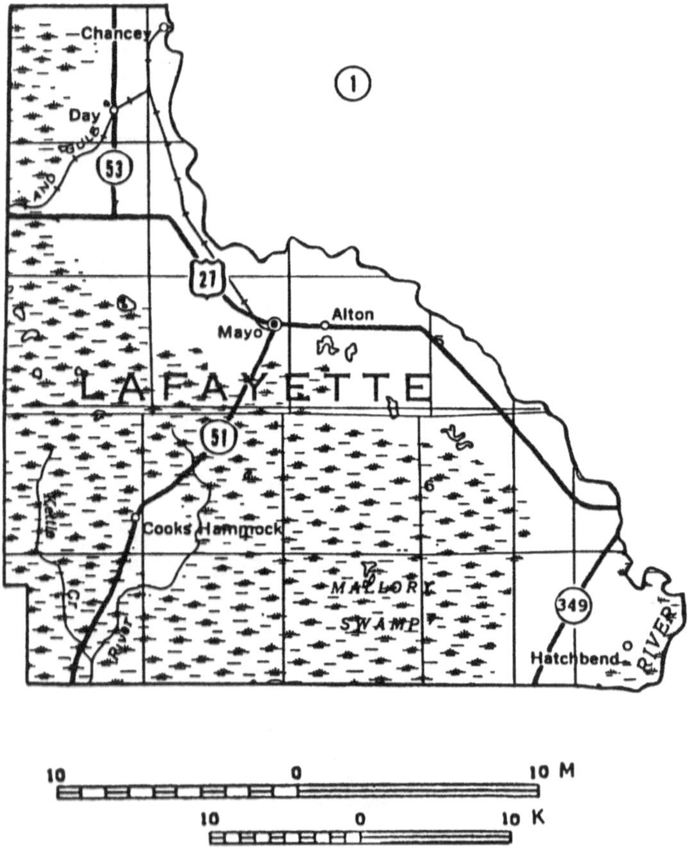

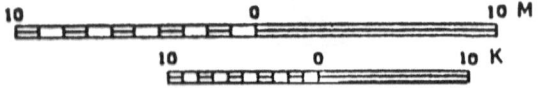

LEVY COUNTY

1. County-wide in quarries.

    Chert. Fossils, marine: coral geodes (Tampa Bay
    Agate var.); other agatized shells. Petrified wood: sili-
    cified. (12)

2. Gulf Hammock area: at the huge Gulf Hammock quarry.

    Fossils, marine: excellent bryozoans, gastropods,
    echinoids, corals, shark teeth; some silicification. (12)

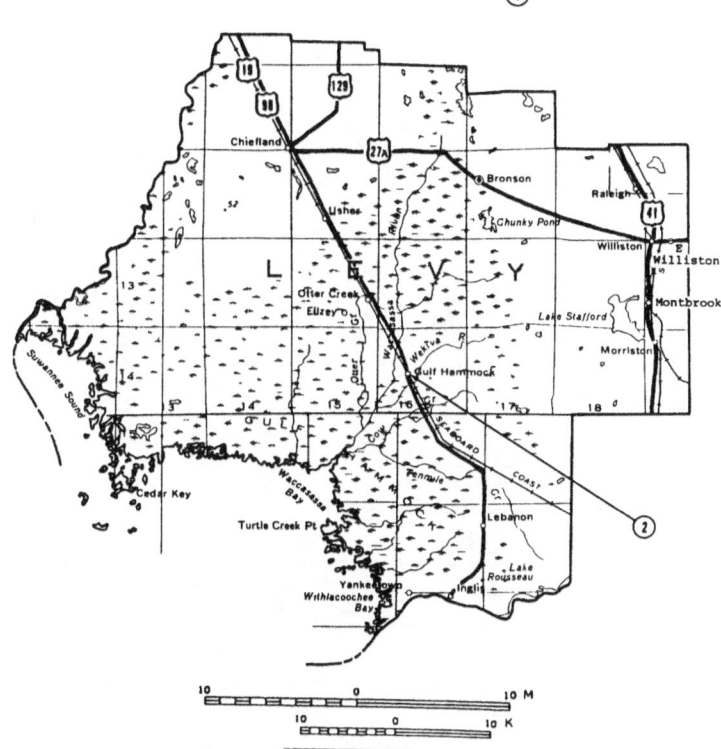

MARION COUNTY

1. County-wide in excavations and in road and RR cuts.

    Chert: good quality. Fossils, vertebrate: ivory (usually
    black). (4)

2. County-wide in limestone quarries.
   Chert: concretions; boulders. (12)

PASCO COUNTY

1. County-wide in lakes and streams.
   Pearl. (7)

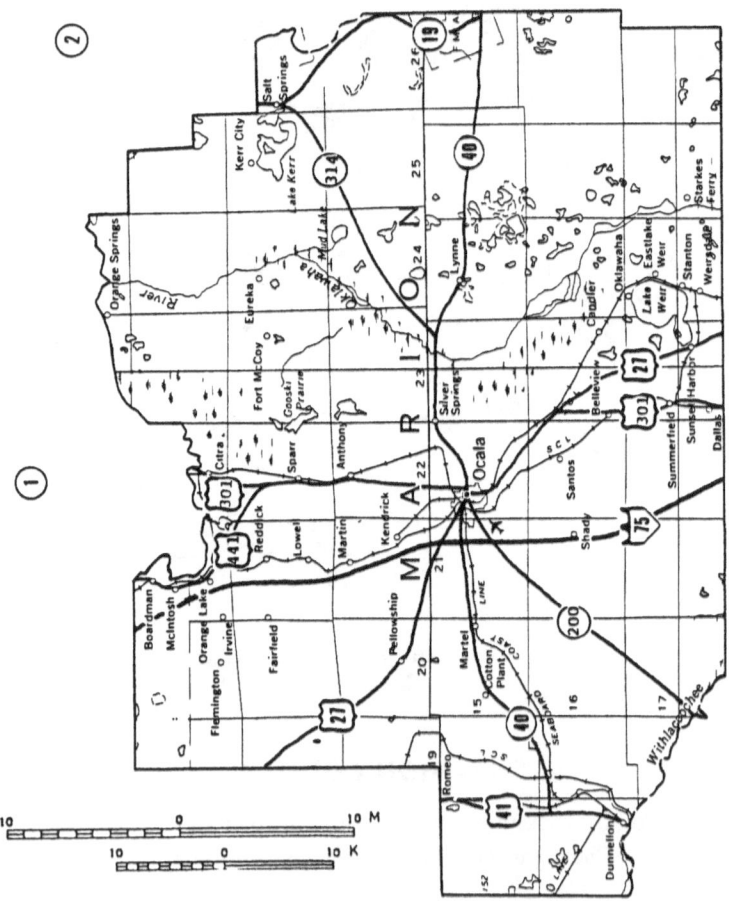

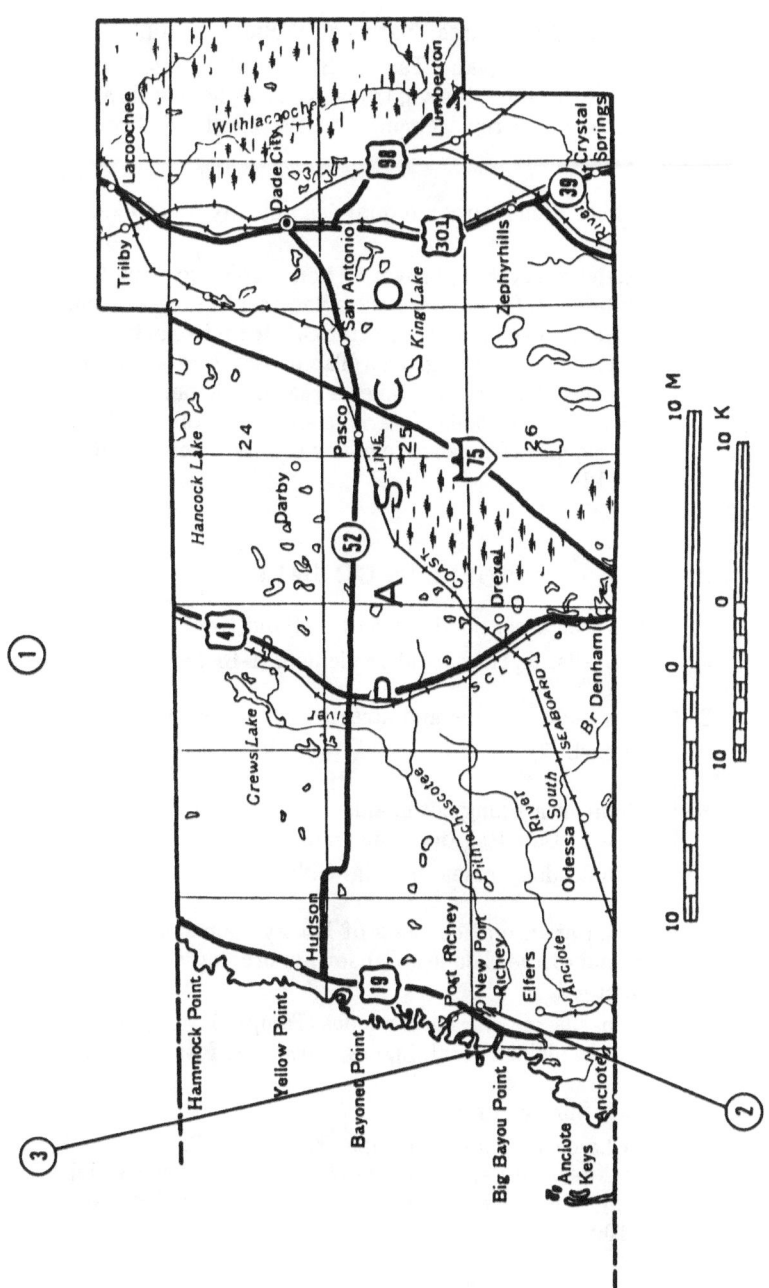

2. New Port Richey area: just west at Bailey's Bluff.

>   Chalcedony (aka "Tampa Bay Agate"): blue, pink, white, orange, brown; translucent; as hollow coral linings and/or replacements. Fossils: coral; silicified. (22-30-37)

3. New Port Richey area: in quarries and any other excavations.

>   Calcite: crystals; brown to honey-colored; usually as fillings in fossil shells or replacements thereof. Chalcedony: pseudomorphs after calcite and selenite. Fossils, marine: coral geodes (Tampa Bay Agate var.); not quality material but takes a reasonable polish; pastel colors; also, shells of various species, many lined with or replaced by brown to honey-colored calcite crystals. (4-12)

PINELLAS COUNTY

1. Clearwater area: northern half of Sand Key.

>   Fossils, marine: corals, silicified. (1-13-20-26)

2. County-wide in lakes and streams.

>   Pearl. (7)

3. Dunedin area: along both banks of the Caladesi Causeway leading to Caladesi and Honeymoon Islands.

>   Chalcedony roses: midnight blue. (37)

4. Dunedin area: on tidal flats of Honeymoon Island at the west end of the Caladesi Causeway; probing is a successful method.

>   Fossils, marine: coral geodes (Tampa Bay Agate var.); also silicified coral fingers; variety of fossil shells. (7)

5. Tarpon Springs area.

>   Chalcedony (aka "Tampa Bay Agate"): blue, pink, white, orange, brown; translucent; as hollow coral linings and/or replacements. Fossils: coral; silicified; to 100 lb. (22-30-37)

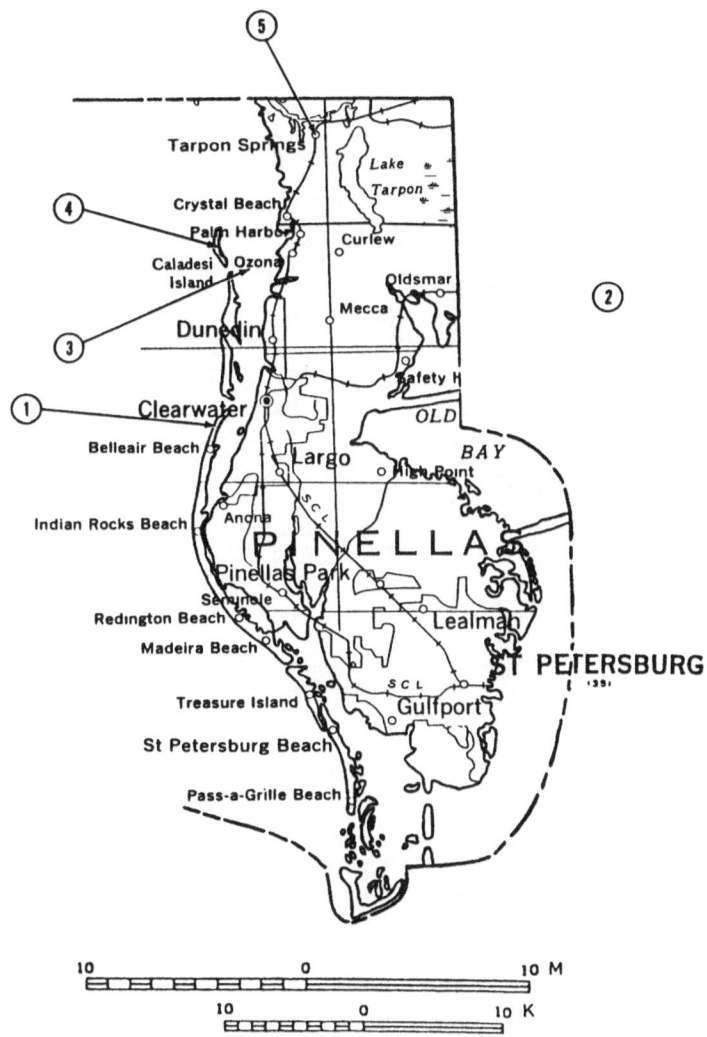

POLK COUNTY

1. Bartow area: at southwest edge of city; at phosphate mine.
    Gypsum: crystals. Vivianite. (2)

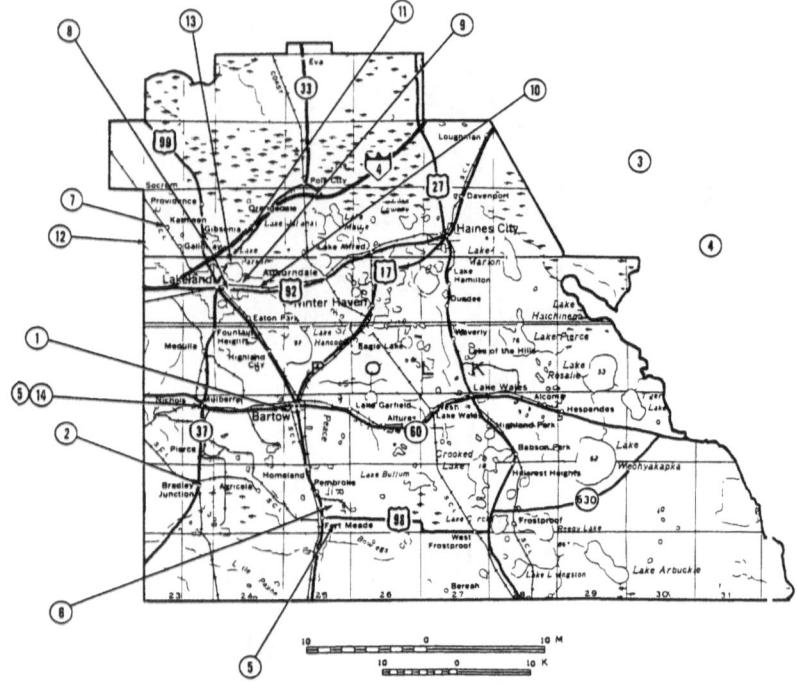

2. Bradley area: at phosphate mines in the Lake Hookers Prairie area.

> Fossils, vertebrate: bones (often silicified); mastodon teeth (usually fragments); ivory (usually black); shark teeth. Gypsum: crystals. Petrified wood: some silicified.

3. County-wide in lakes and streams.

> Pearl. (7)

4. County-wide in phosphate digging operations (quarries, pits, mines, excavations, drillings, dredgings).

> Fossils, vertebrate: bones (often silicified); mastodon teeth (usually fragments); ivory (usually black); shark teeth. Gypsum: crystals. Petrified wood: some silicified. Vivianite. (2)

5. Fort Meade area: 1.5 miles east on SR-630; on north side of road; at phosphate mine.

   Fossils, vertebrate: bones (often silicified); mastodon teeth (usually fragments); ivory (usually black); shark teeth. Gypsum; crystals. (2)

6. Fort Meade area: just north-northeast; between US-17 and the Peace River; at phosphate mine.

   Gypsum; crystals. Vivianite. (2)

7. Kathleen area: on Wear ranch; especially in areas of new diggings of irrigation and drainage canals.

   Chalcedony (aka "Tampa Bay Agate"): blue, pink, white, orange, brown; translucent; as hollow coral linings and/or replacements. Fossils: coral; silicified. (30)

8. Lakeland area: 0.75 mile east on phosphate mine road; then 1 mile south.

   Gypsum: crystals. Vivianite: crystals and crusts. (1-13-30-38)

9. Lakeland area: 1.3 miles east to SR-33; then 6 miles north to area around the I-4 interchange.

   Chalcedony. Fossils, marine: coral; some silicified. (1-4-13-30)

10. Lakeland area: 2 miles east to Saddle Creek; at phosphate mine.

    Fossils, vertebrate: bones (often silicified); mastodon teeth (usually fragments); ivory (usually black); shark teeth. Gypsum: crystals. Petrified wood: some silicified. (2)

11. Lakeland area: 2 miles northeast of Lake Parker; at phosphate mine.

    Fossils, vertebrate: bones (often silicified); mastodon teeth (usually fragments); ivory (usually black); shark teeth. Gypsum: crystals. Petrified wood (some silicified).

12. Lakeland area: 7 miles northwest; on canal near Kathleen.

Chalcedony (aka "Tampa Bay Agate"): blue, pink, white, orange, brown; translucent; as hollow coral linings and/or replacements. Fossils: coral; silicified. (30)

13. Lakeland area: on Lake Parker shoreline.

Chalcedony (aka "Tampa Bay Agate"); blue, pink, white, orange, brown; translucent; as hollow coral linings and/or replacements; nodules and stems; usually less than 5". Fossils: coral; silicified. (20)

14. Mulberry area: east on SR-60 to series of phosphate mines.

Fossils, vertebrate: bones (some silicified); mastodon teeth (usually fragments); ivory (usually black); shark teeth. Petrified wood: some silicified. (2)

15. Mulberry area phosphate mines.

Gypsum: crystals. Vivianite. (2)

ST. JOHNS COUNTY

1. County-wide along ocean beaches.

Calcite, fossiliferous (aka "Coquina Rock"); coquina shells cemented together with calcite. (20)

2. County-wide in quarries.

Calcite, fossiliferous (aka "Coquina Rock"); coquina shells cemented together with calcite. (12)

SARASOTA COUNTY

1. Manasota Key area: along the beach (especially at falling tide to low tide) from Charlotte County border north to Venice Beach.

Fossils, invertebrate and vertebrate: various shells; shark teeth, abundant; antler; bone; ivory; mammal teeth (horse, camel, sloth, mastodon, etc.). (20)

2. Venice area: along both shorelines of the Intracoastal Waterway cut from old US-41 (near Ringling Brothers' winter quarters) west-southwest to mouth at Lemon Bay; probing

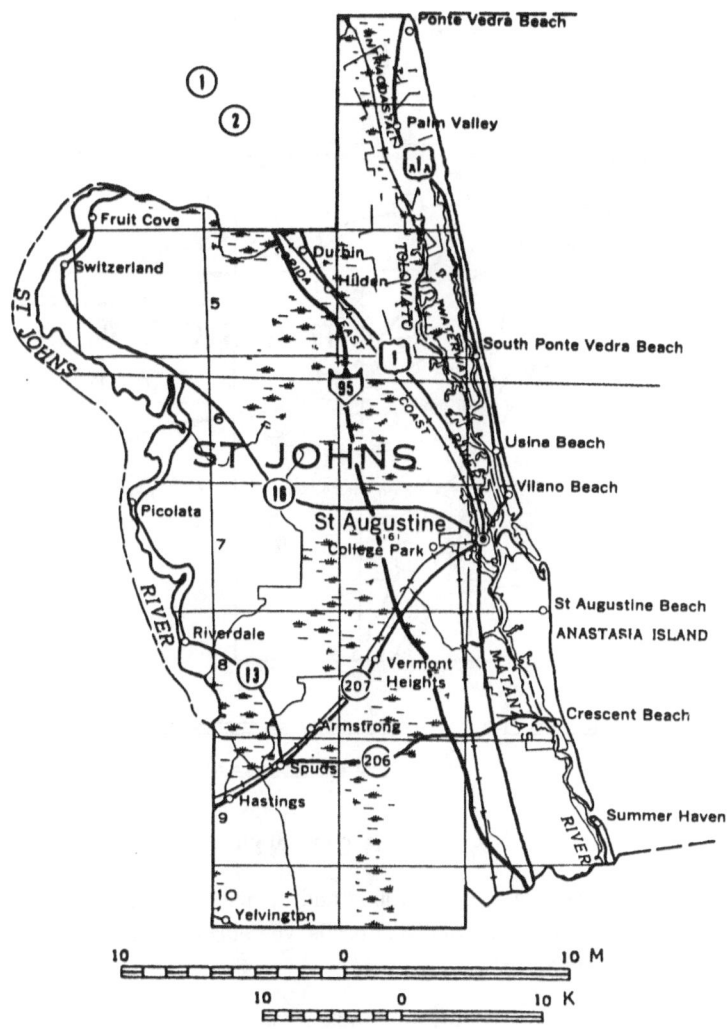

spoil banks is sometimes successful.

Fossils, invertebrate and vertebrate: various shells; shark teeth (very fine; abundant; often not at all waterworn and therefore well defined; some to 5″ high by 4″ wide; light tan through gray to black); antler; bone; ivory; mammal teeth (horse, camel, sloth, elephant, etc.). (20-30)

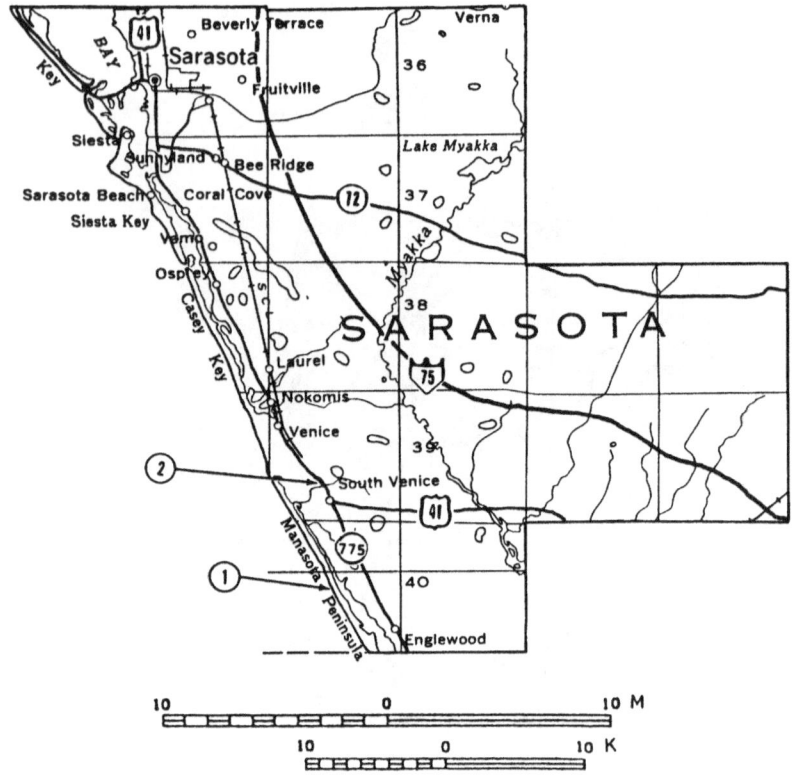

SUMTER COUNTY

1. County-wide in excavations and in road and RR cuts.
   Chert: good quality. Fossils, vertebrate: ivory (usually black). (4)

2. County-wide in limestone quarries.
   Chert: concretions; boulders. (12)

SUWANEE COUNTY

1. Branford area: 2.8 miles east on US-129; quarry on left.
   Fossils, marine: numerous species; many silicified; some shark teeth. (12)

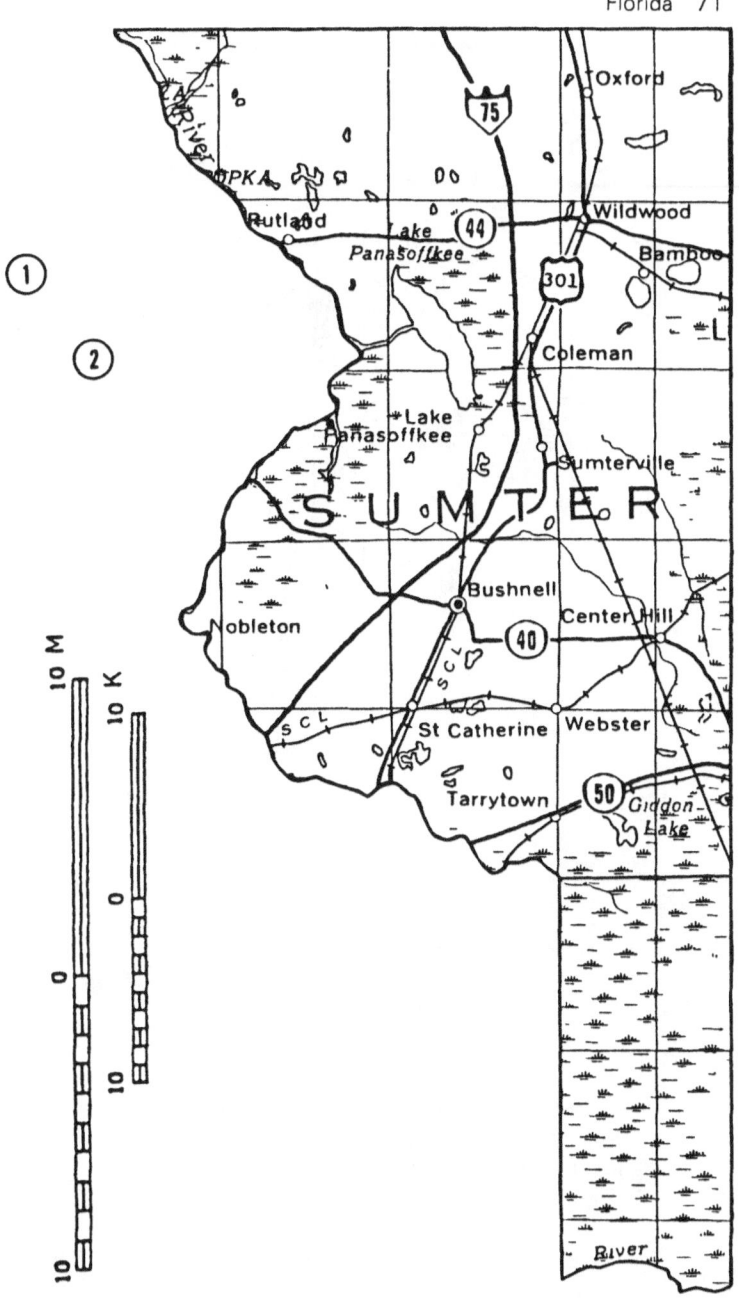

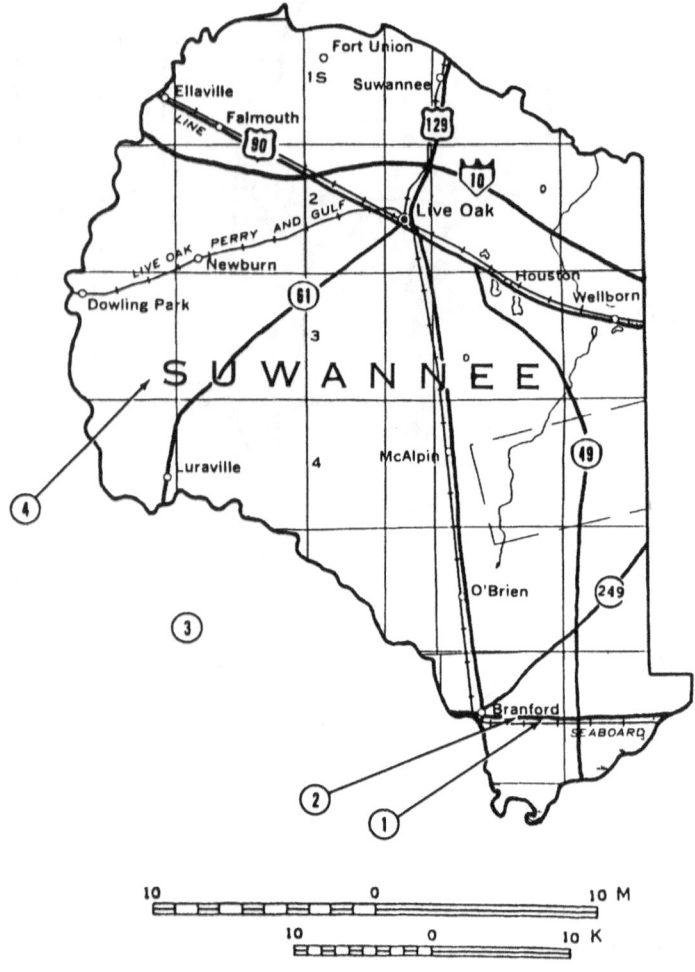

2. Branford area: 1.6 miles east on US-129; quarry on left.
   Fossils, marine: numerous species; many silicified;
   some shark teeth. (12)

3. County-wide in quarries.
   Chert. Fossils, marine: some coral geodes (Tampa Bay
   Agate var.); coral fingers (some silicified). (12)

4. Dowling Park area: 5 miles southeast on SR-252; at mine. Malachite. (2)

TAYLOR COUNTY

1. County-wide in lakes and streams.
   Pearl. (7)

2. Perry area: 20 miles west on US-98 to 0.25 mile past Econfina River bridge; 0.5 mile south on SR-14; then east to parking area; dig there and close to river.
   Fossils: coral; silicified. (6-30)

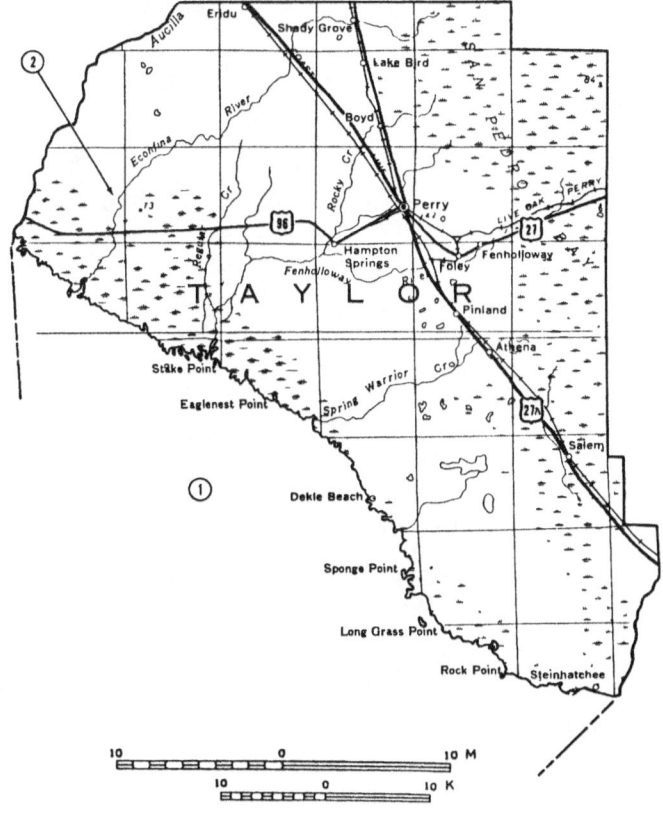

VOLUSIA COUNTY

1. Seville area: 2 miles south in several South Coastal RR cuts.

    Limonite. (4)

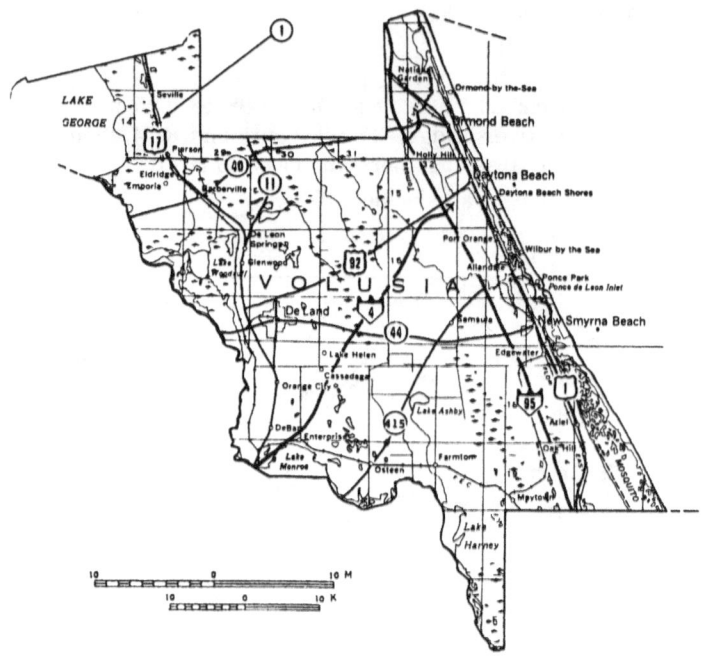

WASHINGTON COUNTY

1. Chipley area: 6 miles southwest on SR-77 to 1.5 miles beyond the Falling Waters Recreational Area; collect along both sides of road.

    Chert, pale blue. (9-19-29)

2. Chipley area.

    Calcite: transparent brown crystals to 1″ long; some in stellate clusters. Chert. (1-4-13)

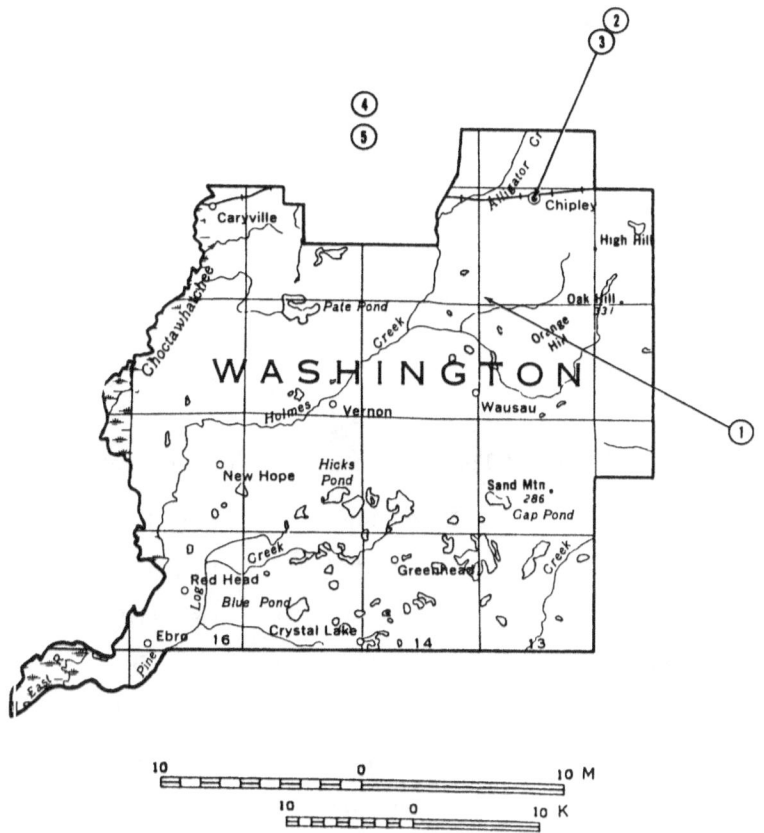

3. Chipley area: in RR cut just south of Chipley air strip.
     Calcite: blocky masses; cream-colored. Fossils: marine; abundant. (4)

4. County-wide in lakes and streams.
     Pearl. (7)

5. County-wide in road and stream cuts.
     Chert: common. Flint: good quality. (4)

GEORGIA

Statewide total of 174 locations in the following counties:

Baldwin (2)	Dodge (4)	Hart (1)
Banks (2)	Dougherty (1)	Heard (1)
Barrow (1)	Elbert (4)	Henry (1)
Bartow (4)	Fannin (7)	Irwin (1)
Bibb (2)	Fayette (1)	Jackson (1)
Burke (2)	Floyd (3)	Jasper (2)
Carroll (3)	Forsyth (1)	Jones (1)
Chattooga (3)	Franklin (1)	Lamar (2)
Cherokee (12)	Fulton (1)	Lee (1)
Clarke (1)	Gilmer (1)	Lincoln (1)
Clayton (1)	Gordon (1)	Lowndes (1)
Cobb (3)	Gwinnett (2)	Lumpkin (2)
Crisp (1)	Habersham (6)	McDuffie (6)
Dade (1)	Hall (4)	Meriwether (2)
Dawson (1)	Hancock (1)	Monroe (4)
De Kalb (2)	Haralson (1)	Morgan (4)

Murray (4)

Muscogee (2)

Oconee (1)

Paulding (3)

Pickens (6)

Polk (1)

Rabun (9)

Rockdale (1)

Screven (1)

Spalding (2)

Stephens (2)

Towns (4)

Troup (5)

Twiggs (1)

Union (4)

Upson (9)

Walton (1)

Washington (1)

White (4)

Whitfield (1)

Wilkes (1)

Wilkinson (2)

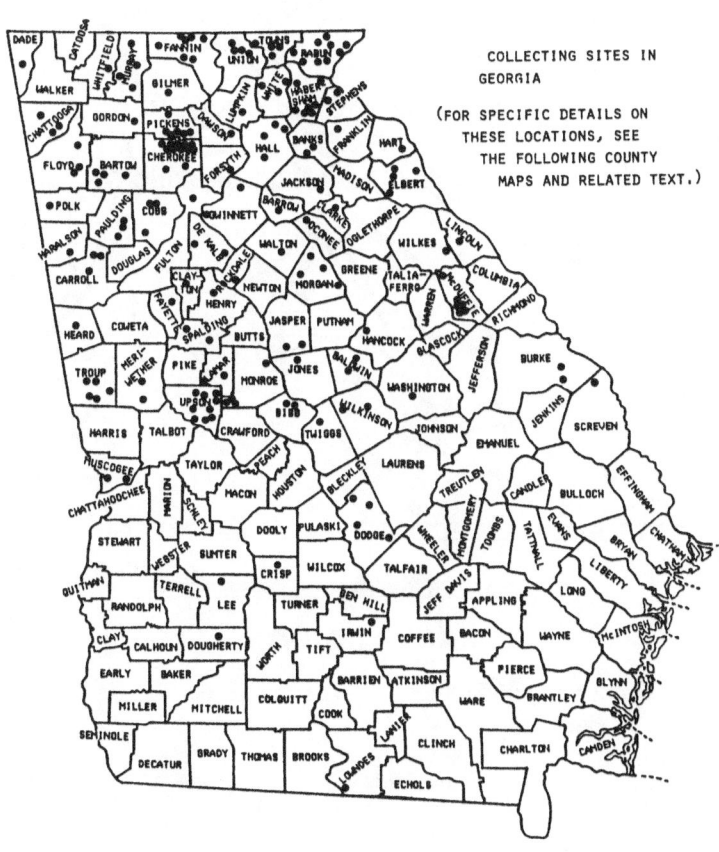

COLLECTING SITES IN
GEORGIA

(FOR SPECIFIC DETAILS ON
THESE LOCATIONS, SEE
THE FOLLOWING COUNTY
MAPS AND RELATED TEXT.)

BALDWIN COUNTY

1. County-wide in lakes (especially Lake Sinclair) and streams.
    Pearl. (7)

2. Milledgeville area: at State farm.
    Jasper. (1)

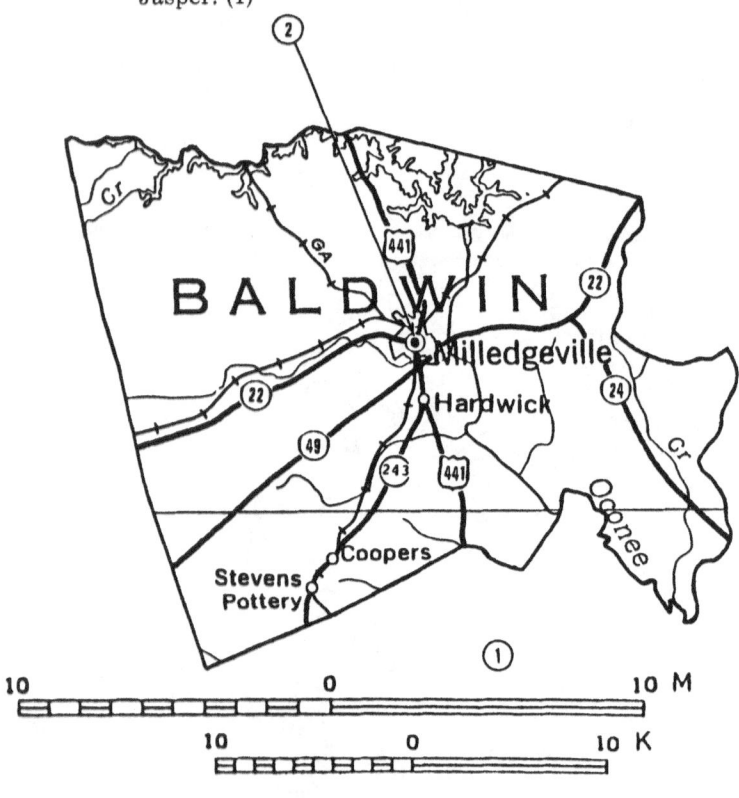

BANKS COUNTY

1. Homer area: 9.5 miles southeast on SR-59.
    Beryl. (1)

2. Maysville area (Jackson County): along RR just north of Jackson County border.
    Quartz, rock. (4)

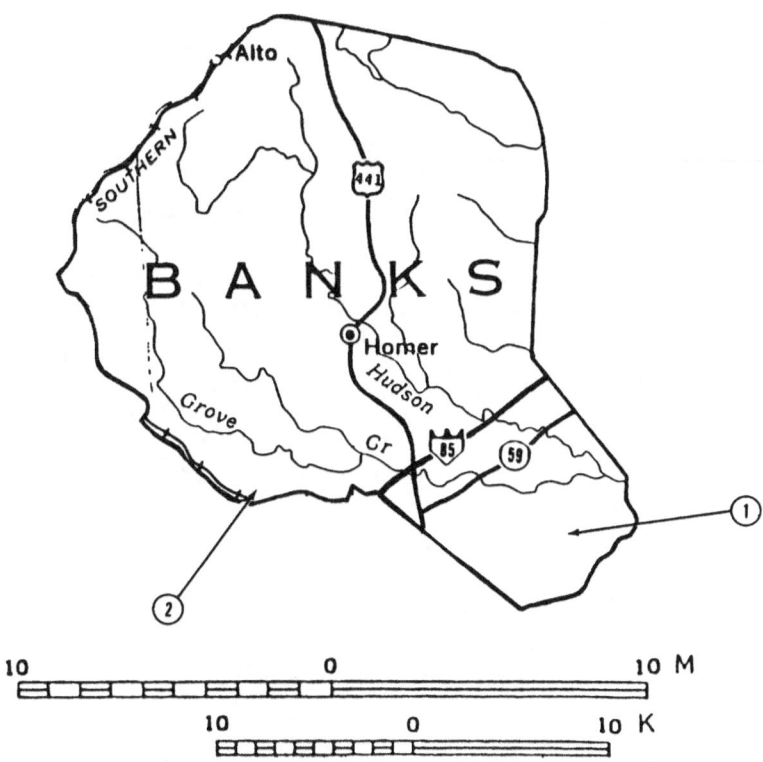

BARROW COUNTY

1. Winder area: northeast on SR-11 to 1.5 miles from the Jackson County border.

    Beryl. Tourmaline, black. (24-hornblende -29-34)

BARTOW COUNTY

1. Cartersville area.

    Manganite. Pyrolusite: crystal-lined geodes. (1-7-13-20)

2. Euharlee area: downstream on the Etowah River.

    Agate, banded: red, yellow, brown, white. (7-20)

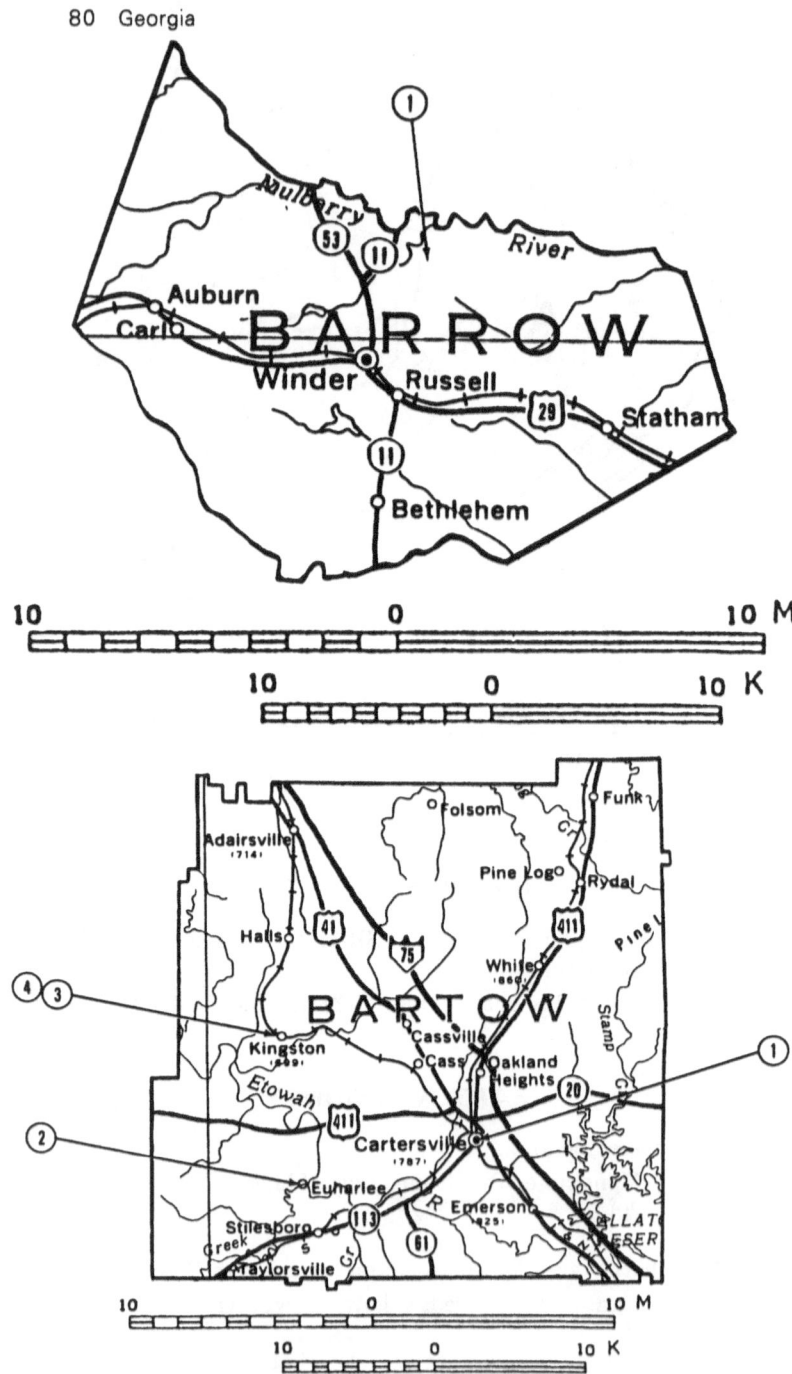

3. Kingston area: south along the Etowah River.
    Agate: brown, red. (20)

4. Kingston area: south; area of Saltpeter Cave.
    Jasper. (1-3)

BIBB COUNTY

1. Macon area: at junction north of Calloway Airport.
    Beryl. (1)

2. Macon area: in the Holton Quarry.
    Agate. (12)

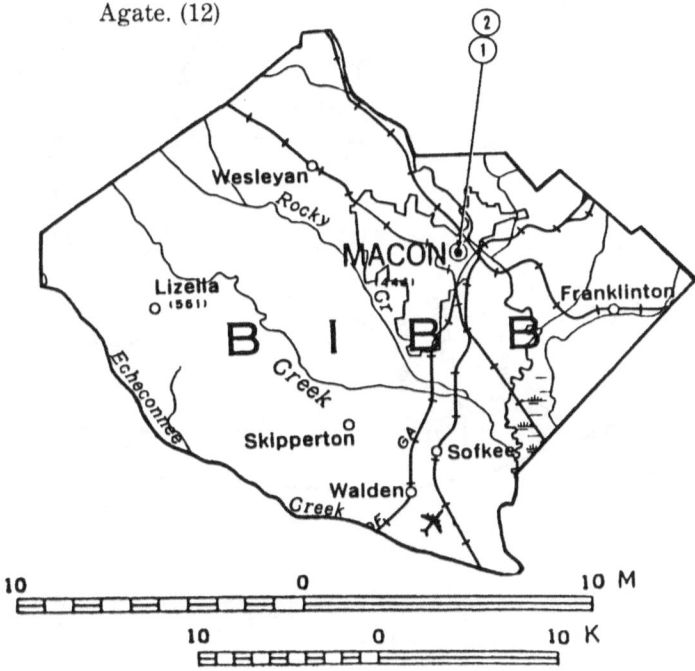

BURKE COUNTY

1. Girard area.
    Agate. (3-26)

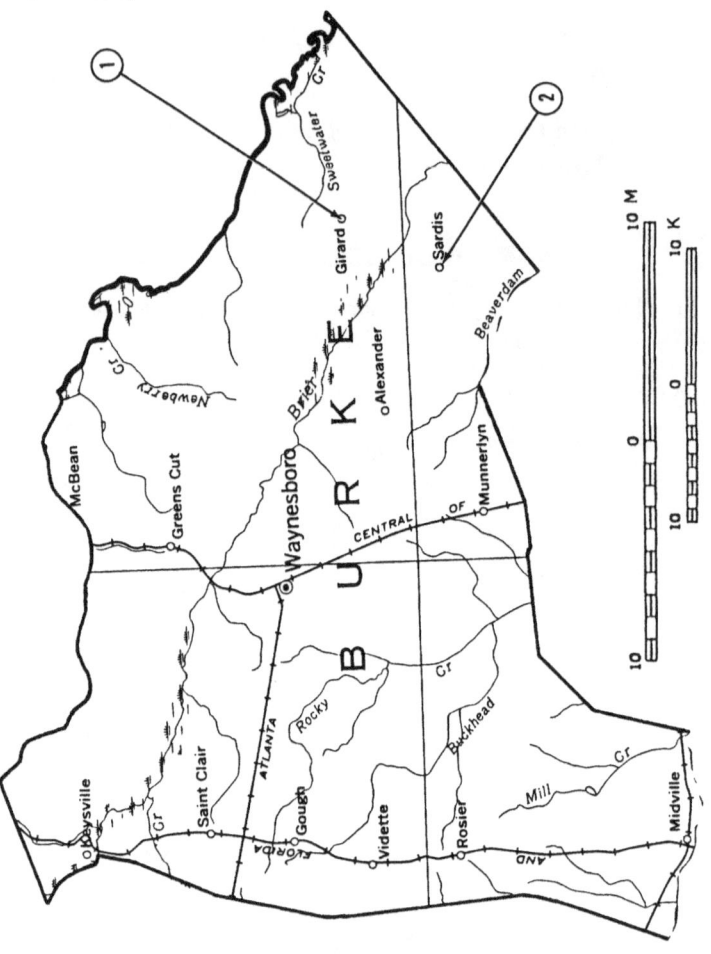

2. Sardis area: in cuts of road leading to Spring Place.
   Oolite: small. (4-9-14-41)

CARROLL COUNTY

1. Carrollton area: upstream and down along both sides of
   the Tallapoosa River.
   Ruby. (3-7-18-20)

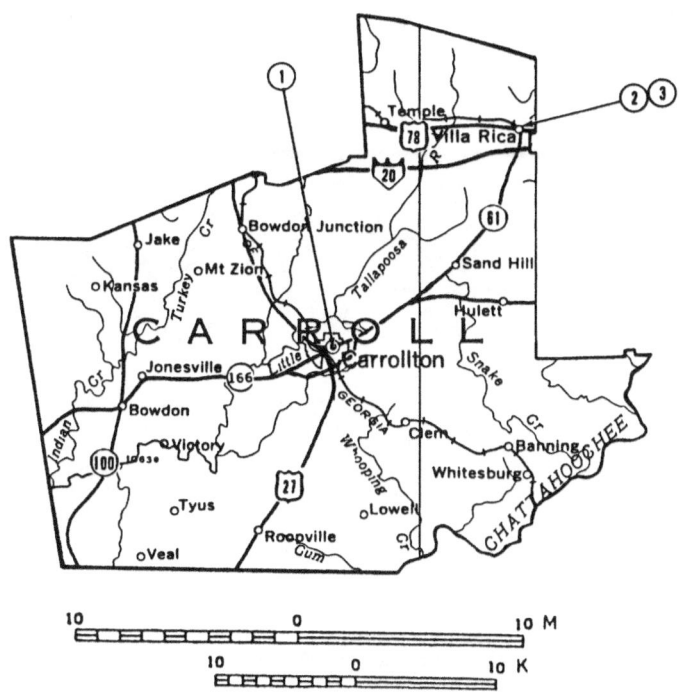

2. Villa Rica area.
   Pyrite. (1-13-14-29)

3. Villa Rica area: on the slopes of Reids Mountain.
   Pyrite: good crystals. (1-13-14-29)

CHATTOOGA COUNTY

1. Summerville area: 3 miles east-southeast on US-27 to two
   stores beyond top of first ridge; turn west on road between
   the two stores; bear left for 2.9 miles; cross bridge and
   immediately turn left through gate; follow right trail 150
   yards; in branch and beyond.
   Chert: agatized; abundant. Serpentine. (1-7-13)

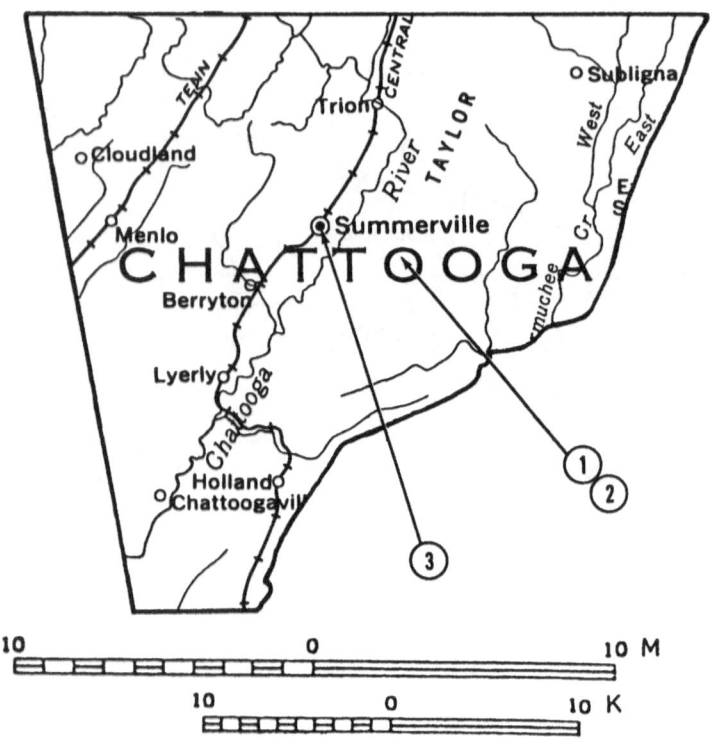

2. Summerville area: 3 miles east-southeast on US-27; go
   over ridge crest and collect on east slope.

   Agate. Chert: agatized; abundant. Serpentine. (1-13-
   29)

3. Summerville area: west on SR-48 to Fish Hatchery sign;
   turn right; go north to sign for the Perennial Springs Baptist Church; turn right; follow narrow road 0.5 mile; collect
   on road, sides of road, and adjacent slopes.

   Agate. Chert: agatized; abundant. (1-13)

CHEROKEE COUNTY

1. Ball Ground area: on the Oscar Robertson farm.
   Staurolite. (1-5-13-27)

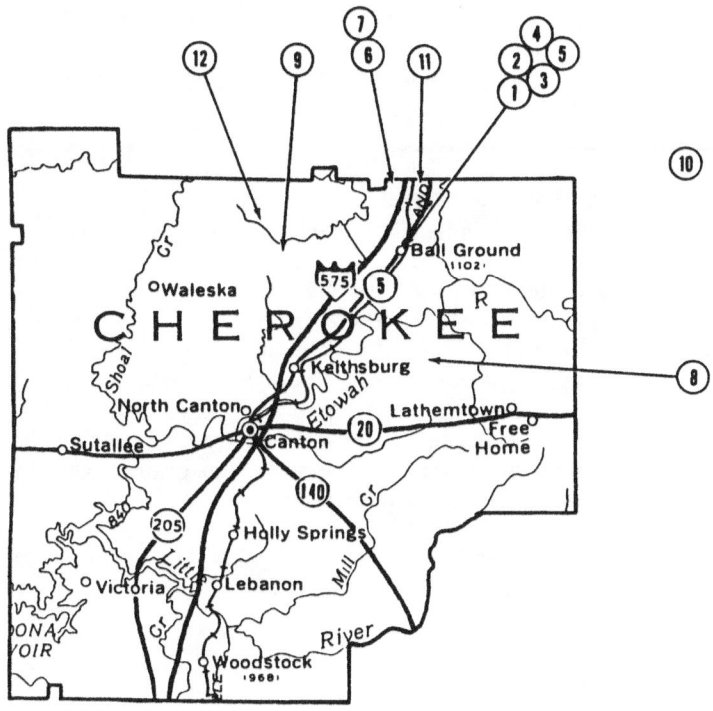

2. Ball Ground area: prospects along SR-5 toward Canton.
   Garnet, almandine: deep red. (21)

3. Ball Ground area: south and west.
   Staurolite: abundant. (1-3-27)

4. Ball Ground area: west on Fairview Church Road; collect
   along Sharp Mountain Creek.
   Staurolite. (1-7)

5. Ball Ground area: west on Fairview Church Road to 1 mile
   west of the Fairview Church; collect along east side of
   Sharp Mountain Creek and upstream in tributaries entering from the east.
   Staurolite: fine crystals. (1-3-27)

6. Ball Ground area: 2.5 miles north, 78 degrees east; at the Cochrane Mine.

> Aquamarine. Beryl. Mica. Quartz, rock: rutilated. (2-19)

7. Ball Ground area: 2.6 miles north; 2 miles west of Centerville; at several prospects and mines.

> Beryl: abundant. Muscovite. Tourmaline. (2-21)

8. Ball Ground area: 4.3 miles south, 86 degrees east; 0.4 mile south and 30 degrees east of Conns Creek Church; at the Amphlett Mica Mine (aka Franklin Mine).

> Apatite. Autinite: small flakes. Beryl: pale chartreuse; crystals to 3″ long. Biotite. Calcite. Chalcopyrite. Columbite: small grains. Garnet. Gummite. Malachite. Mica: good crystals. Muscovite: fine-grained. Quartz, milky: massive. Tourmaline. (2-16)

9. Ball Ground area: 4.5 miles west on CR-1; on north side of Bluff Creek; on the J. M. Spear farm.

> Staurolite: fine; abundant. (1-3-27)

10. County-wide.

> Staurolite. (1-3-27)

11. Nelson area (Pickens County): 1.5 miles south of Bethany Church; at the Bennett Mine (aka the Mica Field Mine).

> Beryl: huge crystals (one weighed 60 lb). Feldspar: kaolinized. Quartz, milky: massive. (2)

12. Tate area (Pickens County): 4.8 miles south-southwest; at the Cagle Mine (aka Dunsmore Mine).

> Garnet: crystals to 2″ diameter. Quartz, rock: fine; clear. Rutile: crystals to 1.5″ thick. (2)

CLARKE COUNTY

1. Athens area: follow Alps Road to fields across from airport.

> Beryl: green. (6)

CLAYTON COUNTY

1. Forest Park area: 4.4 miles northwest; on the Daniel Light farm.

    Diamond: 1 stone; 4.5 ct; yellow; 0.375″ × 0.375″ × 0.3125″; largest Georgia diamond; found in 1887. (1)

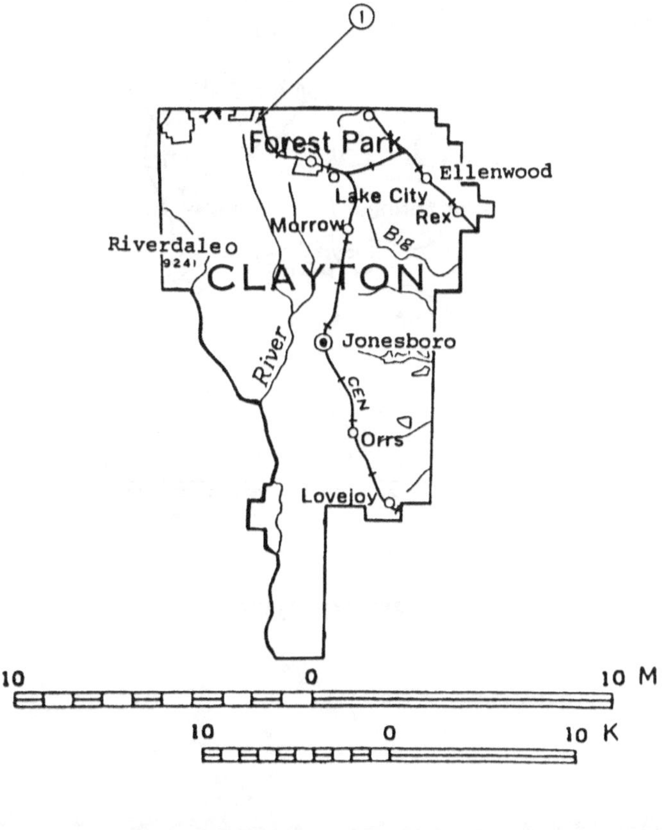

COBB COUNTY

1. Belmont Hills area: 0.25 mile southeast.
    Quartz, rose. Quartz, milky. (19-large)

2. Clarkdale area: 1.2 miles south; on the Turner property.
    Corundum. (1)

3. Marietta area.
    Agate. (7)

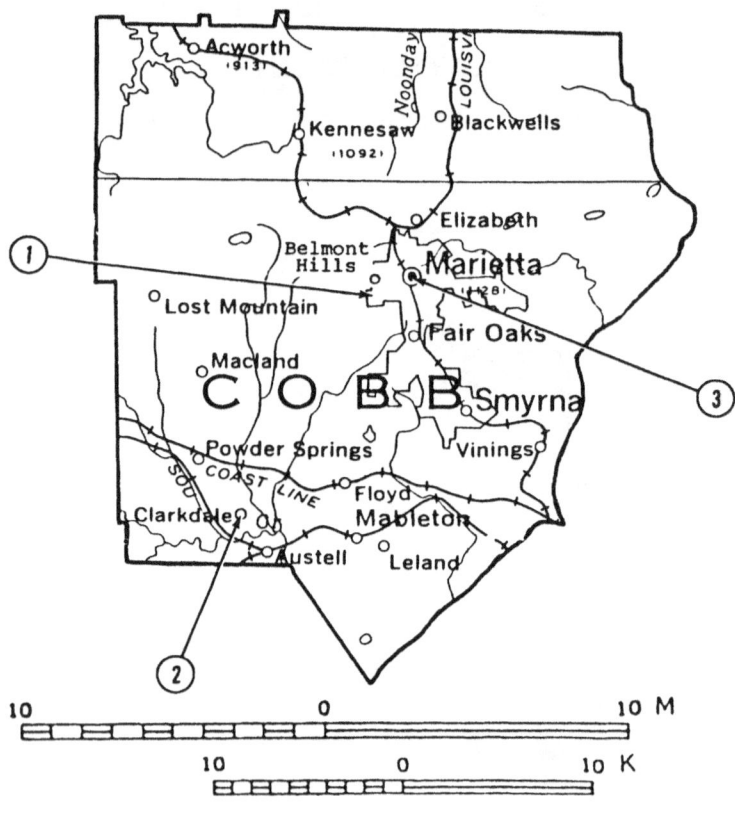

CRISP COUNTY

1. Cordele area.
    Agate, moss. (1-7)

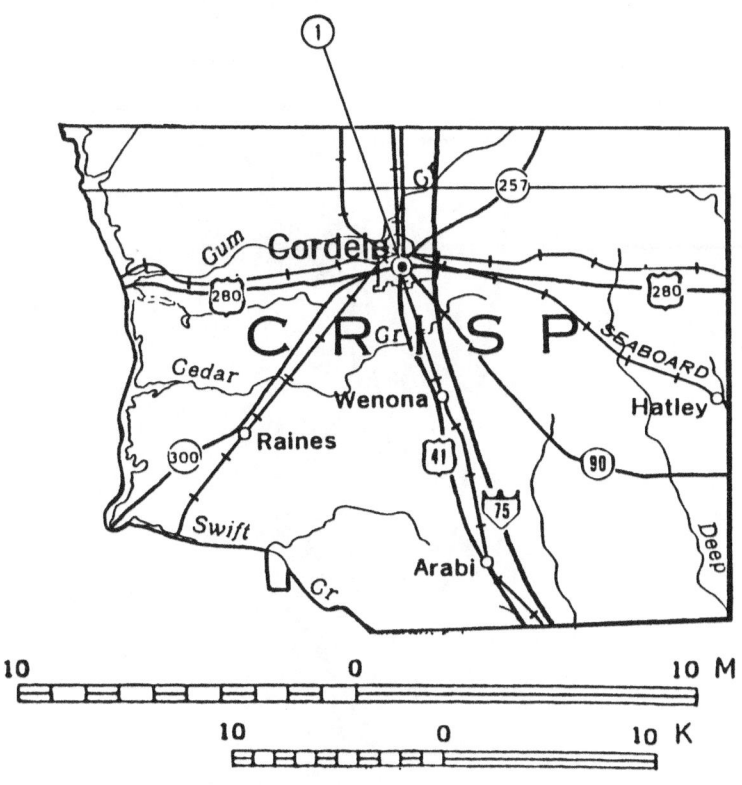

DADE COUNTY

1. Rising Fawn area: 1 mile south to hill; collect on east slope. Halloysite. (1-13-14-29)

DAWSON COUNTY

1. Dawsonville area mines.
   Aquamarine. Beryl. Muscovite. Quartz, rock. Quartz, smoky. Staurolite. Tourmaline, black. (2)

DE KALB COUNTY

1. Atlanta area.
   Diamond: several stones; all badly flawed; found in 1889. (17)

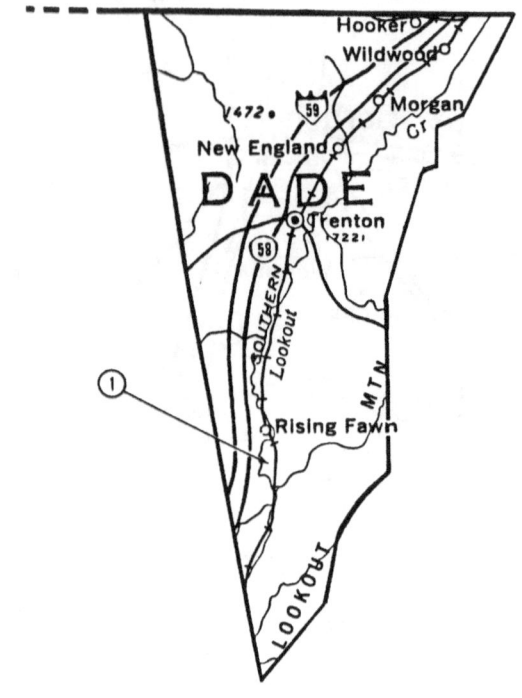

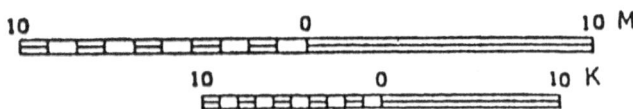

2. Lithonia area: on north side of Arabia Mountain; in the Rock Chapel Quarry.

   Epidote. Thulite. (1-12)

DODGE COUNTY

1. Dubois area: in surrounding fields.

   Tektites: translucent; bottle green. (1-13)

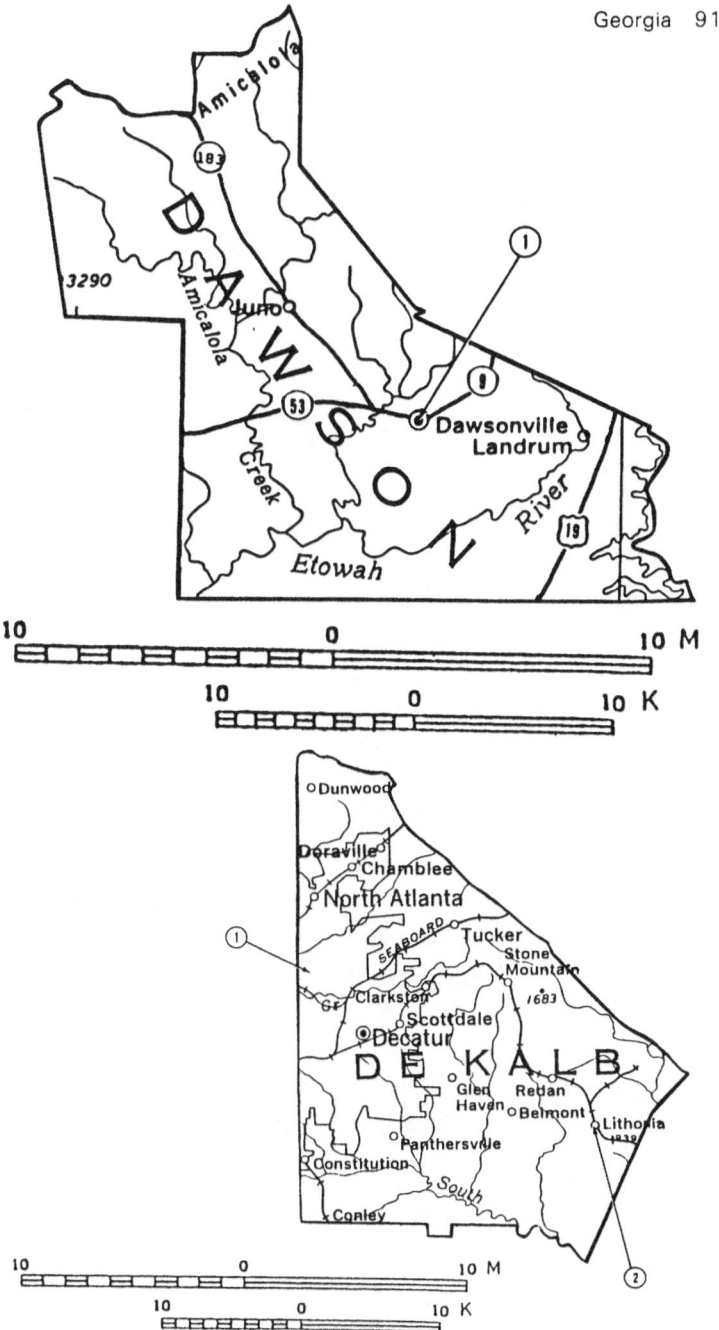

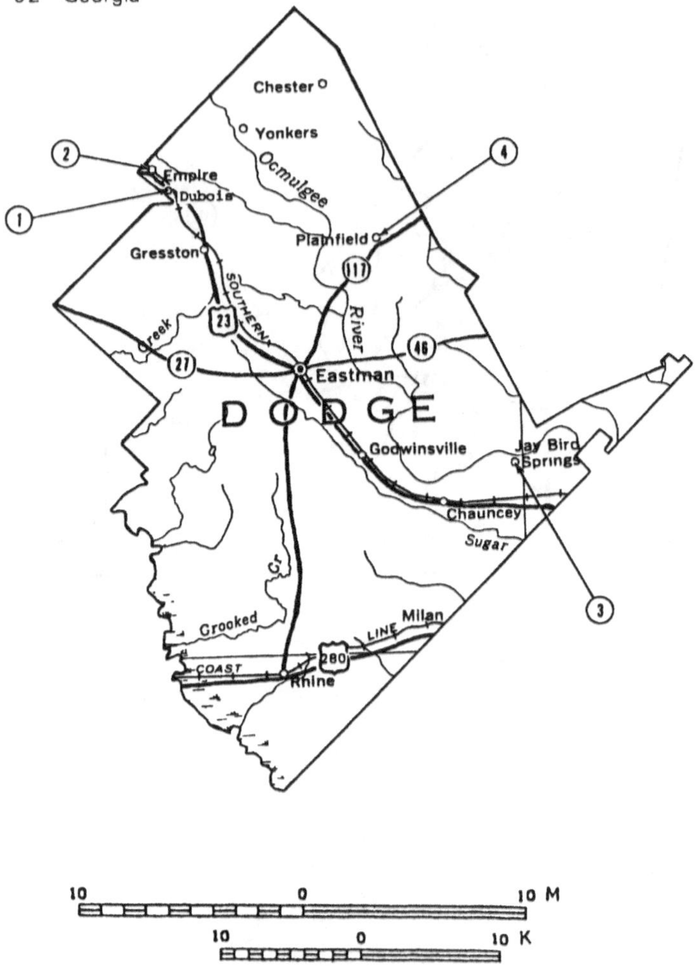

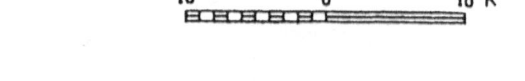

2. Empire area: in surrounding fields.
    Tektites; translucent; bottle green. (1-13)

3. Jay Bird Springs area: in surrounding fields.
    Tektites; translucent; bottle green. (1-13)

4. Plainfield area: in surrounding fields.
    Tektites; translucent; bottle green. (1-13)

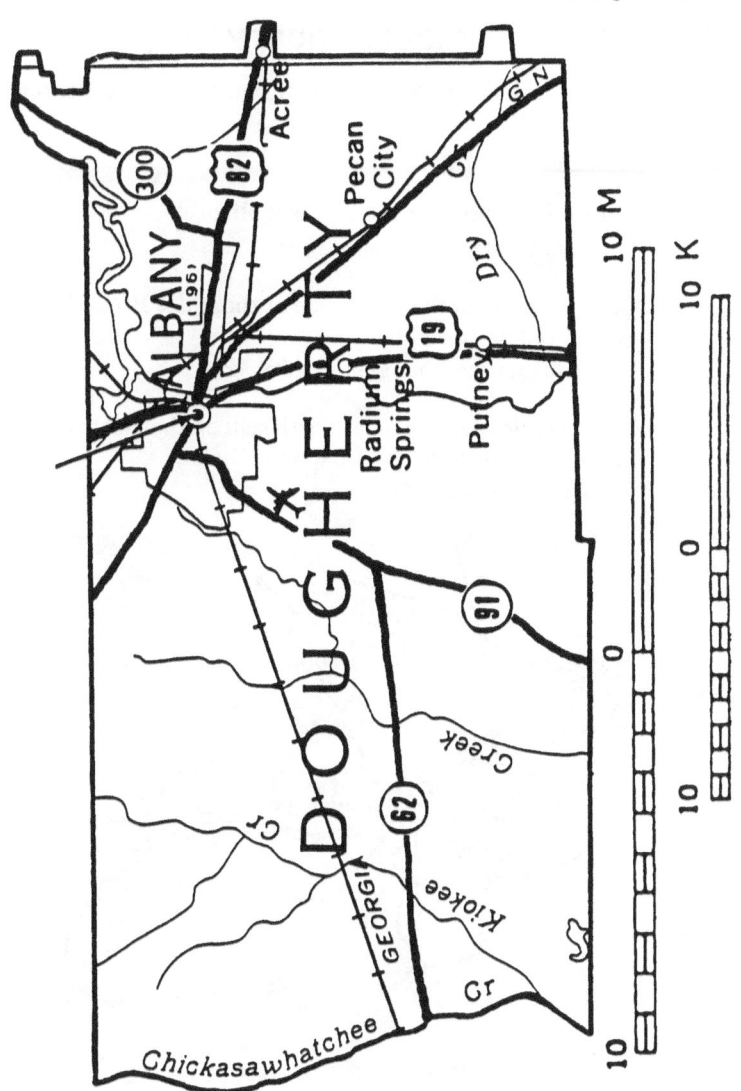

DOUGHERTY COUNTY

1. Albany area: in the Flint River.
   Jasper: red. (7-20)

ELBERT COUNTY

1. Dewey Rose area: 2 miles north; on the W. B. Perkins property.

    Amethyst. (1-2)

2. Dewey Rose area: at the Antioch Mine.

    Amethyst: good color. (2)

3. Oglesby area: on the north side of the Little Broad River; at the Yellow Hill Mine.

    Aquamarine: fine gem quality. (2)

4. Ruckersville area: 4.5 miles northeast; on the north side of Coldwater Creek; at the Chapman Mine.

    Quartz, smoky. (2)

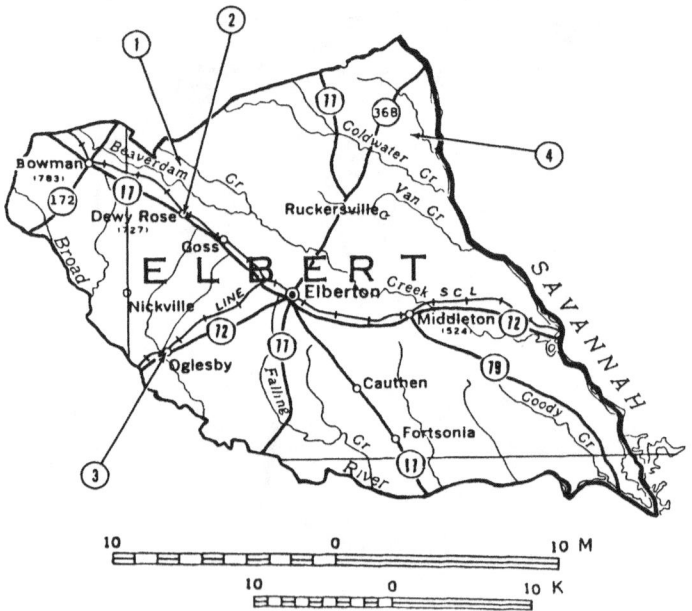

FANNIN COUNTY

1. Blue Ridge area: 1 mile west on SR-2; south on paved road to first gravel road on left; east on gravel road to second house on left; at the J. F. Hackney farm (CF).

   Staurolite: very fine; black; shiny. (1-3-27)

2. Blue Ridge area.

   Staurolite: top quality; black; shiny. (4-27)

3. County-wide.

   Staurolite. (1-3-4-27)

4. Mineral Bluff area: several miles northeast; near Windy Ridge; on the Arp property.

   Staurolite. (1-3-27)

5. Mineral Bluff area: 1 mile northwest to Windy Ridge; then follow paved road to end and collect along dirt road.
    Staurolite. (1)

6. Mineral Bluff area: several miles northwest; near Windy Ridge; at the Richards property.
    Staurolite. (1-3-27)

7. Springer Mountain area: at the Mica Mine dump.
    Emerald: crystals to 0.25" diameter; rare. (2-16)

FAYETTE COUNTY

1. Fayetteville area: 1 mile north; on the Homer Kellin farm.
    Amethyst. Quartz, rock: rutilated. (1-2)

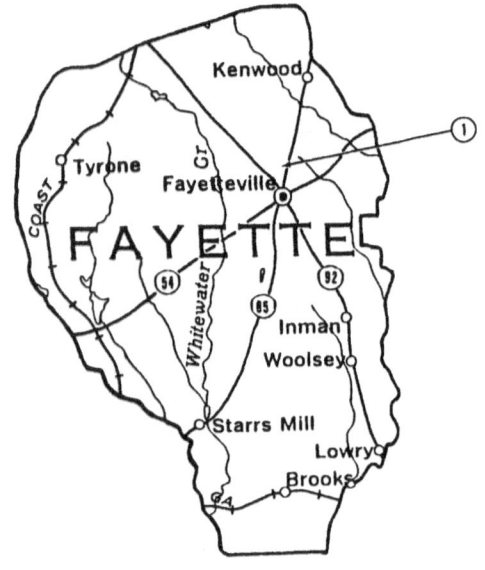

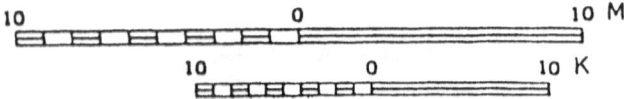

FLOYD COUNTY

1. Hermitage area: 6 miles northeast on SR-53 to right turn at junction; at area bauxite mines.

   Bauxite. Chert. Jasper. Lignite. Marcasite. Pyrite. (2)

2. Rome area: at the Ledbetter Quarry.

   Calcite: good crystals; some with inclusions of brassy pyrite. (9-12)

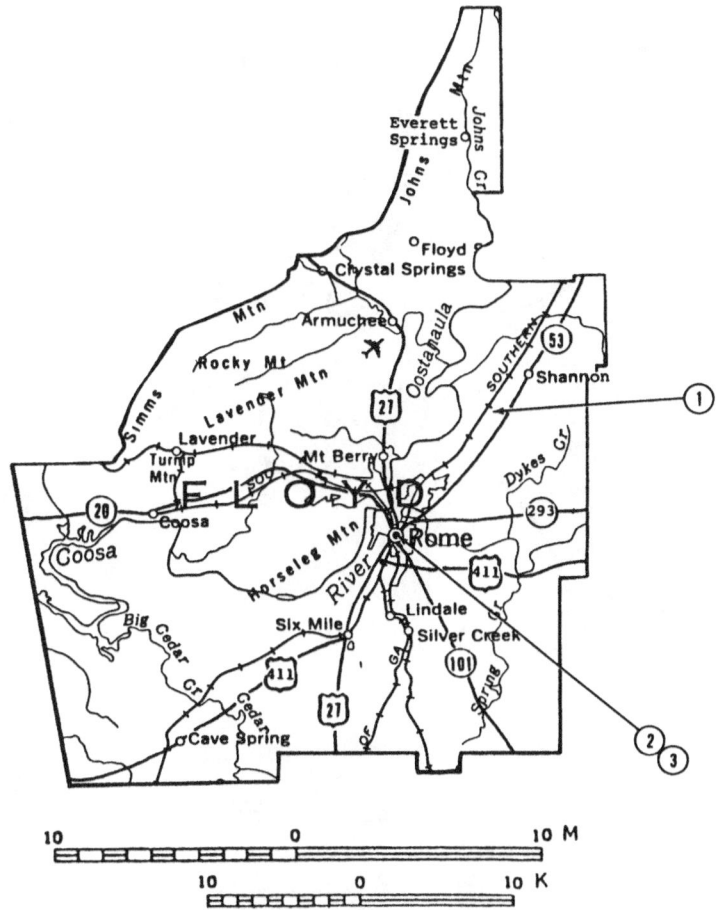

3. Rome area: just west on SR-20; at the Rice Springs Farms.

   Chalcedony: botryoidal clusters. Quartz geodes to 10″ diameter; lined with drusy or crystals; often lined with botryoidal chalcedony; occasionally lined with fine rose quartz. (1-7-13-14-20-29)

FORSYTH COUNTY

1. Cumming area: 6 miles east; on the I. H. Gilbert farm.

   Amethyst. (1-2)

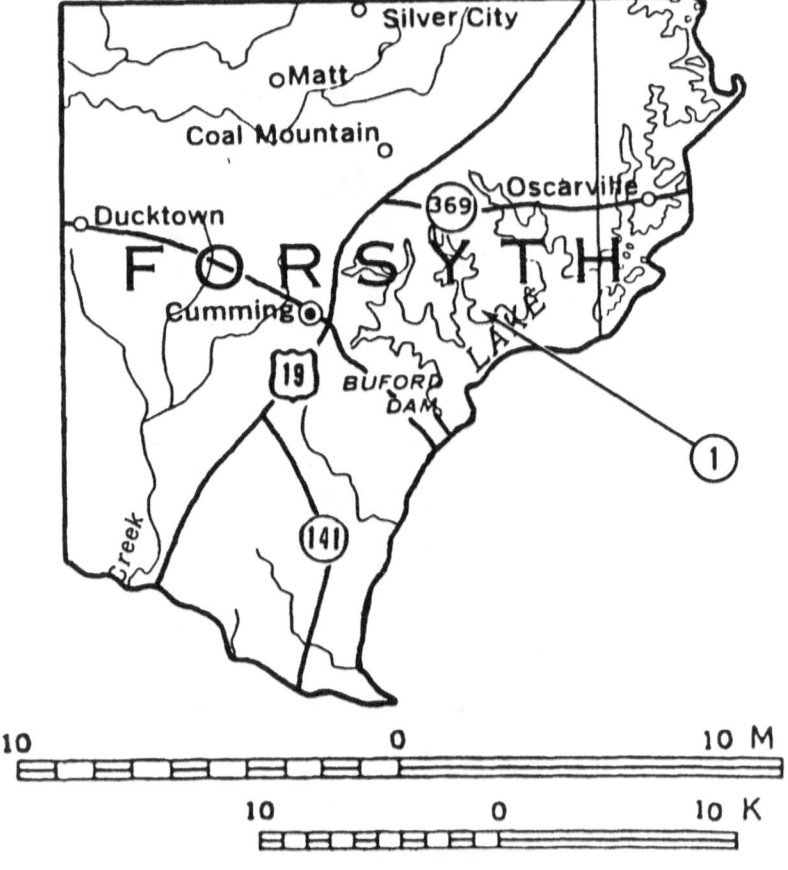

10   0   10 M

10   0   10 K

FRANKLIN COUNTY

1. Carnesville area: 7.5 miles northwest; 2.1 miles south of the Stephens County border.

     Quartz, rock. (1)

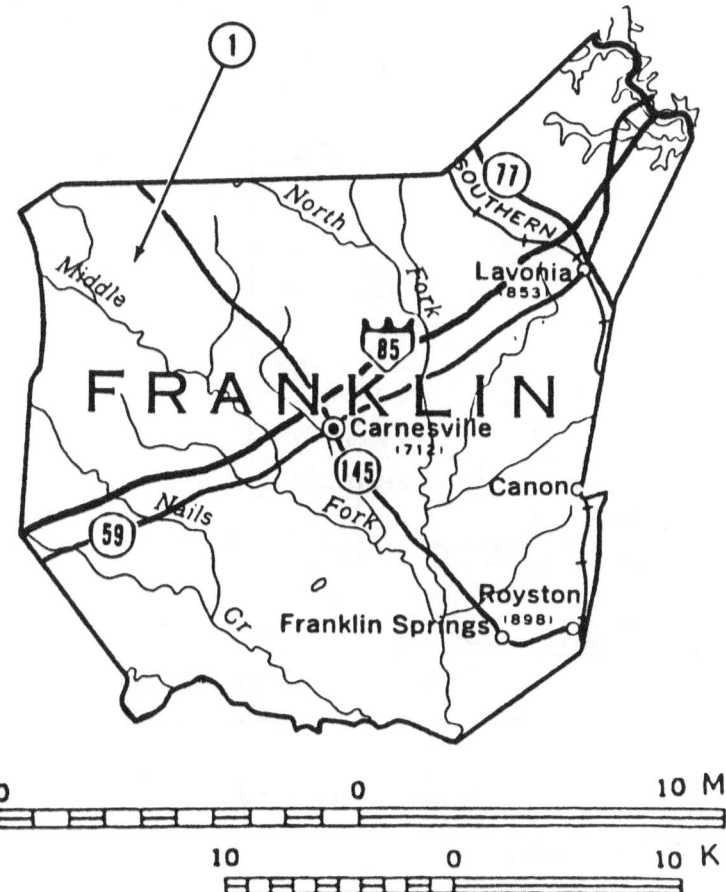

FULTON COUNTY

1. Roswell area: 3.5 miles east.

     Corundum. (1-24)

GILMER COUNTY

1. County-wide.
   Staurolite. (1-3-27)

GORDON COUNTY

1. Ranger area: on US-411; at the Black Marble Quarry.
   Calcite (Marble var.). Fluorite: green; crystals. Pyrite.
   Serpentine. Talc. (12)

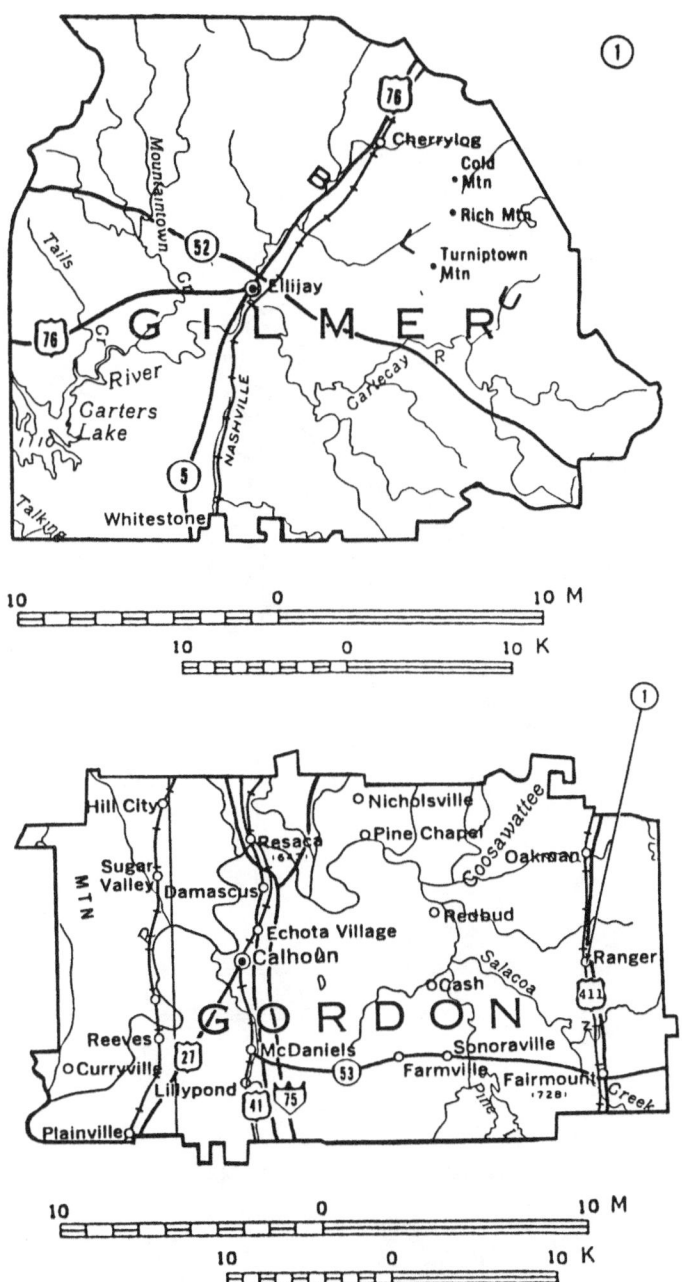

GWINNETT COUNTY

1. Buford area: on the Addison Lowe farm.
     Agate. Moonstone. (1)

2. Norcross area: north on SR-141 to the Chattahoochee
   River crossing; on the Green farm.
     Moonstone. (16)

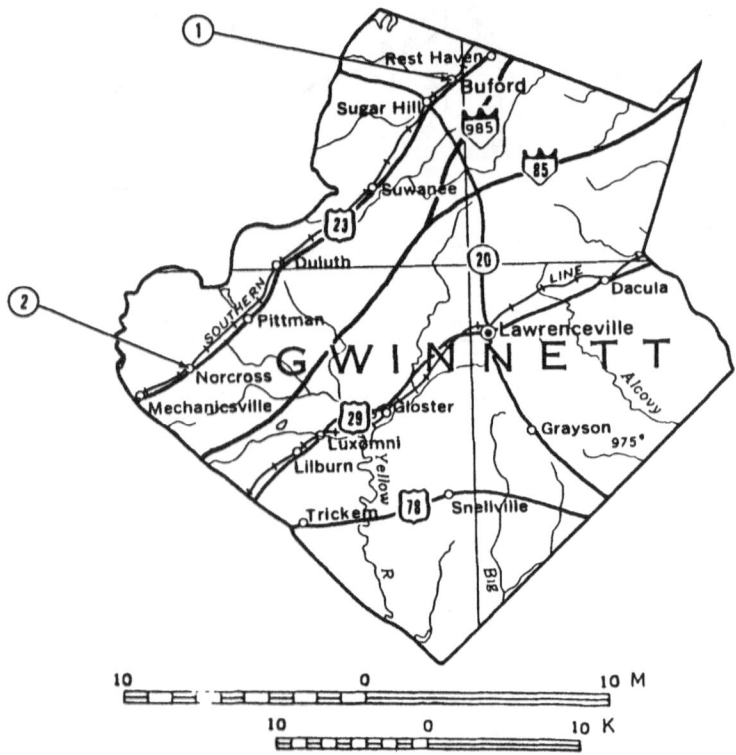

HABERSHAM COUNTY

1. Clarkesville area: 2 miles west; in fields.
     Agate. (1)

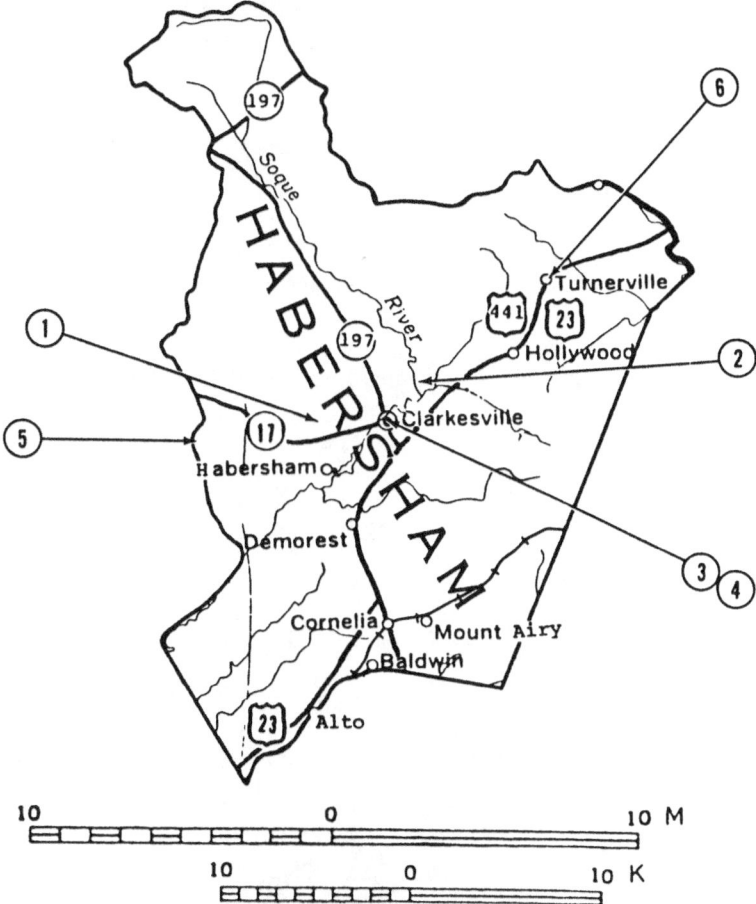

2. Clarkesville area: 0.5 mile east of the grounds of the North Georgia Vocational School; in a small stream and along the bank of the Soque River.

Kyanite. (1-7)

3. Clarkesville area: in area of Stonepile Church.

Margarite. Ruby. (1-27)

4. Clarkesville area: on farms just northeast of town.

Margarite. Ruby. (1-27)

5. Habersham area: 4.9 miles west-northwest, following Beaverdam Creek; on Alec Mountain; in the Piedmont Or-
   chards.

   Margarite. Ruby. (1)

6. Turnerville area: on surrounding farms.
   Margarite. Ruby. (1-27)

HALL COUNTY

1. Gainesville area.
   Diamond: several small crystals found during panning. (17)

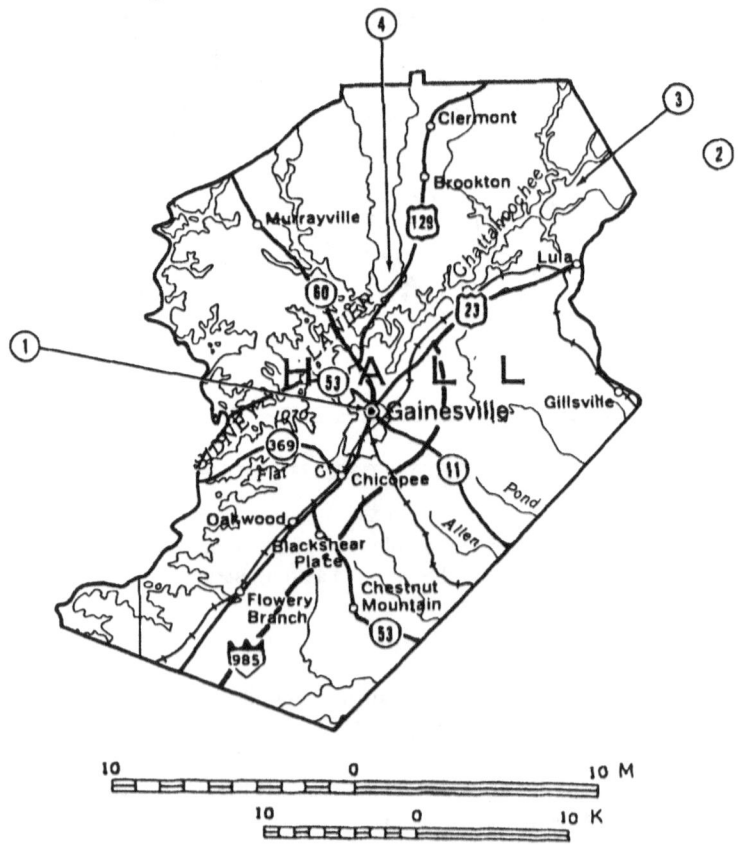

2. Glade Creek area: more precise location not given.

    Diamond: several small crystals found during panning. (17)

3. Lula area: 3.1 miles north-northeast; in Steckeneter Branch at the Glades.

    Diamond: several stones; specifics unknown. (17)

4. Williams Ferry area: in Muddy Creek tributary; more precise location not given.

    Diamond: 1 stone found, described as "large" but further details unknown; found in 1843. (17)

HANCOCK COUNTY

1. Devereux area: 6.4 miles northwest; along shoreline of the Sinclair Reservoir.

    Chert. Jasper. (1-7-13-20)

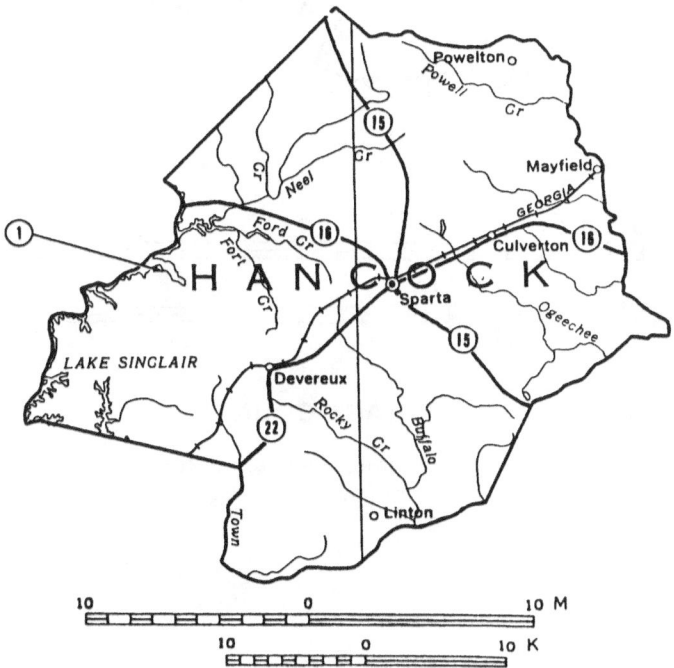

HARALSON COUNTY

1. Bremen area: 1.5 miles west; in fields.
   Quartz, rock. (1)

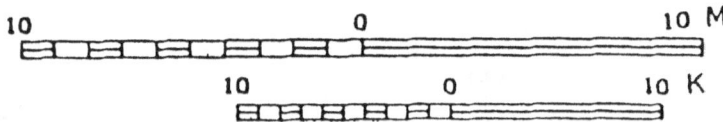

HART COUNTY

1. Hartwell area: east on US-29 to the Hartwell Dam access
   road; quarry on right.
   Beryl. (12)

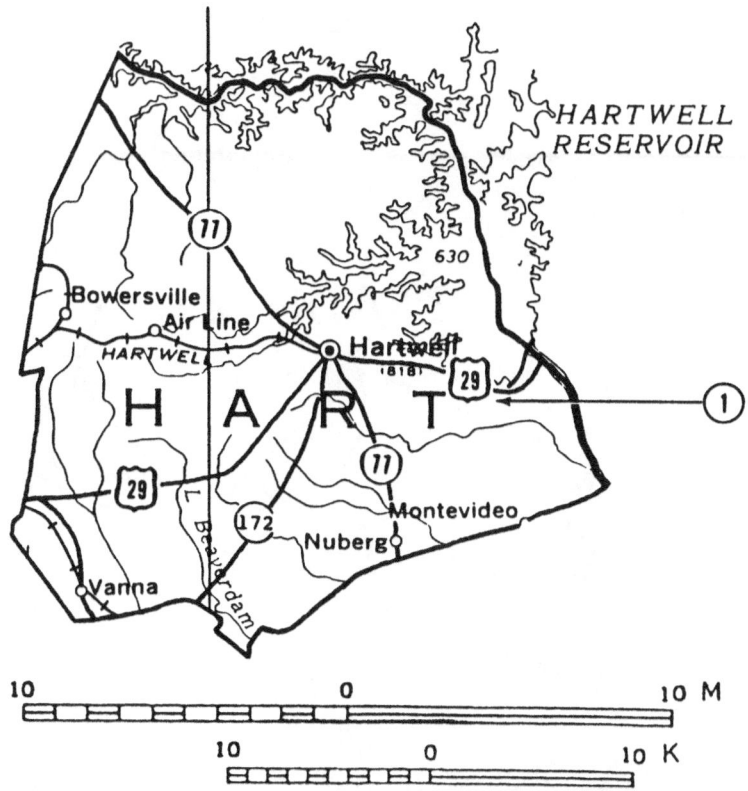

HEARD COUNTY

1. County-wide in stream valleys of Chattahoochee River tributaries.

   > Agate. Chalcedony: varied colors. Chert. Flint. Jasper. Opal, common. Petrified wood: opalized; bright colors. (3-7-18-20)

HENRY COUNTY

1. Stockbridge area: 3.9 miles east; 1 mile west of Millers Mill; at mine.

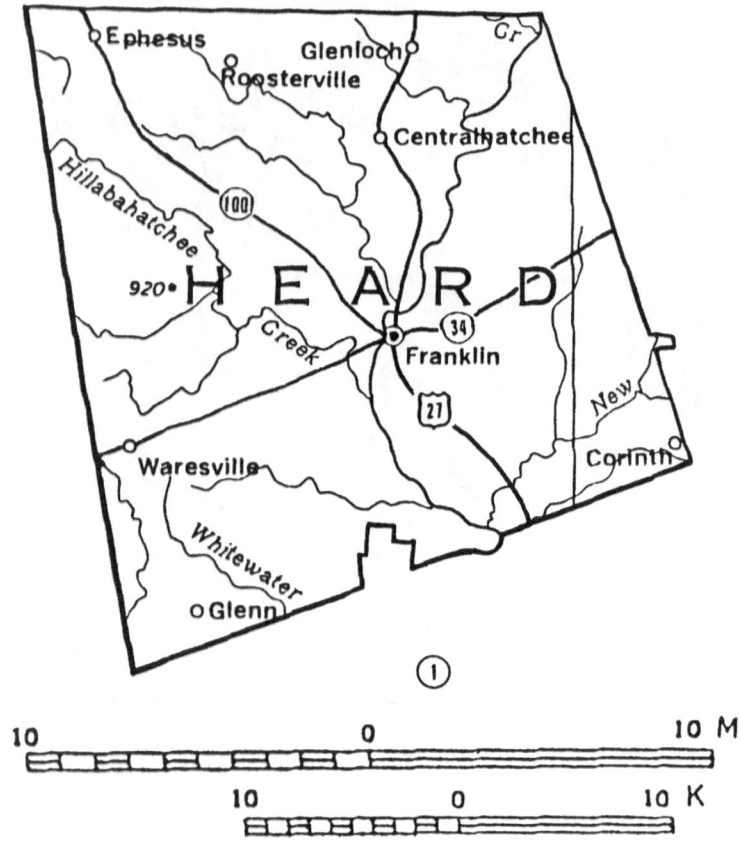

Amethyst. Aquamarine. Beryl. Garnet. Microcline. Muscovite. Tourmaline. (2)

IRWIN COUNTY

1. Osierfield area.
   Tektites: gem quality. (1-13)

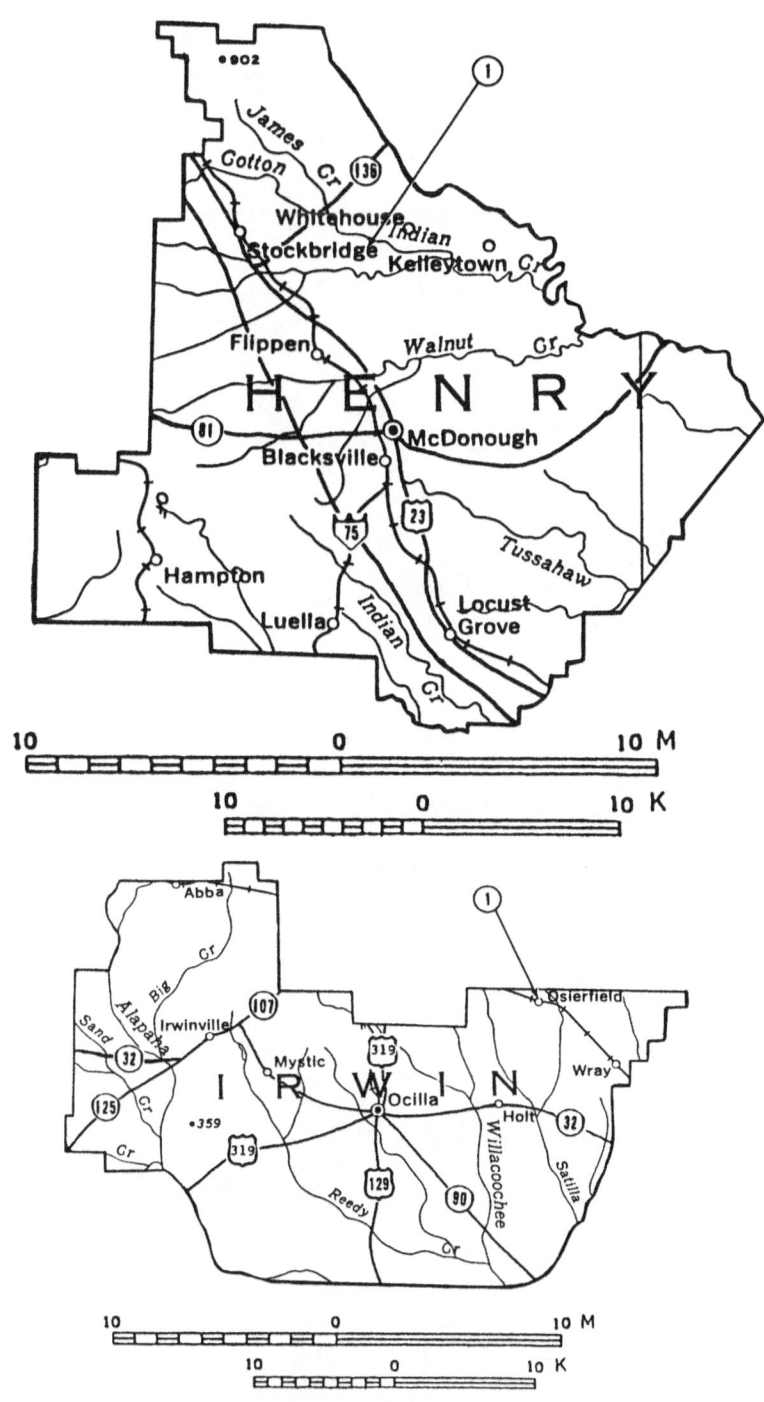

JACKSON COUNTY

1. Nicholson area: just north of town on US-441; on west side of road where dirt road intersects.

    Beryl. (16)

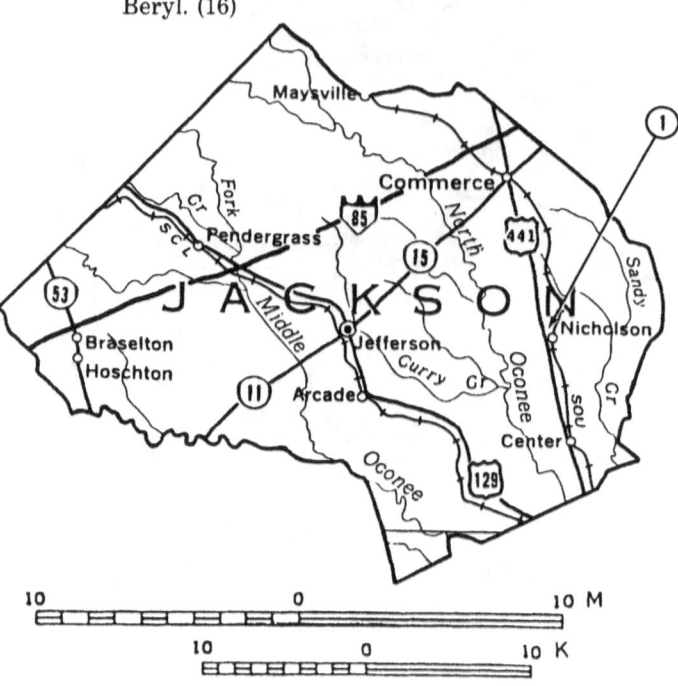

JASPER COUNTY

1. Hillsboro area: 3.3 miles east-northeast; then north on dirt road to Baptist Church; turn right; 3 miles east on dirt road to the Barren Fullerton farm (white farmhouse with large oaks); collect in vicinity of house.

    Amethyst. Beryl. Mica. Quartz, blue: marked asterism. Quartz, smoky: marked asterism. (1-13)

2. Hillsboro area: 2.1 miles west-southwest; on the J. R. Parker property.

    Beryl. (16)

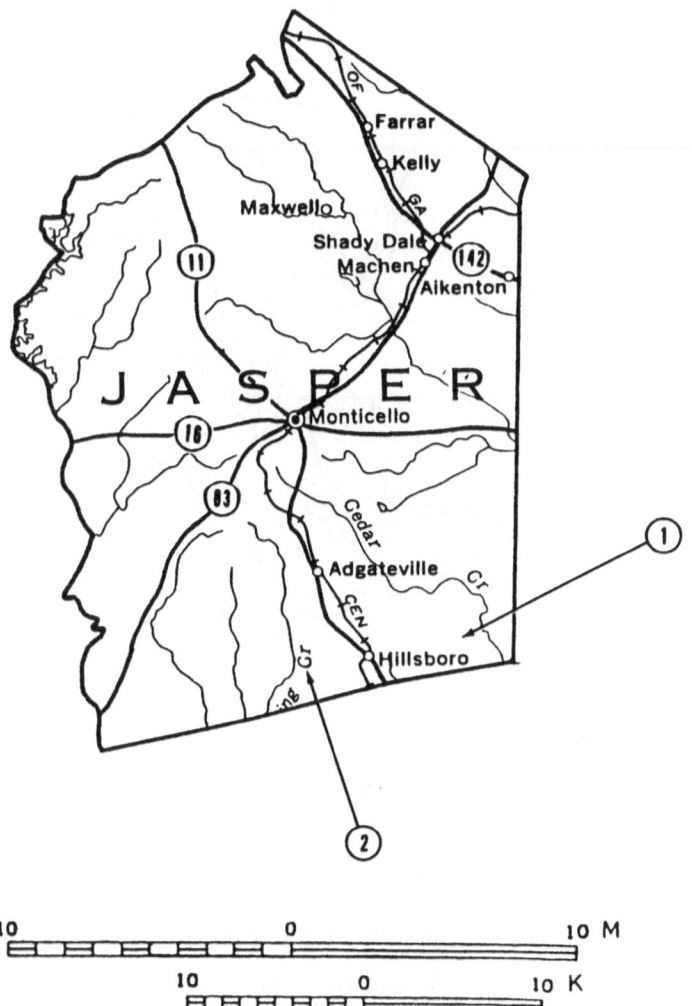

JONES COUNTY

1. Round Oak area.
   Agate. Jasper: red. (1-3)

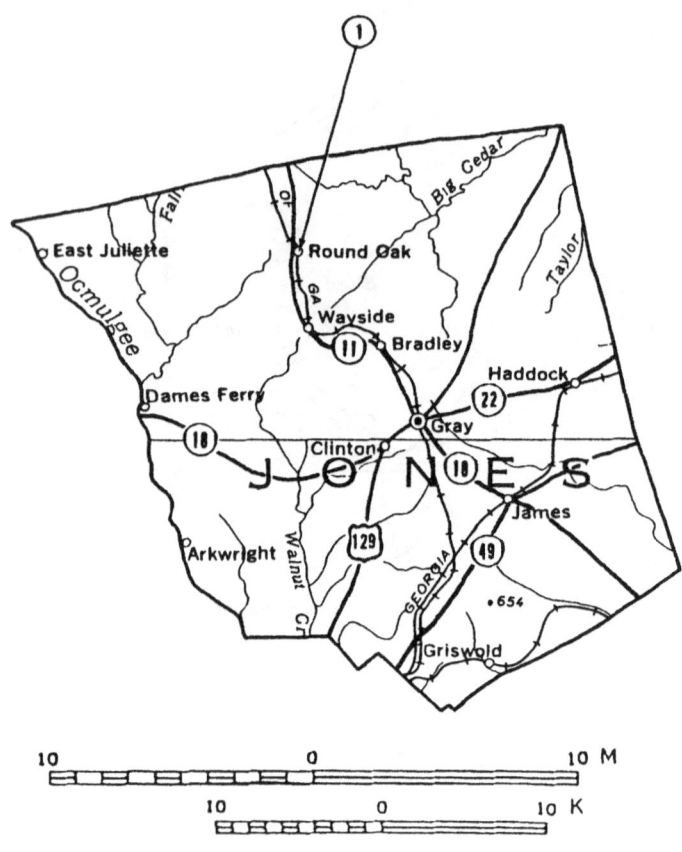

LAMAR COUNTY

1. Barnesville area: east on US-41; at the Early Vaughn Mine.
   Apatite. Beryl. Garnet. Mica: good crystals. Microcline: cleavable. Muscovite. Perthite. Plagioclase: dark green. Pyrite. Quartz, rock. Tourmaline: black. (2)

2. Piedmont area: 1.8 miles southeast; 0.5 mile southwest of the Ramah Church; at the J. T. Means Mine.
   Beryl: bluish green; crystals to 2″ diameter. (2)

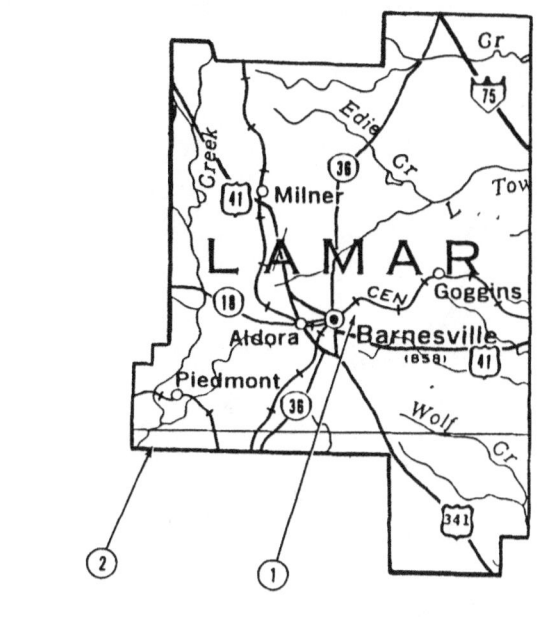

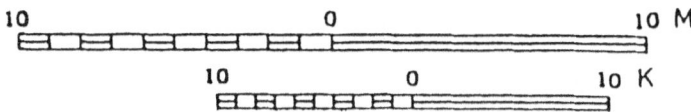

LEE COUNTY

1. Smithville area: 3 miles east on SR-118 to Muckalee Creek bridge; 300 yards south of bridge on west bank.

   Diamond: 1 stone: 3.5 ct; greenish white; flattened octahedron; found in 1901. (20)

LINCOLN COUNTY

1. Lincolnton area: at Graves Mountain.

   Kyanite: spectacular crystals. Lazulite: pale blue; sharply terminated crystals. Pyrophyllite: bladed crystals; silvery green; in stellate clusters to 1″ diameter. Rutile: gem quality. (1-2-5-granular quartz -13-16)

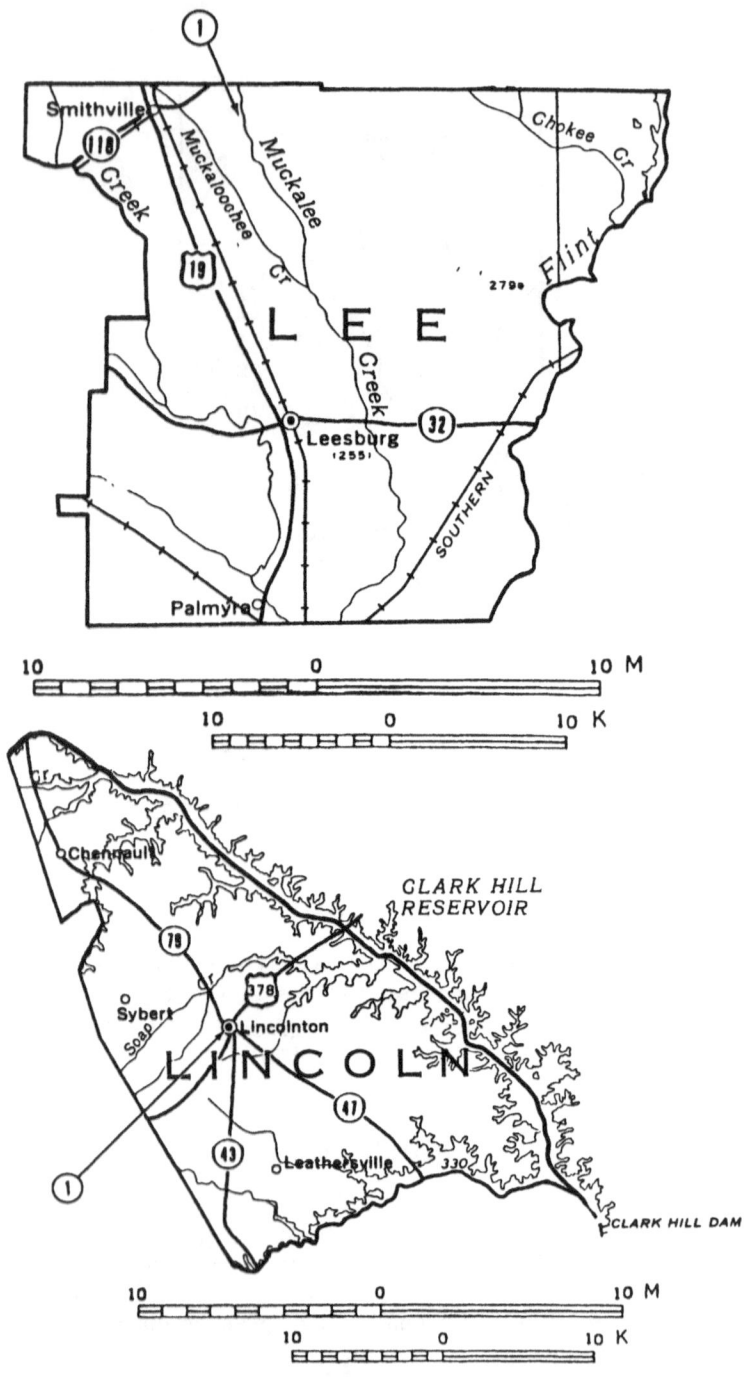

LOWNDES COUNTY

1. Clyattville area: 1.5 miles west on Main Street to left turn
   on Blands Dairy Road; south to the Withalacoochee River;
   best when water level is low.

   Fossils: coral; silicified. (20)

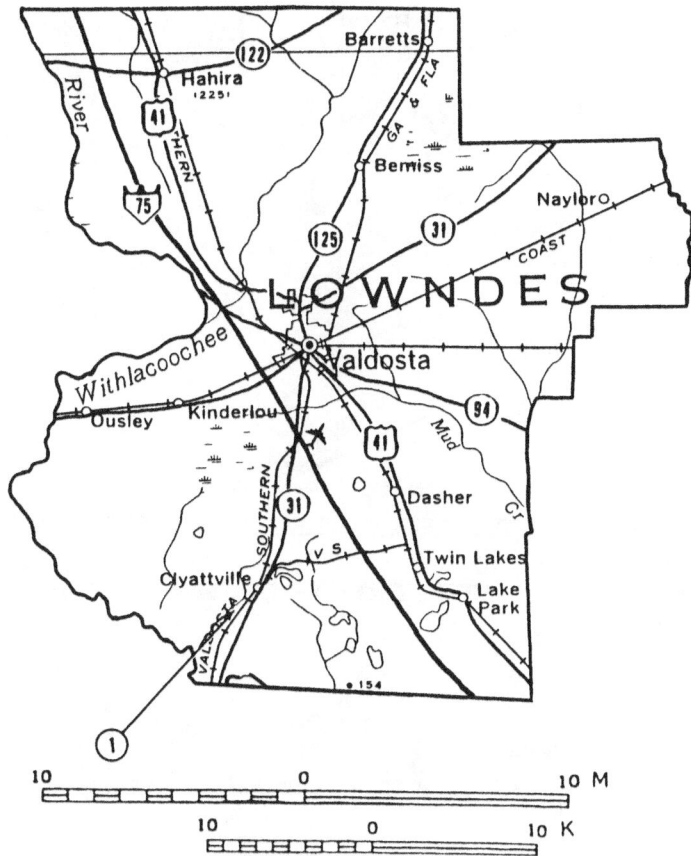

LUMPKIN COUNTY

1. Dahlonega area: 9.5 miles northeast; 2.25 miles southwest
   of Ward Gap; at the Williams Mica Mine.

   Topaz: fine gem quality; colorless. (2)

2. Dahlonega area.
Diamond: 1 stone found during panning; details not specified. (17)

MCDUFFIE COUNTY

1. Thomson area: 12 miles northwest; on bank of the Little River; at quartz mine.
Gold. (2-19-quartz)

2. Thomson area: at the Columbia Mine.
   Gold. (2-19-quartz)

3. Thomson area: at the Hamilton Mine.
   Gold. (2-19-quartz)

4. Thomson area: at the McGruber Mine.
   Gold. (2-19-quartz)

5. Thomson area: at the Park Mine.
   Gold. (2-19-quartz)

6. Thomson area: at the Seminole Mine.
   Gold. (2-19-quartz)

MERIWETHER COUNTY

1. Greenville area: on SR-109 to the first pond; then first road
   left of pond to the fourth house on the left; on the Ernest
   Strozier farm (CF).
   Garnet. Tourmaline. (1-2)

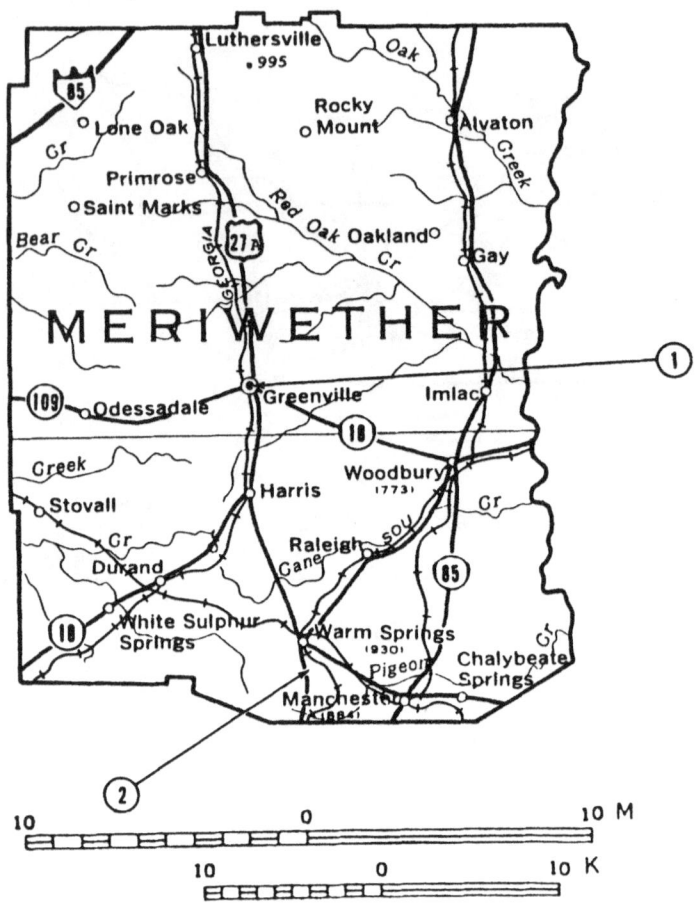

2. Warm Springs area: 1 mile south along the Southern RR tracks.

   Agate. (1-4-13)

MONROE COUNTY

1. Brent area: 3 miles southwest on SR-83; at the Peters Mine.

   Biotite. Feldspar: crystals. Muscovite. Quartz, rose. (2)

2. Culloden area: 4 miles north; at mines and prospects.

   Biotite. Feldspar: crystals. Mica. Muscovite. (2-21)

3. Culloden area: 4.5 miles north; almost on the Lamar County border.

   Feldspar: crystals. Mica. Quartz, milky. Quartz, smoky. (16-24)

4. Juliette area mines and prospects.

   Biotite. Feldspar: crystals. Muscovite. Quartz, milky. Quartz, smoky. (2-21)

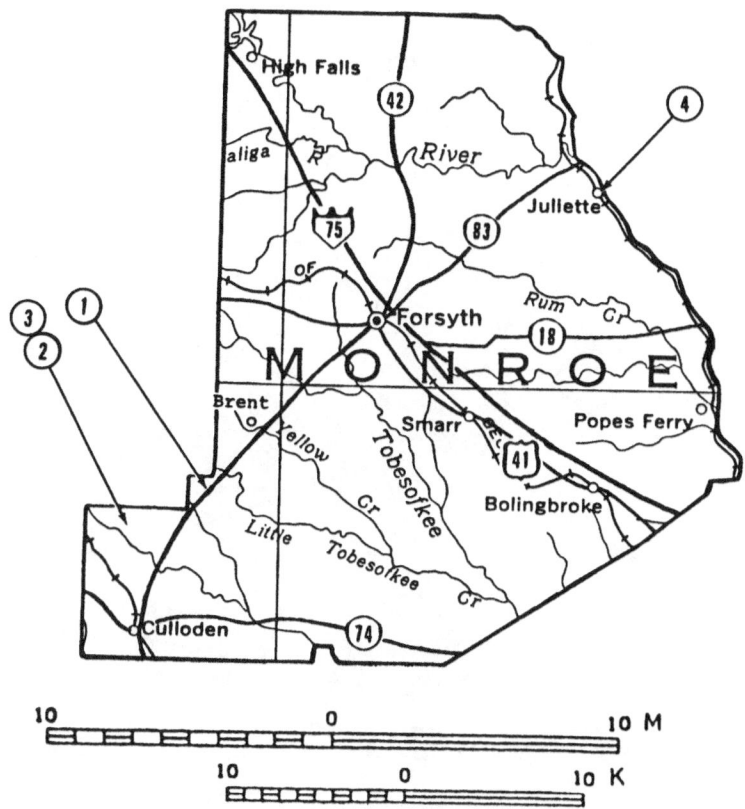

MORGAN COUNTY

1. Apalachee area: west to fields of Adair plantation.
      Aquamarine. (6)

2. Bostwick area: at the Carter Prospect.
      Beryl: abundant small crystals. (2)

3. Buckhead area: 2 miles east; on the Ray farm.
      Amethyst: fine gem quality. (1-2-4-9-quartz)

4. Rutledge area: SR-12 to the Georgia RR; turn onto dirt
   road; 2 miles southwest to the Bill Oxford farm.
      Corundum. (1)

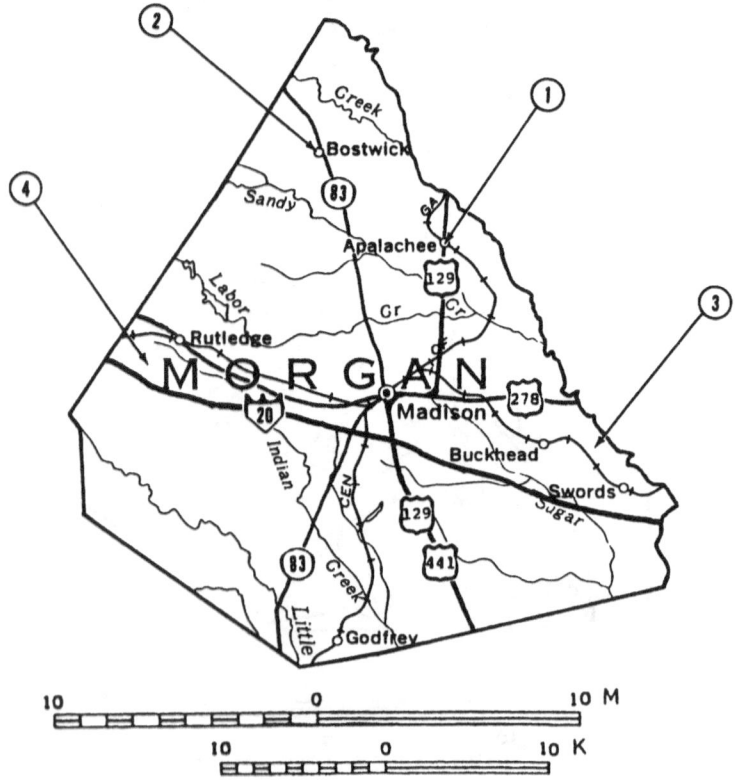

MURRAY COUNTY

1. County-wide in road, RR, and stream cuts.
   Quartz, rock. Quartz, rose: occasional. Quartz, smoky. (4)

2. Eton area mines.
   Barite: crystals. Talc. (2)

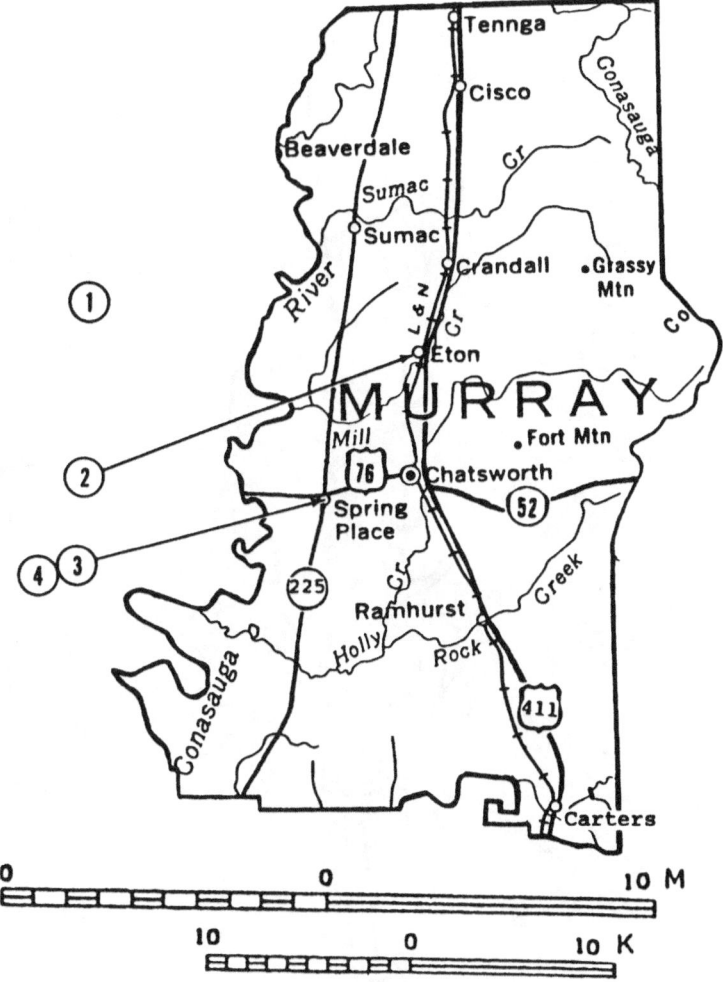

3. Spring Place area: go west on US-76 to the Hooker School area; then north on dirt road to Fincher Bluff.
   Oolite: silicified. (29)

4. Spring Place area: in road cuts for SR-225.
   Agate. Chalcedony. Oolite: silicified. (4)

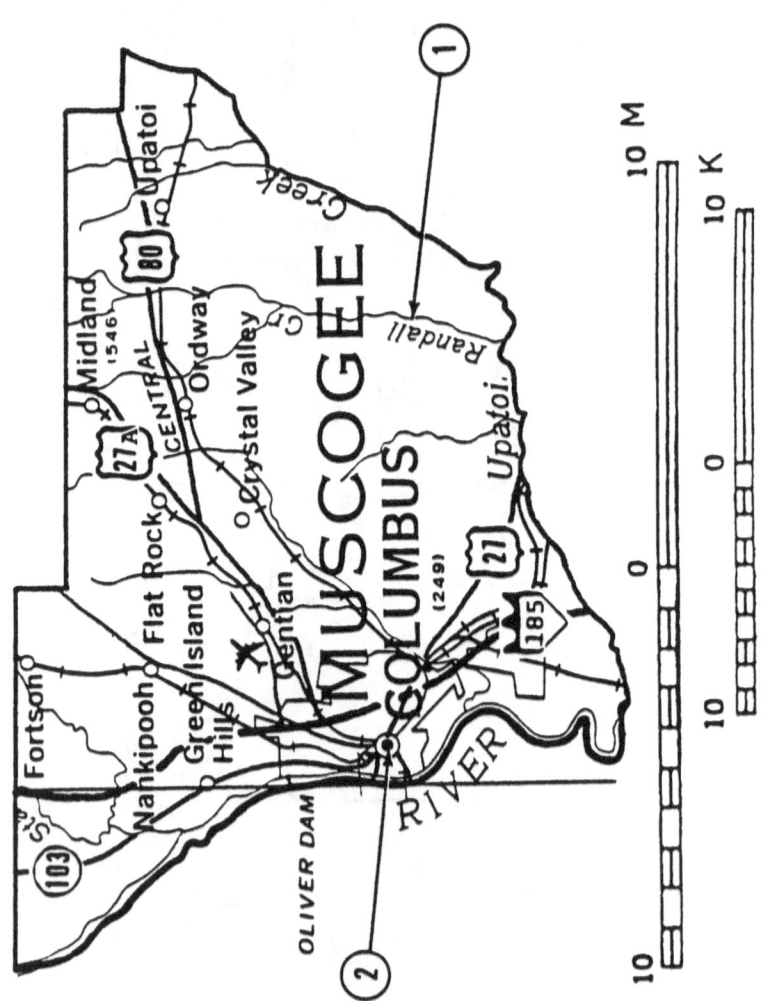

MUSCOGEE COUNTY

1. Columbus area: 10.6 miles east; along Randall Creek.
   Petrified wood. (1-7-20)

2. Columbus area: along Bull Creek.
   Petrified wood. (1-7-20)

OCONEE COUNTY

1. Eastville area: 1.2 miles north to SR-53; 2.5 miles west-northwest to side road at right; follow side road 3 miles to the Dickins farm; at the Dickins Mine.
   Apatite. Biotite. Muscovite. Quartz, rock. Tourmaline. (2)

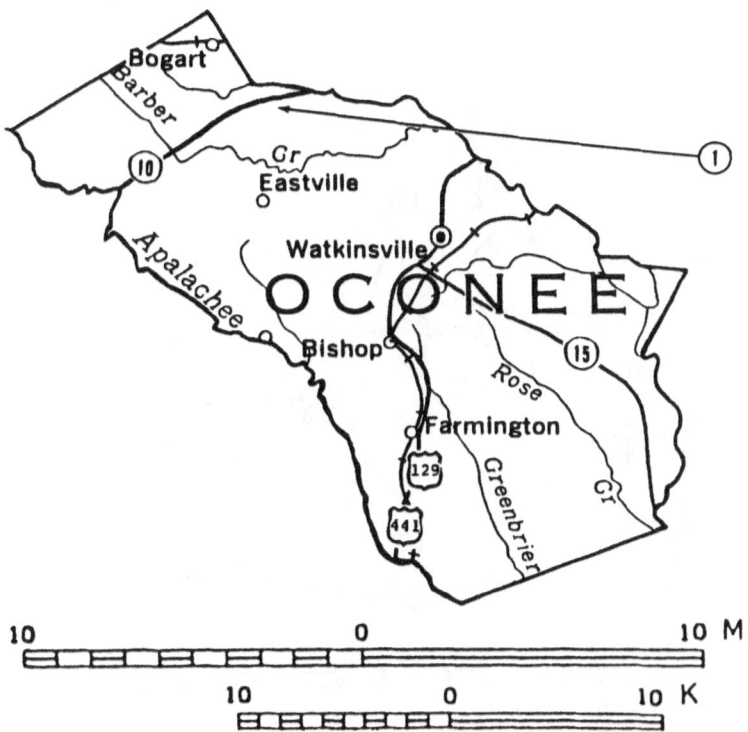

PAULDING COUNTY

1. Hiram area.

    Garnet, almandine: deep red; gem quality. Pyrite. (1-6-13)

2. Hiram area: at the Little Bob Copper Mine.

    Garnet, almandine: deep red; mostly flawless gem quality; crystals to 0.25″ diameter. (1-2)

3. Hiram area: 1 mile southwest; go west to Dunn Store; pavement ends but continue on graded road 0.25 mile to fork; bear right to dirt road leading to Verd Antique Quarry.

    Apatite. Serpentine. (12)

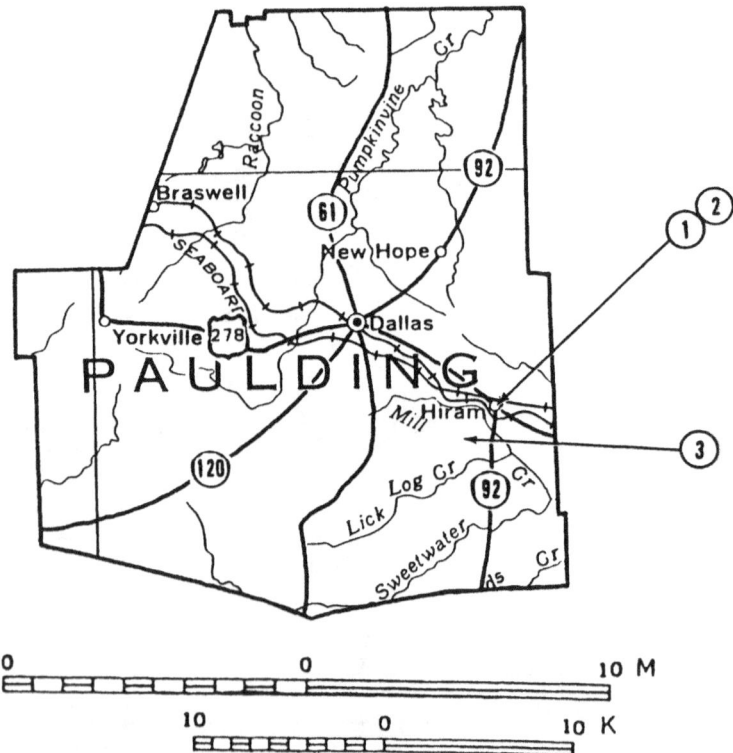

PICKENS COUNTY

1. County-wide.
   Staurolite. (1-3-27)

2. Cagle area: 1.2 miles north-northwest; on the southeast side of Rock Creek.
   Beryl: blue, golden (very deep color), yellow (very bright); fine gem-quality crystals to 6″ long. (16-quartz core-29)

3. Marblehill area: 3 miles southeast.
   Garnet. Kyanite. Staurolite. (21)

4. Tate area: 5.5 miles east; in the Partain Prospects.
   Garnet. Kyanite. Staurolite. (21)

5. Tate area quarries.
   Calcite (Marble var.): good quality. (12)

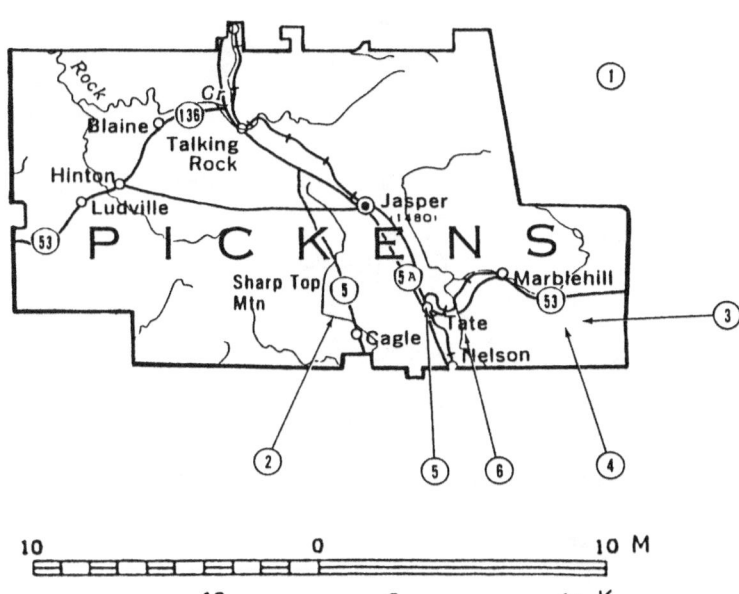

6. Tate area: southeast of Rock Creek near Refuge Church; on the Ralph Cook property.

> Aquamarine. Beryl, golden. (16)

POLK COUNTY

1. Cedartown area iron mines.

> Hematite: some botryoidal crystals. Limonite. Rhodochrosite. (2)

RABUN COUNTY

1. Clayton area: 8.8 miles west on US-76 to the Tallulah River; follow the river road into Towns County; turn right at Upper Charlie Creek and come back into Rabun County to mines.

> Amethyst. (2)

2. Clayton area: 7 miles east; several miles south of War Woman Creek; at the Beck Beryl Mine.

> Beryl. Quartz, rock. (1-2)

3. Clayton area: 4 miles southeast; at the John A. Wilson Prospect.

> Amethyst: fine gem quality. (9-19-quartz-21)

4. Clayton area: southeast to the junction of US-23 and US-76; at the W. T. Smith Mine.

> Amethyst. (2)

5. Clayton area: south to Walnut Fork crossing; at the Kell Mica Mine.

> Quartz, rose: very fine. (2-16)

6. Dillard area: 1 mile east; at elevation 2000'; on southeast side of Black Creek; at the Ledbetter Mine and Prospects.

> Amethyst: gem quality. (2-9-quartz -19-21)

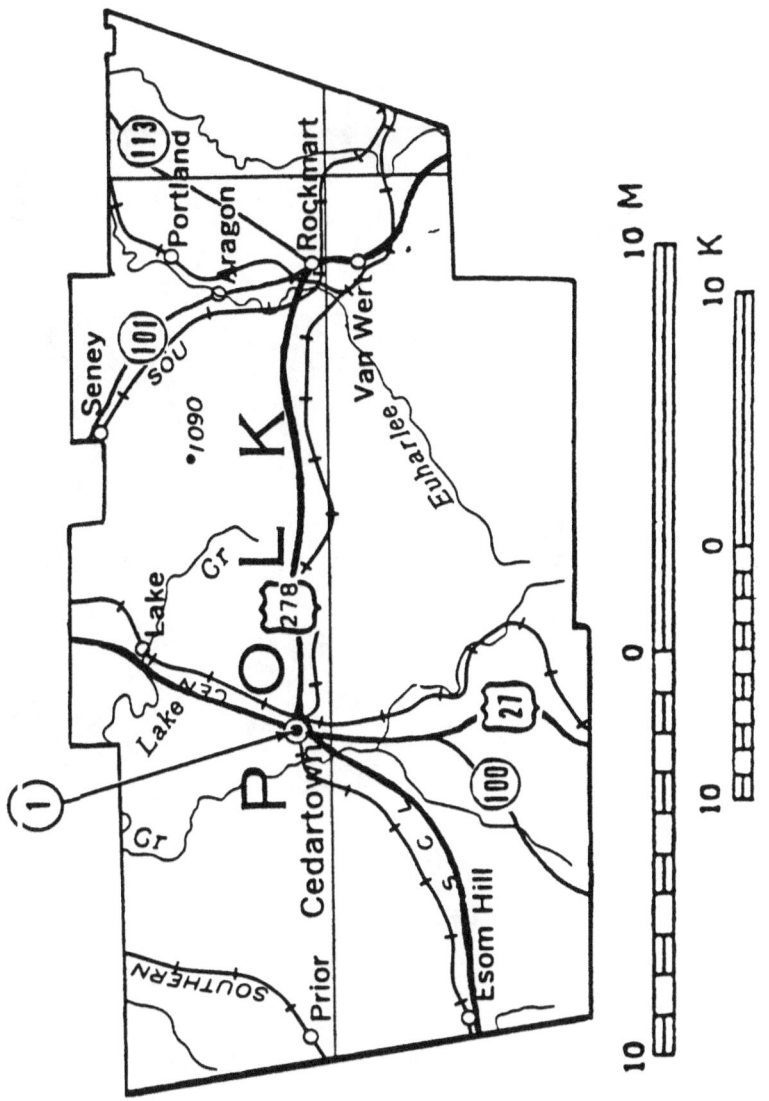

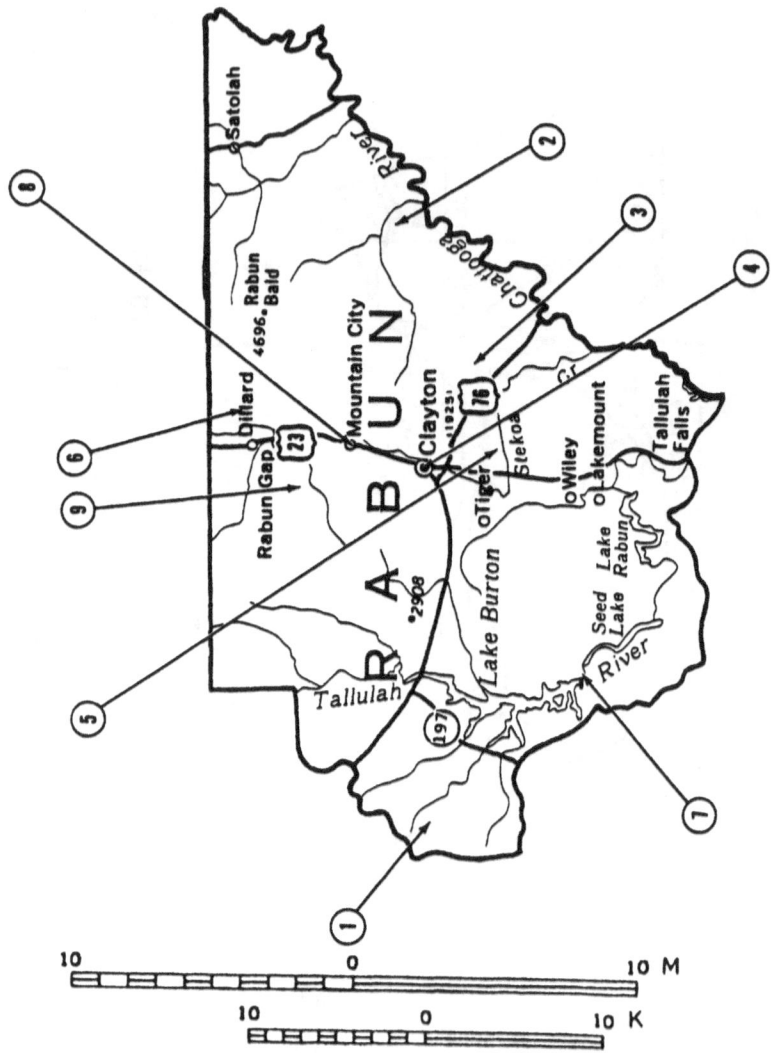

10   0   10 M

10   0   10 K

7. Lakemount area: 6.2 miles west; in diggings at the lower
end of Lake Burton near the dam.
   Kyanite. (1-2)

8. Mountain City area: at mine dumps.
   Amethyst. (2)

9. Rabun Gap: 1.5 miles southwest; at the North Georgia Company Mine.

> Amethyst. (2)

ROCKDALE COUNTY

1. Magnet area: 2 miles northwest (1 mile north of Ocmulgee River).

> Feldspar: crystals. Muscovite: crystals. Quartz, milky. (14-24)

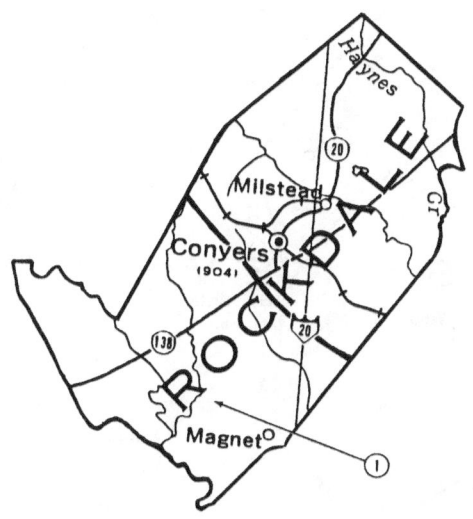

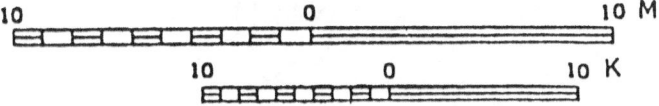

SCREVEN COUNTY

1. Sylvania area: 14.8 miles northeast on US-301 to paved side road on left leading toward Girard (in Burke County); follow this side road north for 8 miles to where it turns sharply left; follow the dirt road from this point for 0.5 mile.

Agate. Fossils: marine; agatized. (1-13)

SPALDING COUNTY

1. Griffin area: 2.5 miles southeast on River Road.

    Beryl, blue: gem quality. Quartz, rose. Tourmaline. (24)

2. Vaughn area: 2 miles north on unmarked paved road to dirt road right; 0.15 mile east on dirt road to the T. J. Allen Prospect.

    Beryl: fine; clear; light blue (one crystal cut to an excellent 2-ct gem). Quartz, rose: massive. Quartz, smoky: good crystals to 1.5″ diameter. Tourmaline: blue; crystals to 1″ long. (6-9-granite 16-21-24)

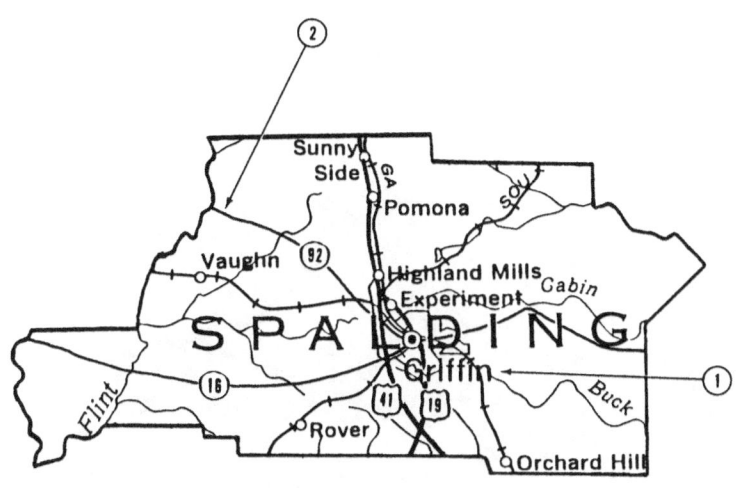

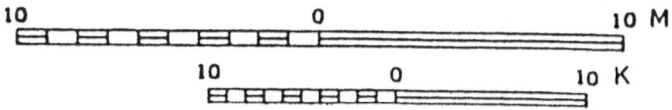

STATEWIDE

1. In such rivers and creeks as the Etowah, Oostanaula, Appalachicola, Suwannee, Johns, Flint, Oconee, Ocmulgee, Ogeechee, Connesauga, Chattanooga, and Spanish and their tributaries; also in Altamaha and Bulltown swamps.

    Pearl. (7)

STEPHENS COUNTY

1. Toccoa area: 5.9 miles north-northeast; just above the Tugaloo River; at the Tugaloo Overlook Mine.

    Corundum: tiny transparent areas in massive corundum. Ruby: red; fine gem quality. (2)

2. Toccoa area: 2.1 miles northwest to Dicks Creek Road; follow it to the Marck Beck property.

    Beryl. Quartz, rock. (1-2)

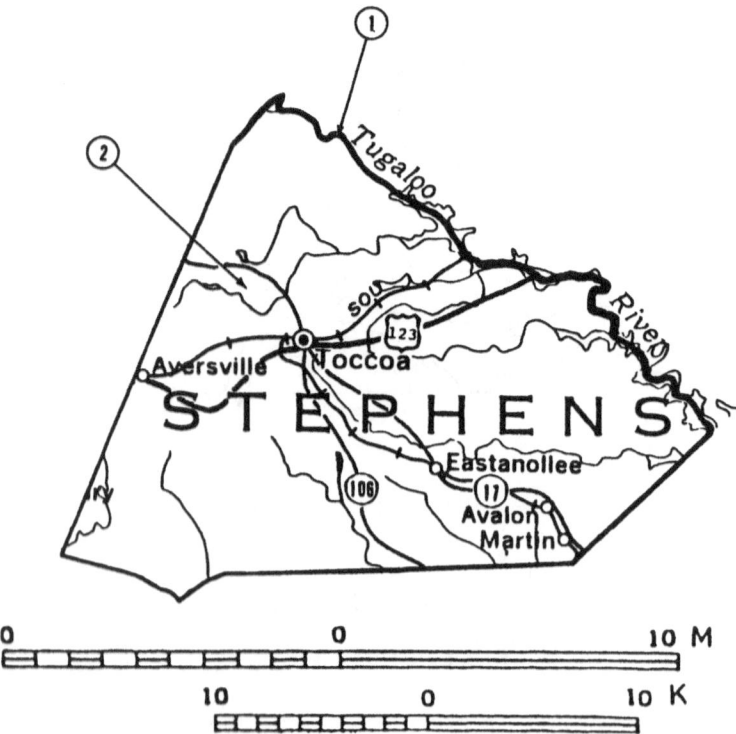

TOWNS COUNTY

1. Hiawassee area: 2 miles southwest; at the Hog Creek
   Mine.
   >Corundum: pink. Smaragdite (Amphibolite, green). (2)

2. Hiawassee area: north to Elf area; in the area of Chatuge
   Lake.
   >Corundum: pink. (4-20)

3. Hiawassee area: at Charlies Creek.
   >Amethyst: crystals to 4″ diameter. (7)

4. Young Harris area: 2.5 miles west on US-76; south to
   Track Rock Gap; on south side of slope.
   >Corundum. (1)

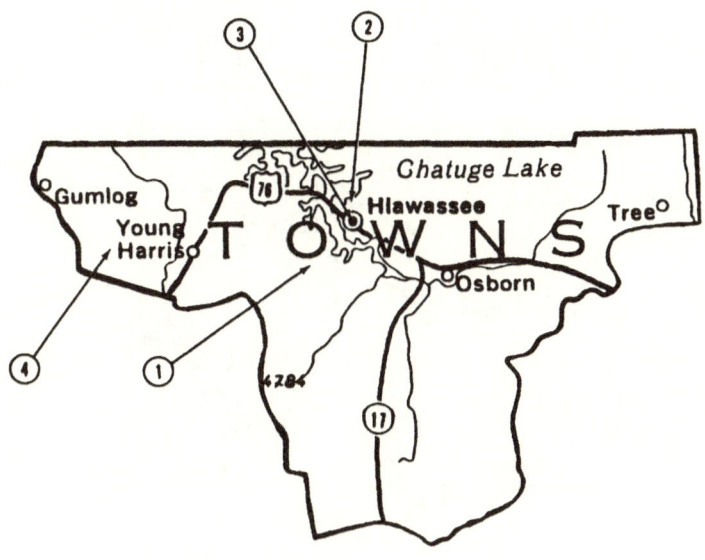

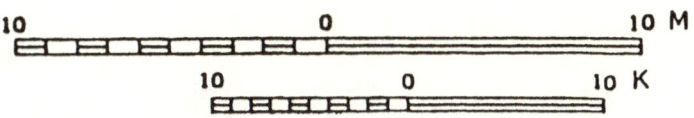

TROUP COUNTY

1. La Grange area: 10.4 miles southeast on US-27; then 0.75 mile south; in Pine Mountain Valley.

   Beryl. Quartz, rose. (1)

2. La Grange area: 8.6 miles south on SR-219; 1.3 miles south of Smiths Store crossroads.

   Beryl: pale blue. Beryl: pale green; highly translucent; crystals to 18″ thick. Garnet: good crystals to 2″ diameter. Tourmaline: black; crystals to 1″ diameter and 6″ long. (6)

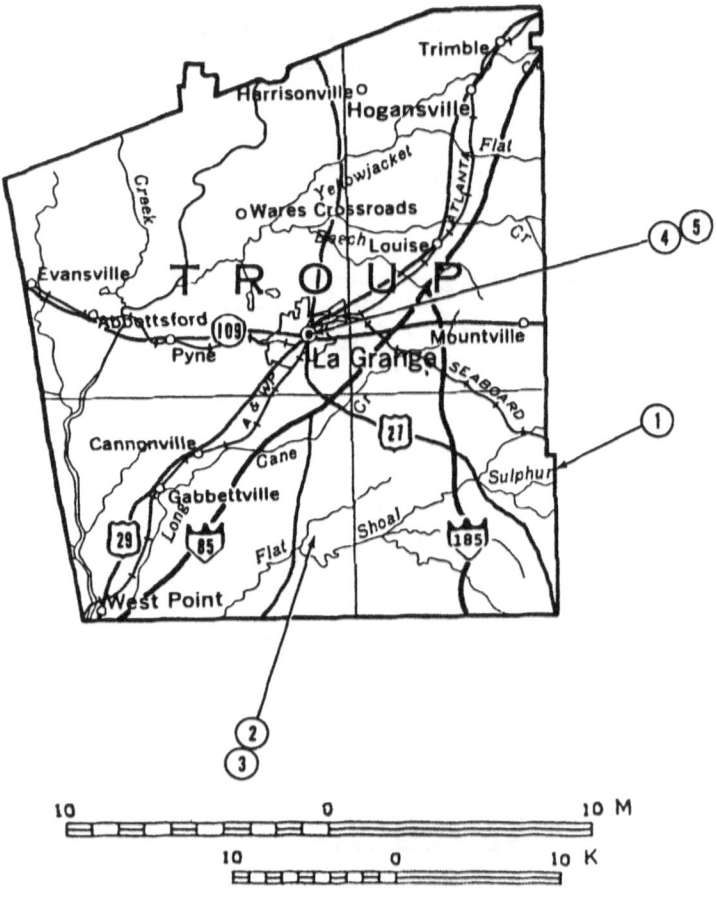

3. La Grange area: 8.6 miles south on SR-219; 1 mile south of Smiths Store crossroads; west side of SR-219; at the Minerals Processing Company Mine.

> Aquamarine. Beryl. Kaolinite. Mica. Muscovite. Quartz, rose: fine crystals; some flawless gem quality. Tourmaline: black. (2-9-milky quartz -16-19)

4. La Grange area: just north of the airport.

> Beryl: clear; fragments. (1)

5. La Grange area: south on SR-219 to Cleveland Crossroads; follow signs to Big Beryl Mine (aka Foley Mine, Hogg Mine) (CF).

> Beryl. Quartz, rose. (2)

TWIGGS COUNTY

1. Huber area: 1.5 miles northeast; in Ocmulgee River tributary.

> Diamond: several stones; specifics unknown. (17)

UNION COUNTY

1. Blairsville area: 5 miles northwest; then 1 mile below road in the Teece Creek valley.

> Kyanite. (1-7)

2. Blairsville area: Hightower Bald region; 1 mile south; on a ridge separating Shoal Branch and Jacks Branch; at the Garrett Mine.

> Amethyst. (2-5-milky quartz -9)

3. Blairsville area: in fields 0.5 mile east of Akin Mountain.

> Kyanite. (1)

4. Blairsville area: south side of Track Rock Gap.

> Ruby. (1)

1
57
Dry Branch
Big
80
18
Huber
Creek
Jeffersonville
96
231
SEABOARD
River
TWIGGS
SOUTHERN
Bullard
16
Flat
Cr
Adams Park
96
Savage
129A

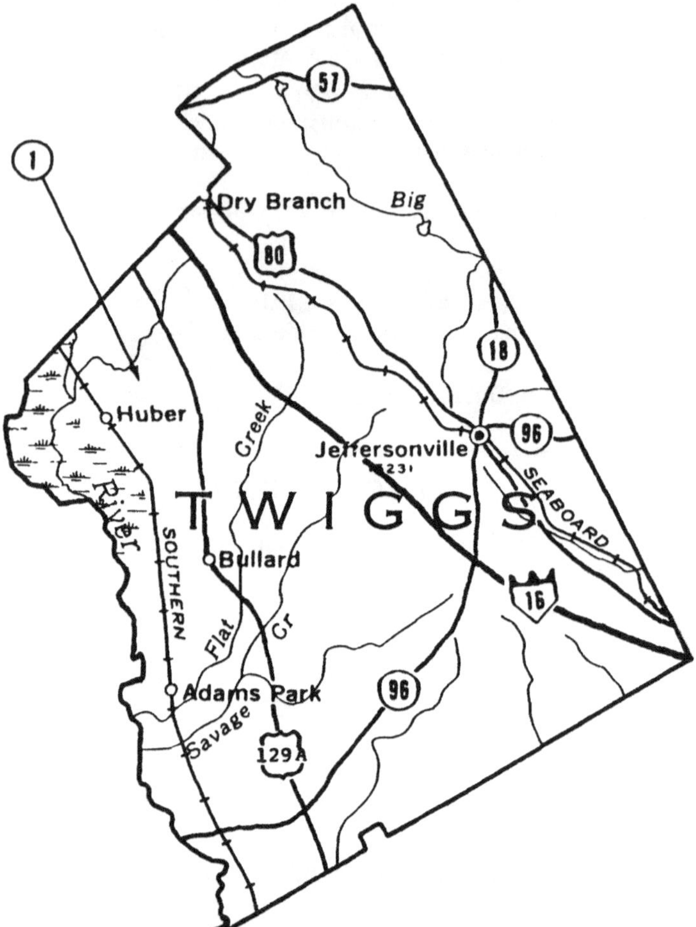

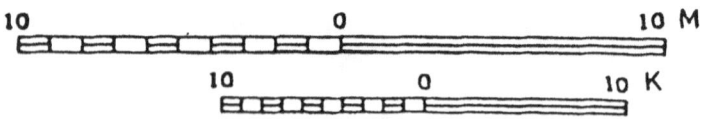

10
0
10 M
10
0
10 K

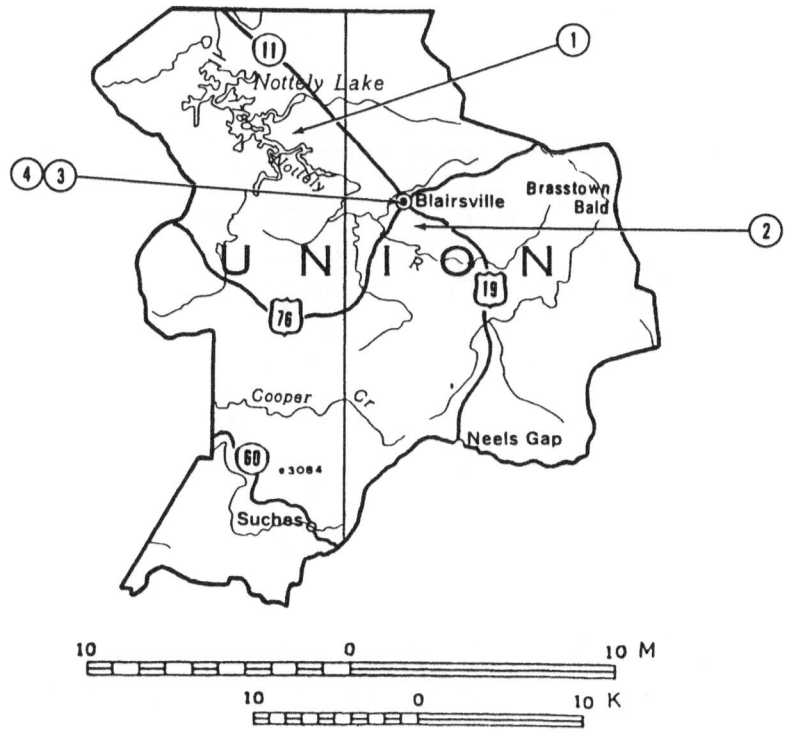

10   0   10 M

10   0   10 K

UPSON COUNTY

1. Lincoln Park area: 1 mile south on US-19 to Shepherd School; turn right on dirt road and go 1.75 miles to mine road between Bell Creek and Potato Creek; on that road to the Dolly Cherry Mine.

    Garnet, almandine: crystals to 1″ diameter. Kyanite: crystals. Staurolite. (1-2)

2. The Rock area: 1.4 miles south; in Wilmots Ravine.
    Agate. (1)

3. The Rock area: at the Kelly property.
    Corundum. (1-24)

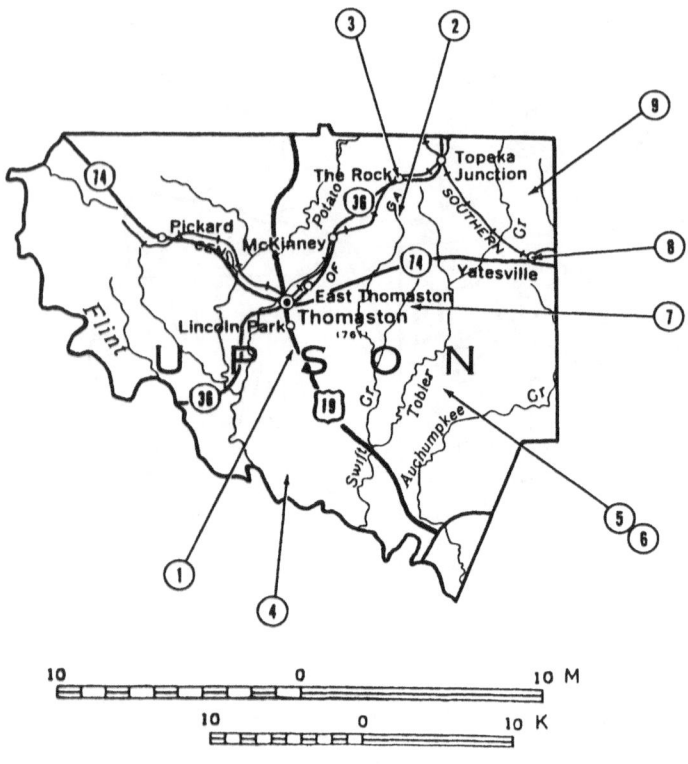

4. Thomaston area: 7.3 miles south, 76 degrees east; 0.2 mile from the Mitchell Creek Prospect, on opposite side of the road.

Apatite: pale green; glassy. Garnet: purplish; small crystals. (2)

5. Thomaston area: 7.25 miles southeast to Waymansville; then 1 mile northeast on Mitchell Creek tributary of Tobler Creek; at the Joe Persons Mine.

Apatite. Moonstone. (2)

6. Thomaston area: 7.25 miles southeast to Waymansville; then 1 mile northeast on Mitchell Creek tributary of Tobler Creek; at the Mitchell Creek Mica Mine.

Apatite: chartreuse to dark green; some pale green; excellent crystals to 0.5″ in diameter and 2″ long. Biotite: well-developed crystals. Feldspar, plagioclase:

translucent; very cleavable grains to 3″ diameter.
Garnet. Mica: excellent crystals. Moonstone. Musco-
vite: good crystals with brilliant luster. Pyrite.
Quartz, rock. Tourmaline. (2)

7. Thomaston area: 5 miles east; at the Blount No. 1 Mine.

Apatite: pale green. Beryl: blue-green; crystals to
1.5″ diameter and 4″ long. Garnet. Moonstone. Tourmaline: black. (2)

8. Yatesville area: 0.25 mile east; 300′ south; at the Herron
Mine.

Beryl: light blue to olive green; gem-quality crystals
to 7″ diameter. (2)

9. Yatesville area: 2.5 miles north; at the Adams Mine.

Beryl: blue to dark green; crystals to 1.5″ diameter
and 15″ long. (2)

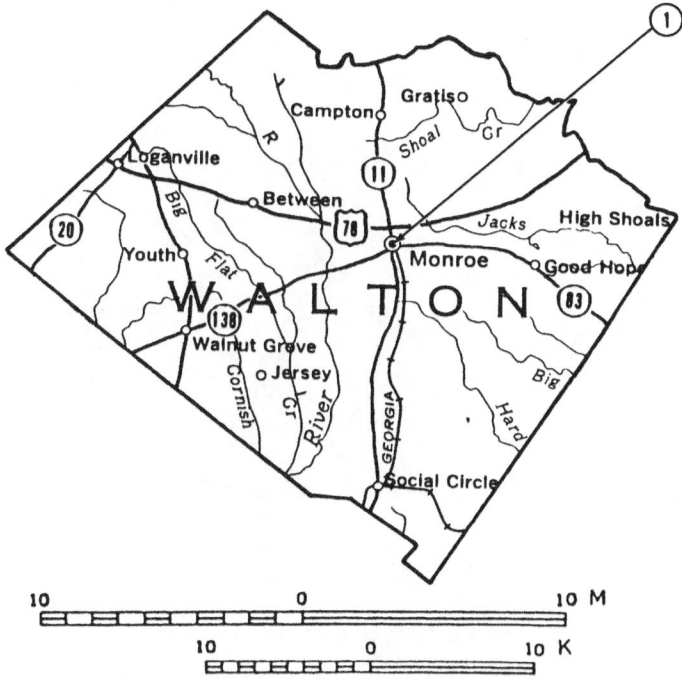

WALTON COUNTY

1. Monroe area: southeast on SR-83 to Blasingame; take dirt
   road from there 6.5 miles to the Malcolm farm.
   Aquamarine. (1)

WASHINGTON COUNTY

1. Tennille area: 1 mile south on SR-15; collect on the Hugh
   Taubutton property.
   Quartz, cryptocrystalline: gem quality. (1-13-14-29)

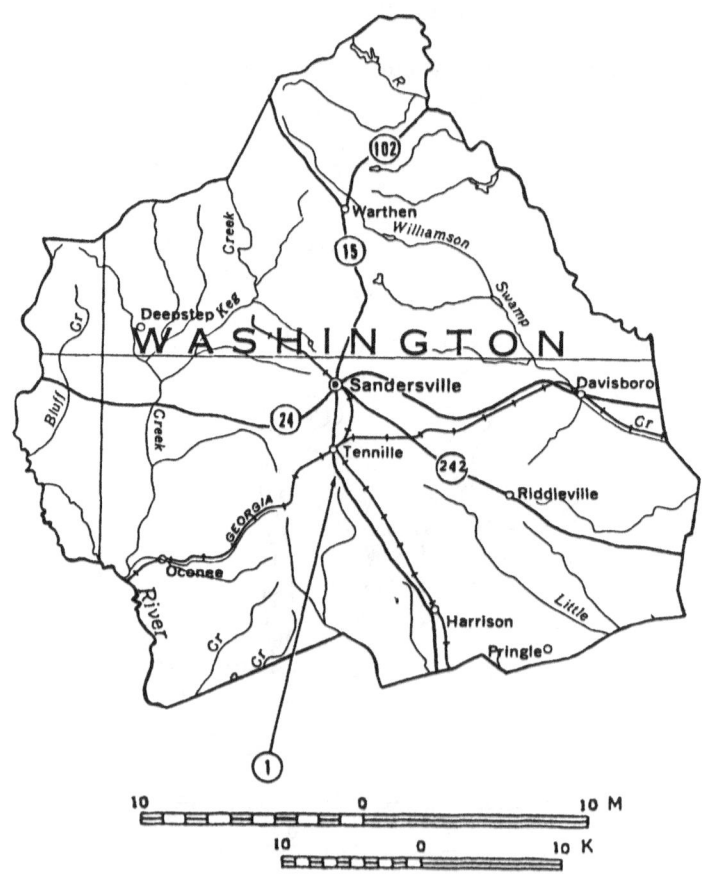

WHITE COUNTY

1. Nacoochee area: 3 miles east on SR-17 to cut on north side of road.

    Apatite: green. (4)

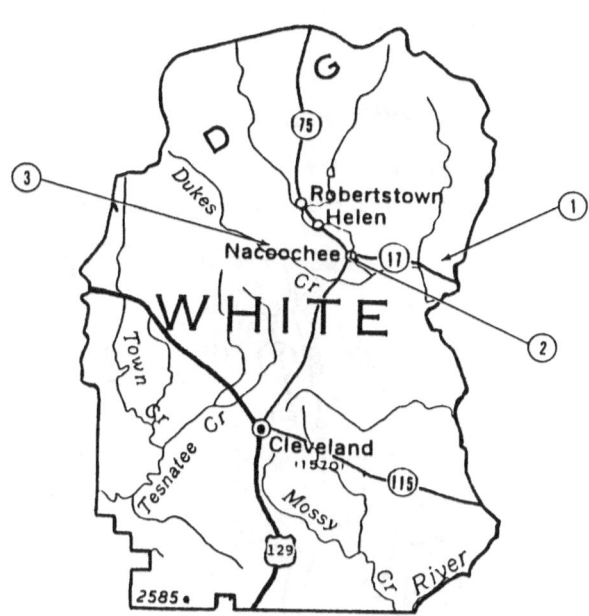

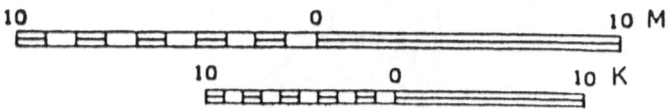

2. Nacoochee Valley: at the Horshaw Mine.

    Diamond: 1 stone; under 0.125 ct; opaque; found in 1866. (2)

3. Robertstown area: 2.4 miles south-southwest; on north bank of Dukes Creek.

> Diamond: 1 stone; 1.3 ct; greenish; fine; found in 1882. (17)

4. Unspecified location.

> Diamond: 1 stone; very fine; 1 ct; specifics unknown. (17)

WHITFIELD COUNTY

1. Dalton area: along Tarr Creek.

> Jasper, oolitic: red. Oolite: black; siliceous. (4)

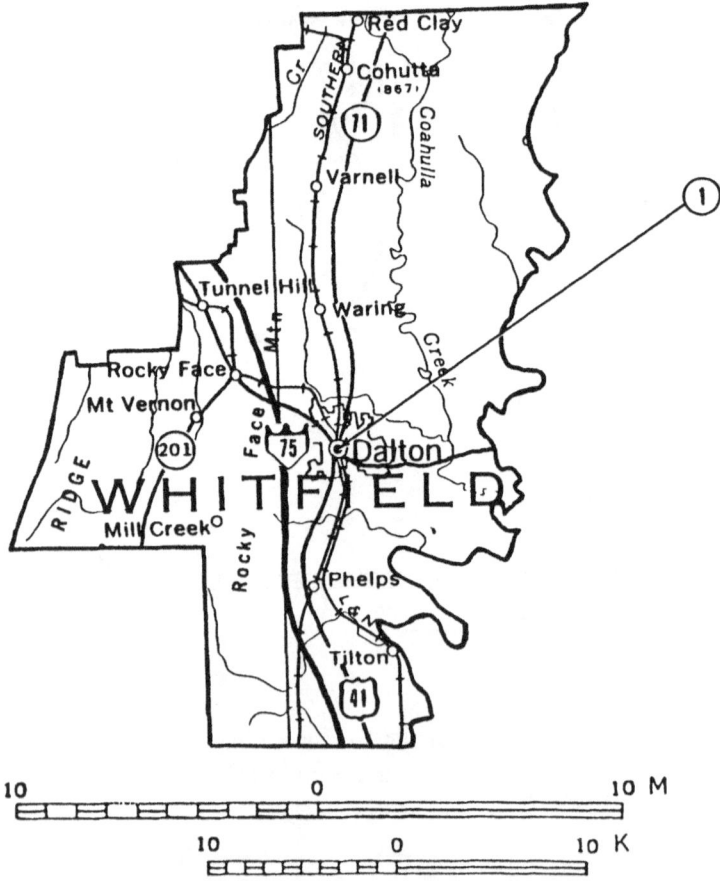

WILKES COUNTY

1. Metasville area: just east, on Lincoln County border; at the old Magruder Mine.

    Azurite. Barite. Chalcocite. Chalcopyrite. Galena. Garnet. Malachite. Mica. Pyrite. Sphalerite. Spinel. (2)

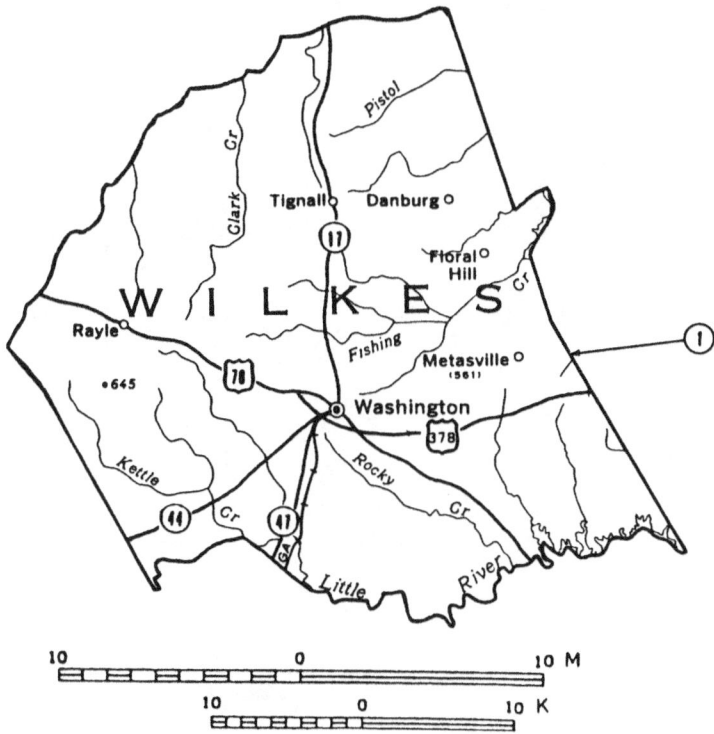

WILKINSON COUNTY

1. Gordon area: along the shores of area lake.
    Opalite. Quartz, rock. Quartz, smoky. (3-20)

2. McIntyre area mines.
    Bauxite: crystals. (2)

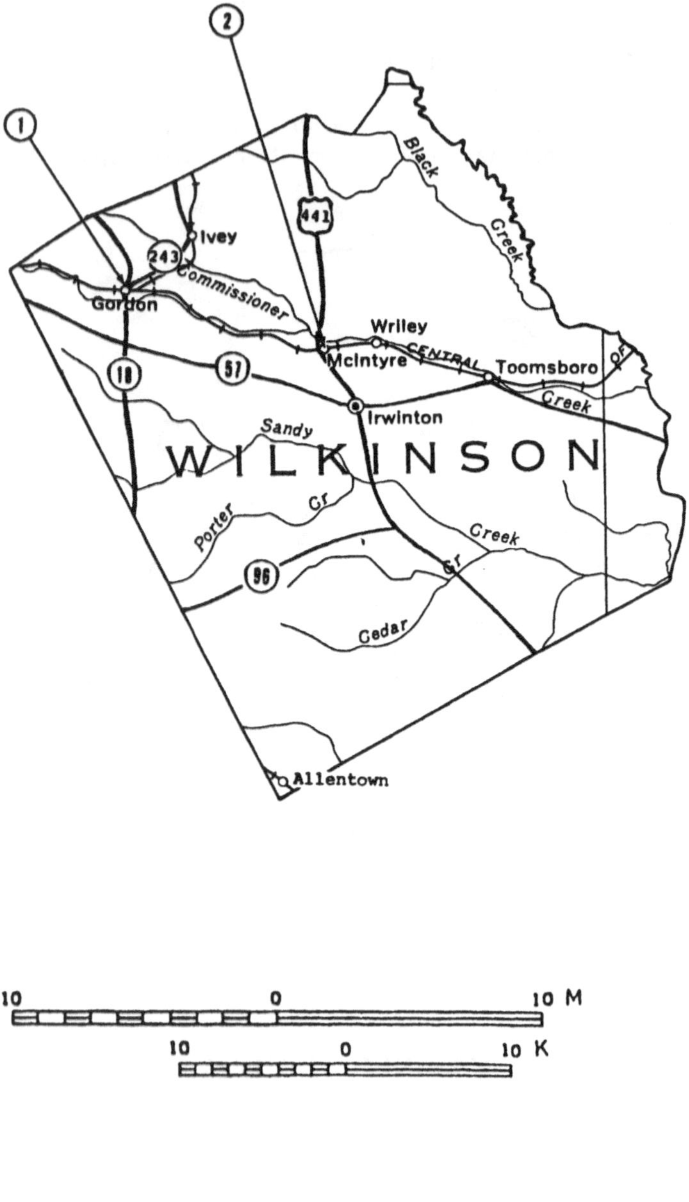

①

②

Black Creek

441

Ivey

243

Commissioner

Gordon

18

57

Wriley

McIntyre

CENTRAL

Toomsboro

Cr

Creek

Irwinton

Sandy

W I L K I N S O N

Porter

Cr

Creek

Cr

96

Cedar

Allentown

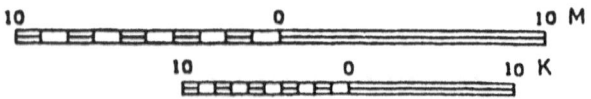

10   0   10 M

10   0   10 K

KENTUCKY

Statewide total of 103 locations in the following counties:

Adair (5)	Elliott (4)	Livingston (9)
Allen (1)	Estill (1)	Lyon (2)
Anderson (1)	Fayette (2)	Madison (2)
Ballard (1)	Franklin (1)	Mercer (2)
Barren (2)	Garrard (3)	Monroe (1)
Bath (3)	Graves (6)	Owen (1)
Bourbon (2)	Hardin (2)	Powell (1)
Boyd (1)	Harrison (2)	Rockcastle (6)
Boyle (3)	Hart (1)	Rowan (4)
Caldwell (3)	Jefferson (5)	Russell (1)
Clark (1)	Jessamine (2)	Trigg (1)
Crittenden (13)	Lincoln (6)	Woodford (2)

COLLECTING SITES IN KENTUCKY

(FOR SPECIFIC DETAILS ON THESE LOCATIONS,
SEE THE FOLLOWING COUNTY MAPS AND RELATED
TEXT.)

ADAIR COUNTY

1. Chance area: near Cabin Fork Creek; at the Henry Burris farm.

    Diamond: 1 stone; 0.78 ct; fine; pale yellow; found in 1888. (1)

2. Coburg area: northeast; upstream in the Green River and its tributaries.

    Pearl. (7)

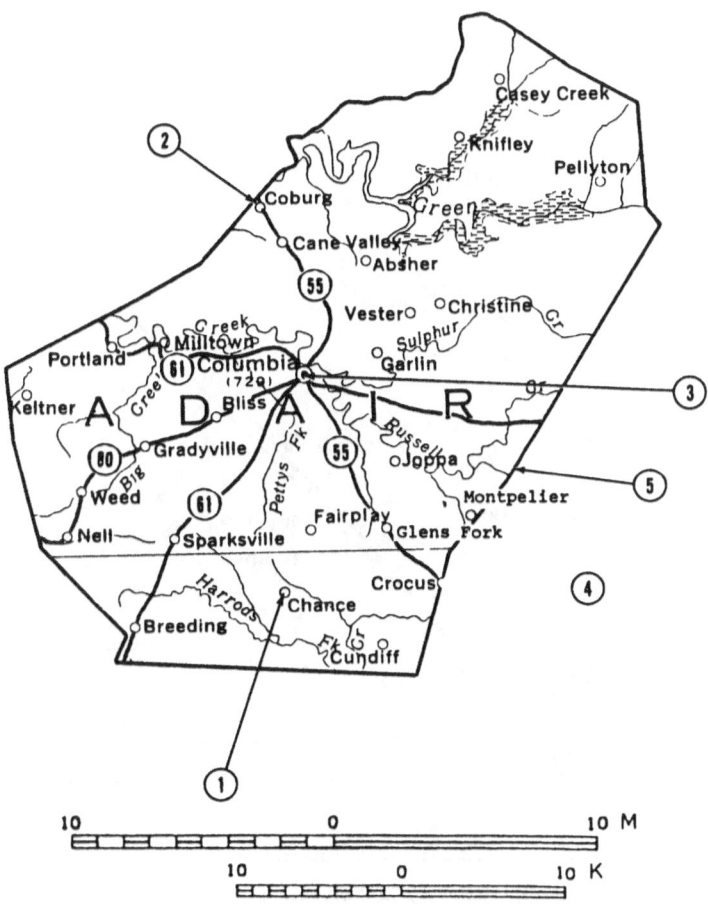

3. Columbia area: at the Shamrock Stone Company Quarry.
   Quartz: geodes to 16″ diameter. (12)

4. County-wide.
   Quartz geodes to 16″ diameter, containing variously
   barite crystals, pink calcite crystals, pale blue celestite crystals, banded chalcedony, goethite crystals,
   and rock quartz in both bright drusy and sharply terminated crystals. (1-4-7-13-20-29)

5. Montpelier area: Russell Creek tributary.
   Diamond: 1 stone; 0.766 ct; yellow; 0.375″ × 0.125″
   × 0.125″. (7)

ALLEN COUNTY

1. Scottsville area quarry.
   Quartz: geodes to 16″ diameter. (12)

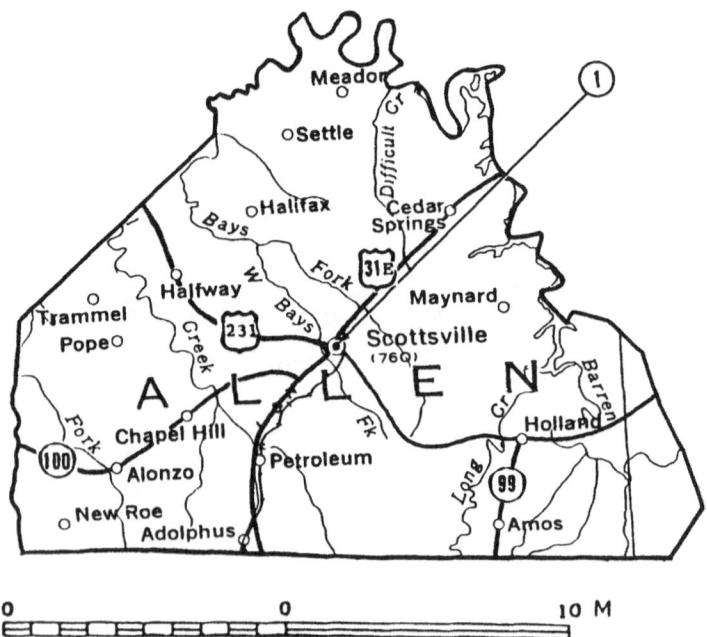

ANDERSON COUNTY

1. County-wide.

Quartz: geodes to 16″ diameter variously containing barite crystals, pink calcite crystals, pale blue celestite crystals, banded chalcedony, goethite, and rock quartz in sparkling drusy and sharply terminated crystals. (1-4-7-13-20-29)

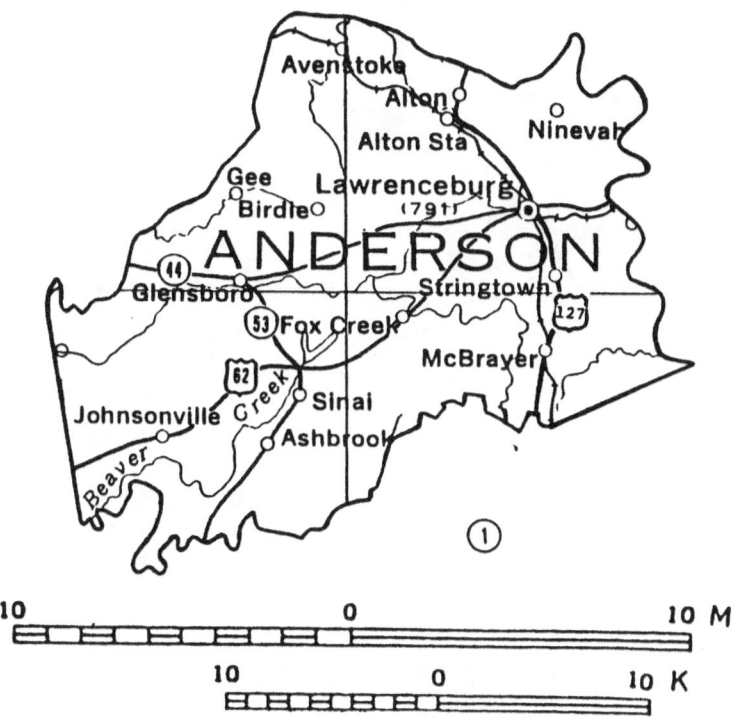

BALLARD COUNTY

1. Wickliffe area.

Jasper. (10)

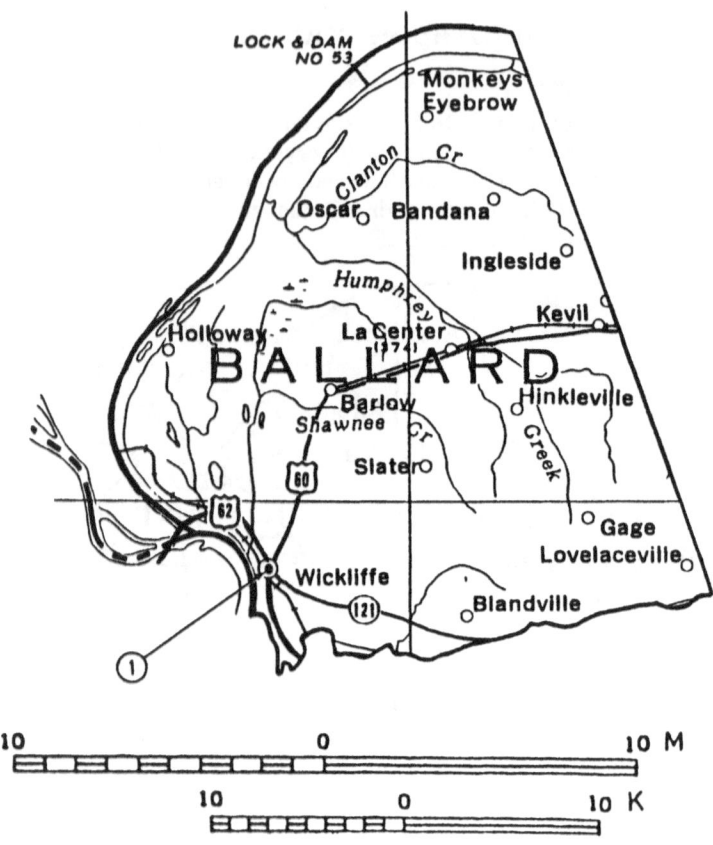

BARREN COUNTY

1. Cave City area quarries.

    Calcite (Marble var.): pink, yellow; some gemmy material. (12)

2. Glasgow area: in streams to east.

    Quartz: geodes to 16″ diameter variously containing barite crystals, pink calcite crystals, pale blue celestite crystals, banded chalcedony, goethite, and rock quartz in sparkling drusy and sharply terminated crystals. (7-20)

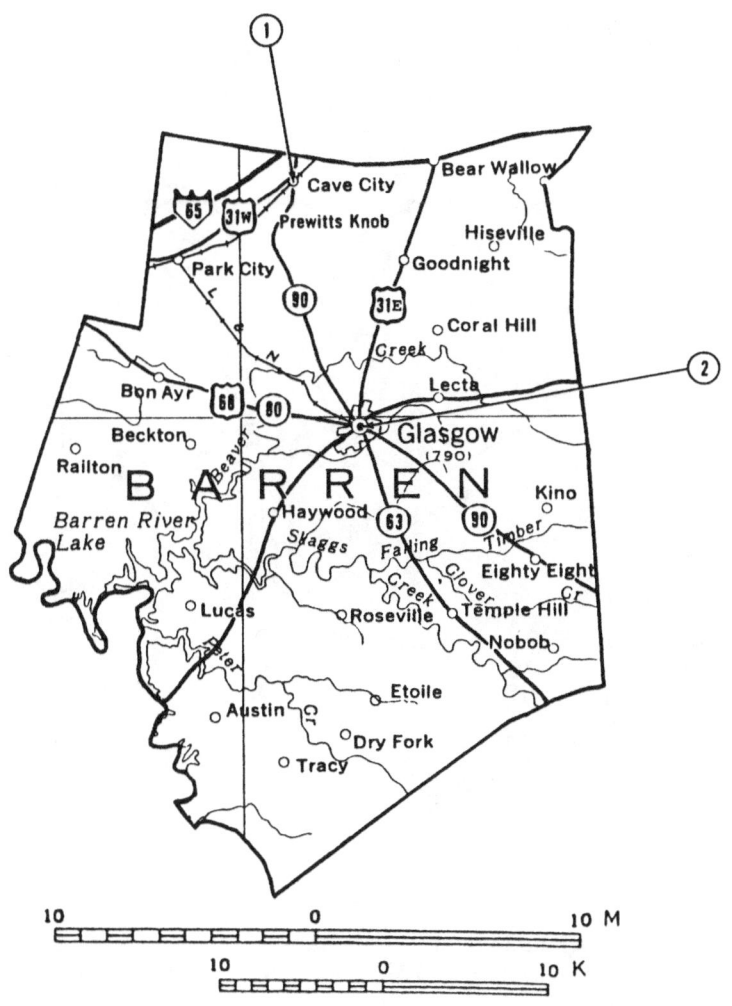

BATH COUNTY

1. Olympia Springs area: 0.75 mile west; 3 miles north of US-60; at the old Rose River Iron Strip Mines.

    Hematite: nodules to 3" diameter. Jasper: red. (20)

2. Owingsville area iron mines.

    Hematite: some good botryoidal clusters. (2)

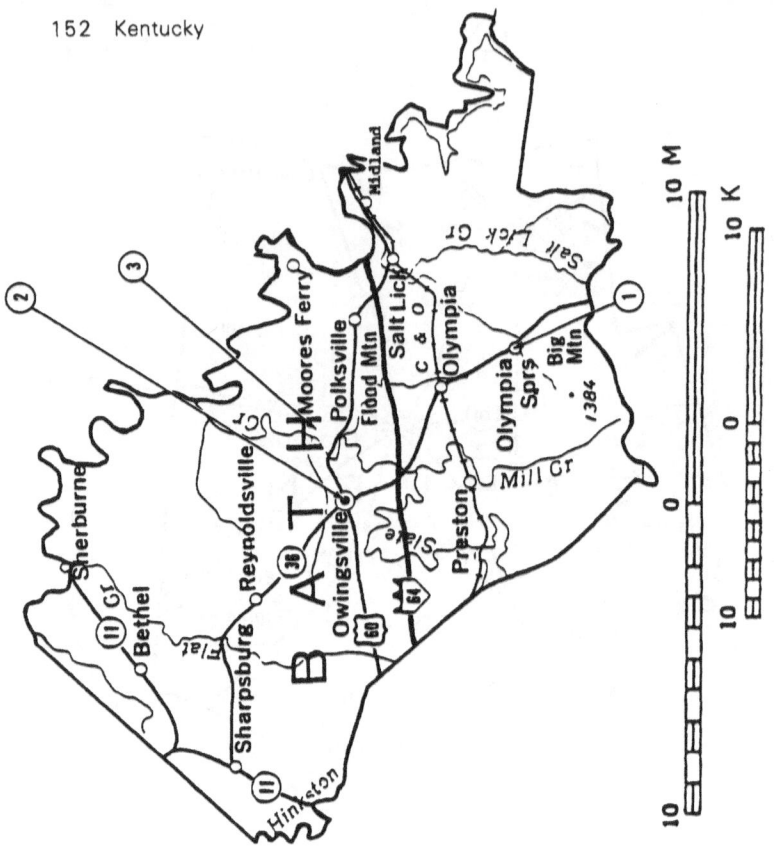

3. Salt Lick area: 5 miles west on SR-60 to side road right; then 3 miles north to abandoned Rose River Iron Mines (strip).

Hematite: nodules; botryoidal. Jasper, red. (2)

BOURBON COUNTY

1. Millersburg area mines.
   Barite: crystals. Galena. (2)

2. Paris area mines.
   Barite: crystals. Galena. (2)

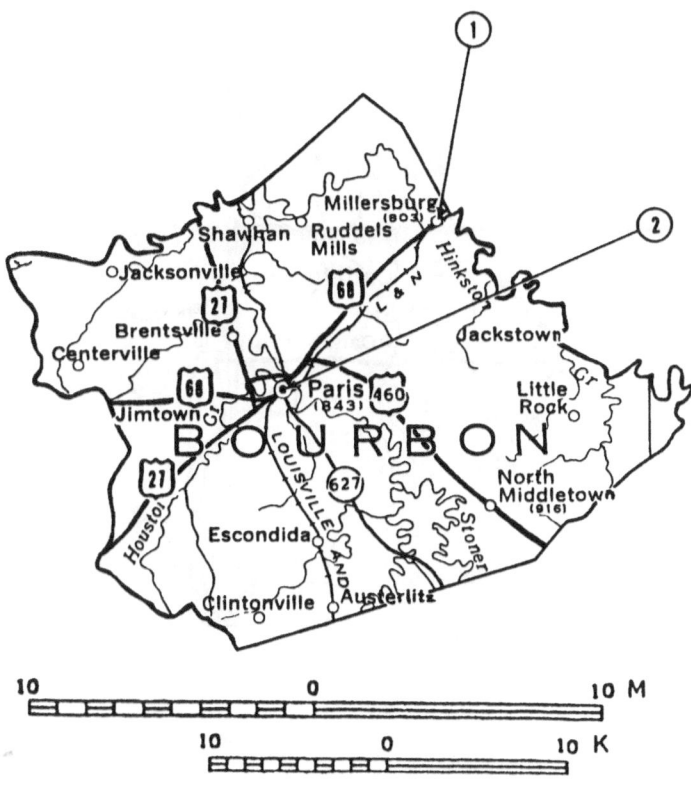

```
10          0          10 M
├┤┤┤┤┤┤┤┤┤┤┤┤┤┤┤┤┤┤┤┤┤┤┤┤┤┤┤┤┤┤┤┤┤┤┤┤┤┤┤┤┤┤┤┤┤┤┤┤┤┤┤┤┤┤┤┤┤┤┤┤┤┤┤┤

        10      0      10 K
    ├┤┤┤┤┤┤┤┤┤┤┤┤┤┤┤┤┤┤┤┤┤┤┤┤┤┤┤┤┤┤┤┤┤┤┤┤┤┤
```

BOYD COUNTY

1. Ashland area iron mines.
   Siderite. (2)

BOYLE COUNTY

1. County-wide.
   Quartz: geodes to 16″ diameter variously containing
   barite crystals, pink calcite crystals, pale blue celestite crystals, banded chalcedony, goethite, and rock
   quartz in sparkling drusy and sharply terminated
   crystals. (1-4-7-13-20-29)

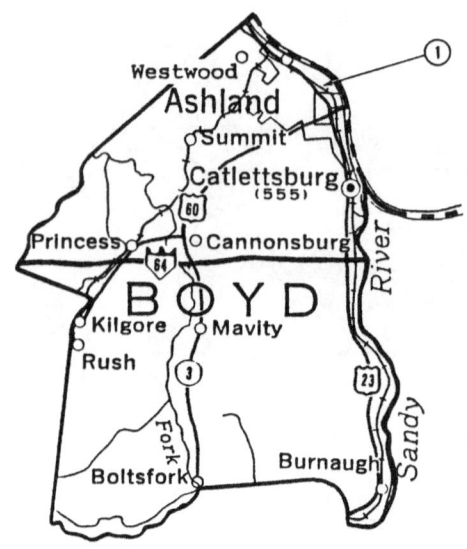

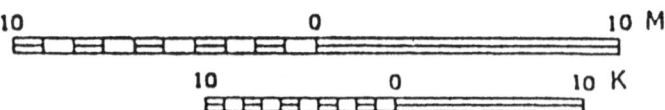

2. Danville area: 3.5 miles east on SR-52 to bridge; just south of the confluence of Hanging Rock Creek and the Dix River; at barite pit.

>Barite: tabular crystals. Calcite. Fluorite. Sphalerite. (10)

3. Danville area mines.

>Barite: crystals. (2)

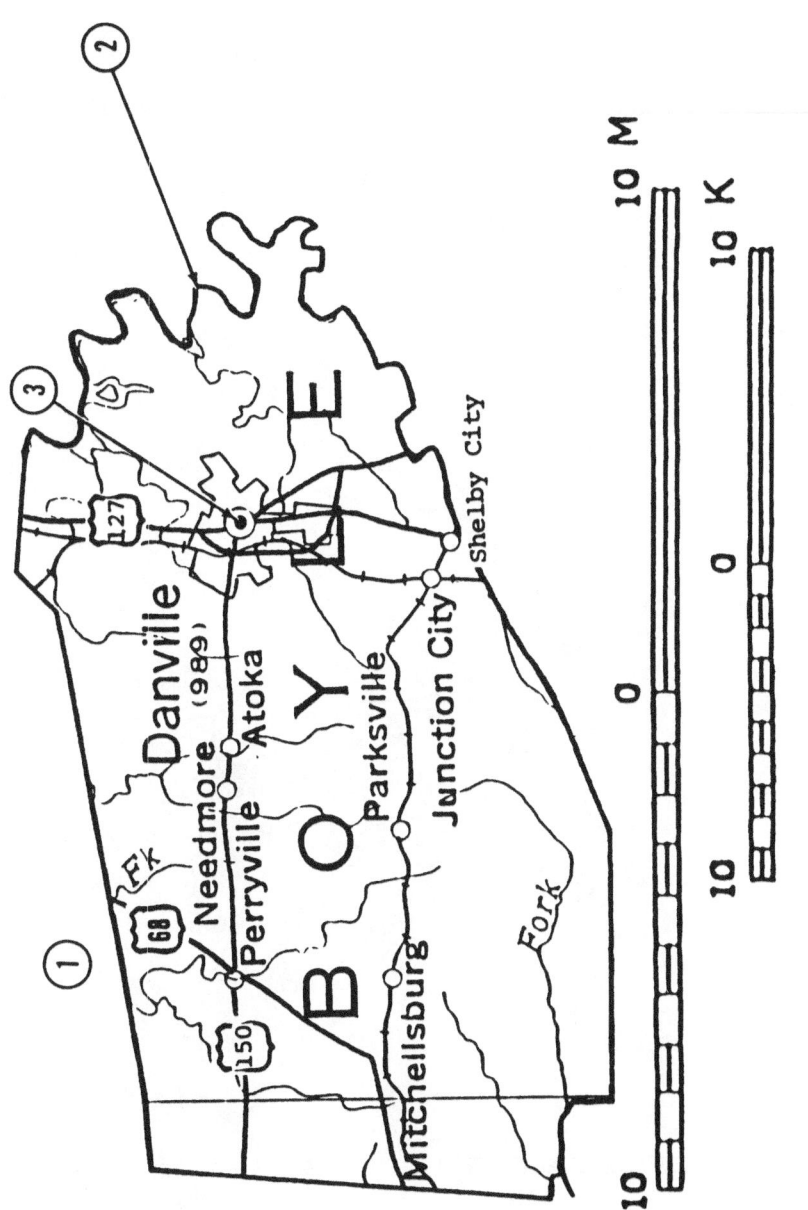

CALDWELL COUNTY

1. Crider area: just north.
   Fluorite. (29-faults)

2. Fredonia area mines.
   Barite: crystals. (2)

3. Princeton area quarries.
   Calcite. Fluorite. (12)

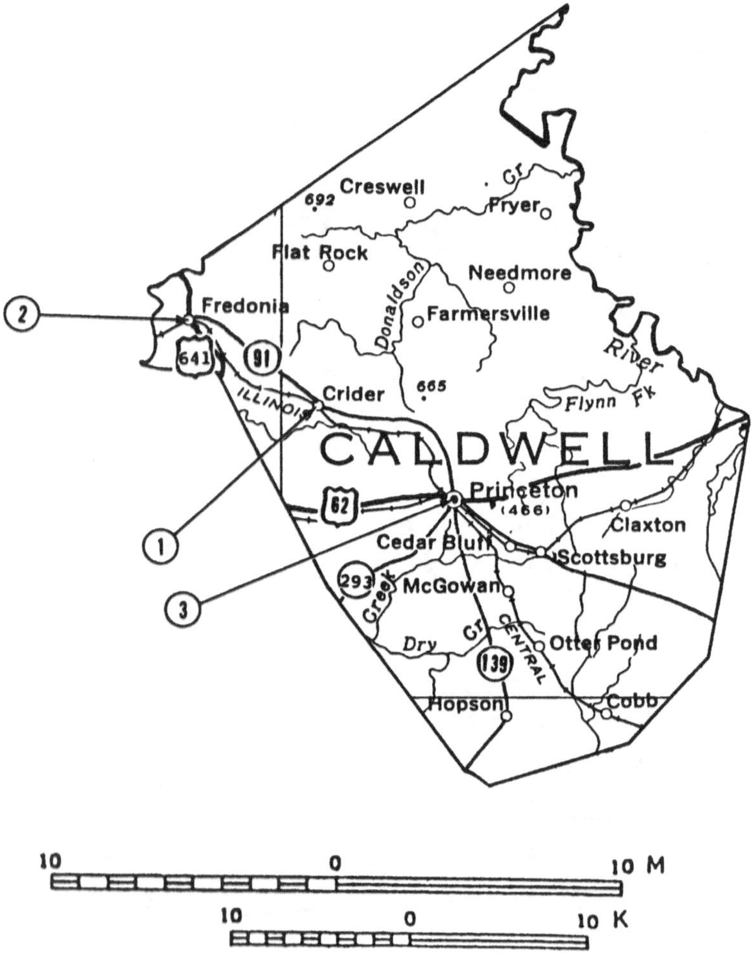

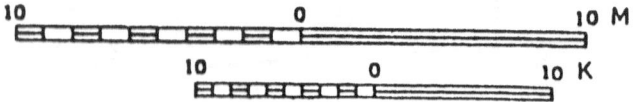

CLARK COUNTY

1. County-wide.

Quartz: geodes to 16″ diameter variously containing barite crystals, pink calcite crystals, pale blue celestite crystals, banded chalcedony, goethite, and rock quartz in sparkling drusy and sharply terminated crystals. (1-4-7-13-20-29)

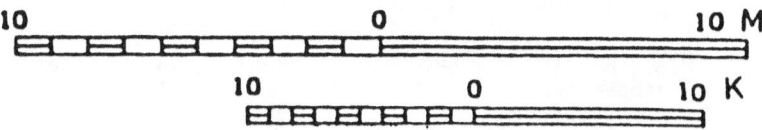

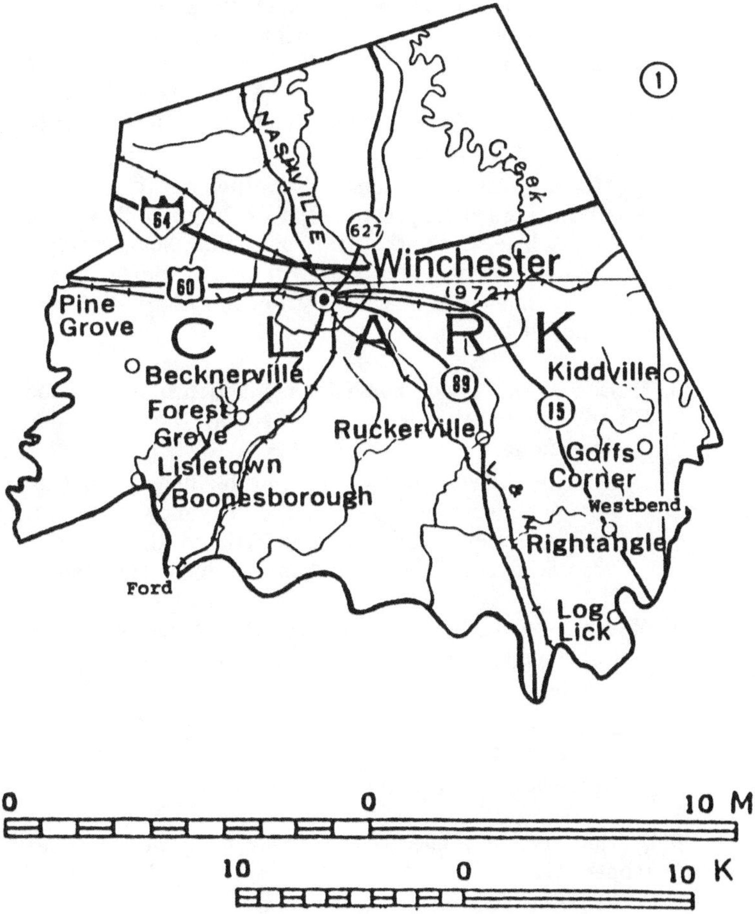

CRITTENDEN COUNTY

1. County-wide in fluorite mines.

   Fluorite: beautiful individual crystals to 5″ wide; some single crystals, some twinning, abundant and spectacular stacked clusters; some striated and etched; many clusters sprinkled with drusy quartz, calcite, marcasite, and pyrite; both transparent and translucent; predominantly blue, yellow, and clear, but also honey-colored, rich hyacinth, cerise, lettuce green, deep purple, lavender, apple green, bright canary, pink, brown, and black. (2)

2. Marion area: 1 mile west from junction of US-60 and SR-855; at the Kirk Mine.

   Barite. Calcite. Dolomite. Fluorite: fine crystals to 5″ on edge; pink, green, deep rose, bright yellow, deep blue, lilac, hyacinth, clear, pale blue, some orange; striated, etched, sprinkled with pyrite and/or marcasite. Galena. Marcasite. Pyrite. Silver. Smithsonite. Sphalerite. (2)

3. Marion area: 3 miles north from junction of US-60 and CR-1688; on east side of the road; at the Crittenden Springs fault.

   Barite. Calcite. Dolomite. Fluorite: fine crystals to 5″ on edge; color-zoned; pink, blue, yellow, lavender, green, hyacinth, clear; etched, striated. Galena. Marcasite. Pyrite. Silver. Smithsonite. Sphalerite. (29)

4. Marion area: 3 miles northwest on CR-1688; at the Crittenden Springs Quarry.

   Calcite. (12)

5. Marion area mines.

   Fluorite: crystals and massive. (2)

6. Marion area: in mines west of town; many of the mines inoperative.

   Fluorite. Quartz, rock. (2)

7. Marion area: near the Glendale Baptist Church; 1 mile

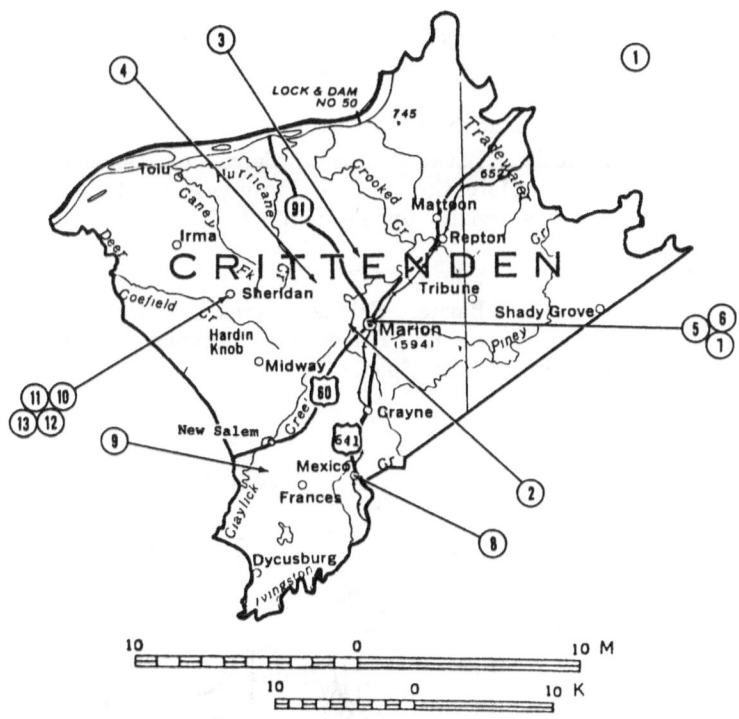

south of CR-1668 on the Glendale Church Road; west of the road and upstream on Caney Creek a short distance; at the Hickory Cane Mine.

Fluorite. Marcasite. Peridotite. (2)

8. Mexico area: 0.8 mile southwest on US-70 to the Crider Fluorspar Company sign; turn left and go 0.3 mile to turn left again across RR tracks; at the Pigmy Fluorspar Mine.

Fluorite. (2)

9. New Salem area: 1 mile south on CR-855; at the inoperative Kirk Mine dump.

Fluorite. Marcasite. Pyrite. (2)

10. Sheridan area: at the Cartwright Mine.

Anglesite. Barite. Calcite. Cerrusite. Dolomite.

Galena. Marcasite. Pyrite. Pyromorphite. Quartz, rock. Smithsonite. Sphalerite. (2)

11. Sheridan area: at the LaRue Mine.

Anglesite. Barite. Calcite. Cerrusite. Dolomite. Galena. Marcasite. Pyrite. Pyromorphite. Quartz, rock. Smithsonite. Sphalerite. (2)

12. Sheridan area: at the Lead Mine.

Anglesite. Barite. Calcite. Cerrusite. Dolomite. Galena. Marcasite. Pyrite. Pyromorphite. Quartz, rock. Smithsonite. Sphalerite. (2)

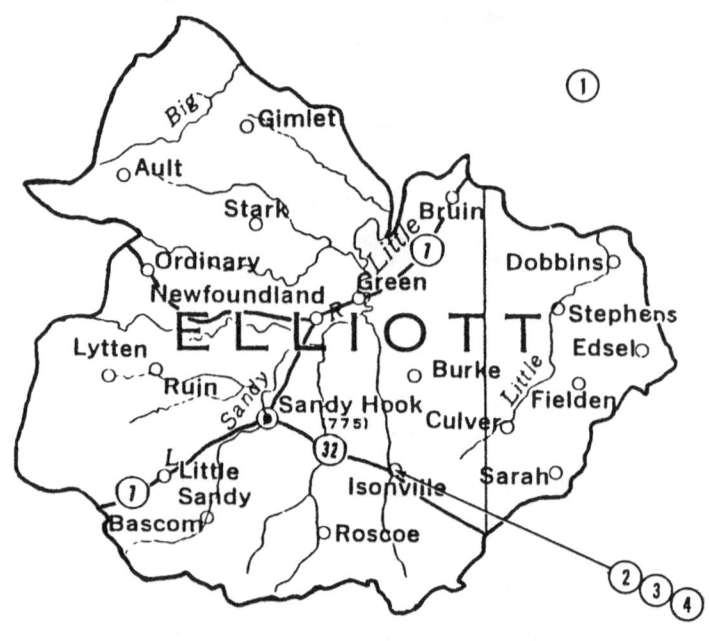

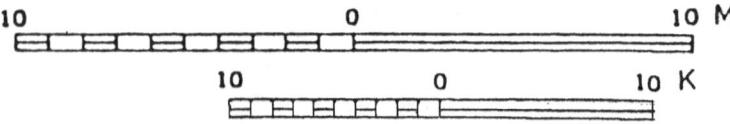

13. Sheridan area: at the Macer Mine.

> Anglesite. Barite. Calcite. Cerrusite. Dolomite. Galena. Marcasite. Pyrite. Pyromorphite. Quartz, rock. Smithsonite. Sphalerite. (2)

ELLIOTT COUNTY

1. County-wide in igneous dikes.

> Apatite. Chromite. Diopside. Enstatite. Feldspar. Garnet, almandine: fine color. Peridotite. Quartz, rock. Serpentine. (1-13)

2. Isonville area.

> Garnet, pyrope. (1)

3. Isonville area: across from Ison Johnson School; on south bank of Ison Creek above its confluence with Johnson Creek.

> Garnet, almandine. Ilmenite. Magnetite. Peridotite (Olivine). (1-13-20-29-igneous)

4. Isonville area: in streambed of Isons Creek.

> Enstatite: fine gem quality. Garnet, pyrope: gem quality. Ilmenite. Mica. Peridotite (Olivine). (7)

ESTILL COUNTY

1. County-wide.

> Quartz: geodes to 16″ diameter variously containing barite crystals, pink calcite crystals, pale blue celestite crystals, banded chalcedony, goethite, and rock quartz in sparkling drusy and sharply terminated crystals. (1-4-7-13-20-29)

FAYETTE COUNTY

1. County-wide.

> Quartz: geodes to 16″ diameter variously containing barite crystals, pink calcite crystals, pale blue celes-

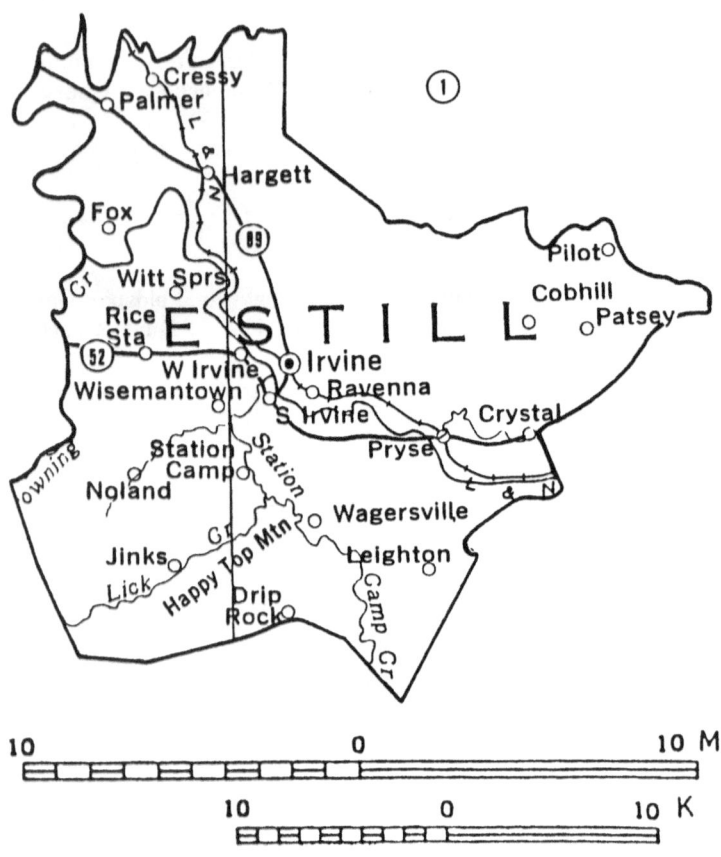

tite crystals, banded chalcedony, goethite, and rock quartz in sparkling drusy and sharply terminated crystals. (1-4-7-13-20-29)

2. Lexington area: in the Elk Lick Falls area.
   Travertine (aka "Cave Onyx"): banded; yellow, brown. (32-39)

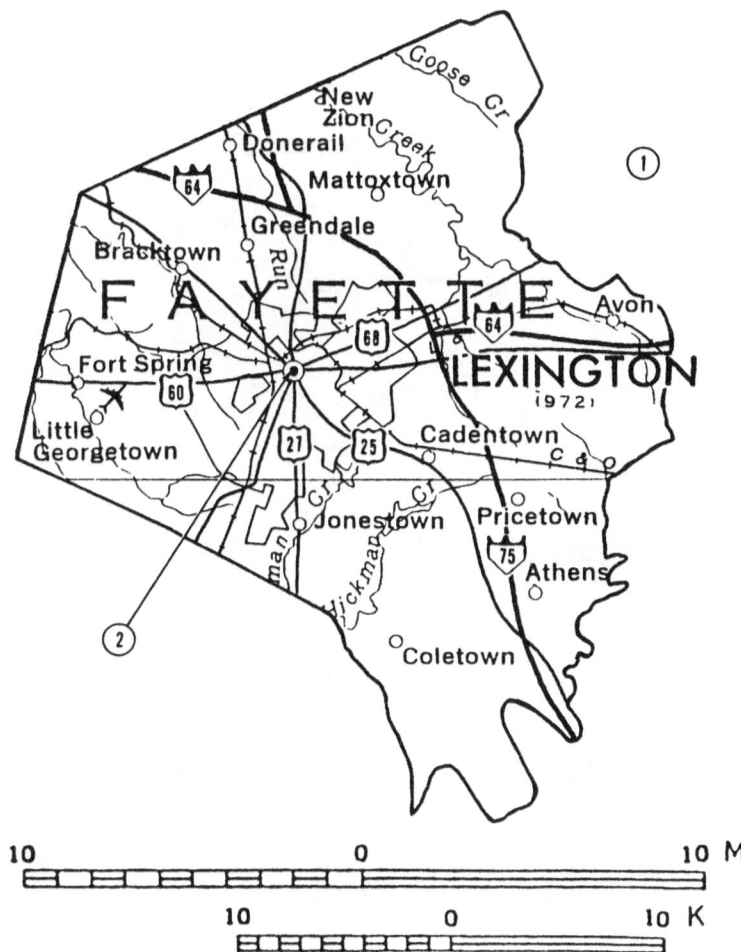

FRANKLIN COUNTY

1. Frankfort area: mines in the vicinity of Kissinger.
   Barite. Galena. (2)

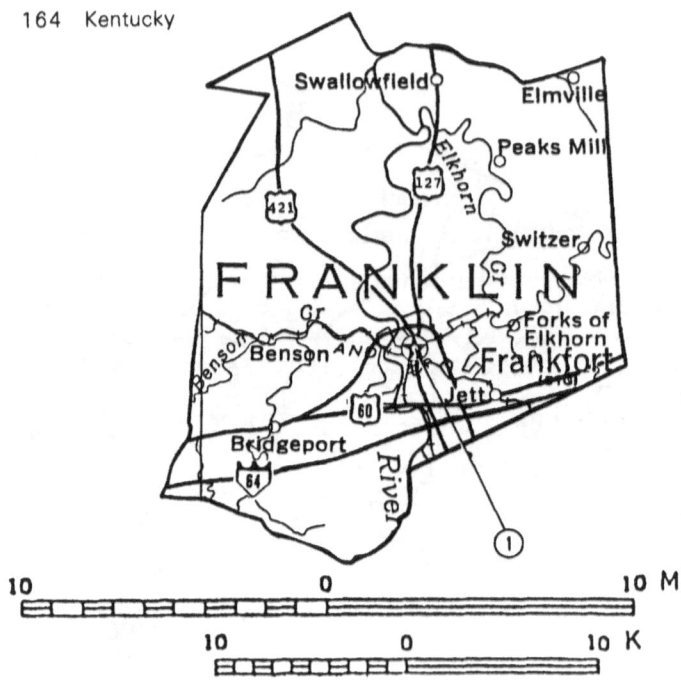

GARRARD COUNTY

1. County-wide.

    Quartz: geodes to 16" diameter variously containing barite crystals, pink calcite crystals, pale blue celestite crystals, banded chalcedony, goethite, and rock quartz in sparkling drusy and sharply terminated crystals. (1-4-7-13-20-29)

2. Hyattsville area: north of SR-52; along Boone Creek; in barite pit.

    Barite: tabular crystals. Calcite. Quartz: geodes to 11" diameter variously containing barite crystals, pink calcite crystals, pale blue celestite crystals, banded chalcedony, goethite, and rock quartz in sparkling drusy and sharply terminated crystals. (1-4-7-10-13-20-29)

3. Lancaster area: 4 miles west.

    Barite. (1-13)

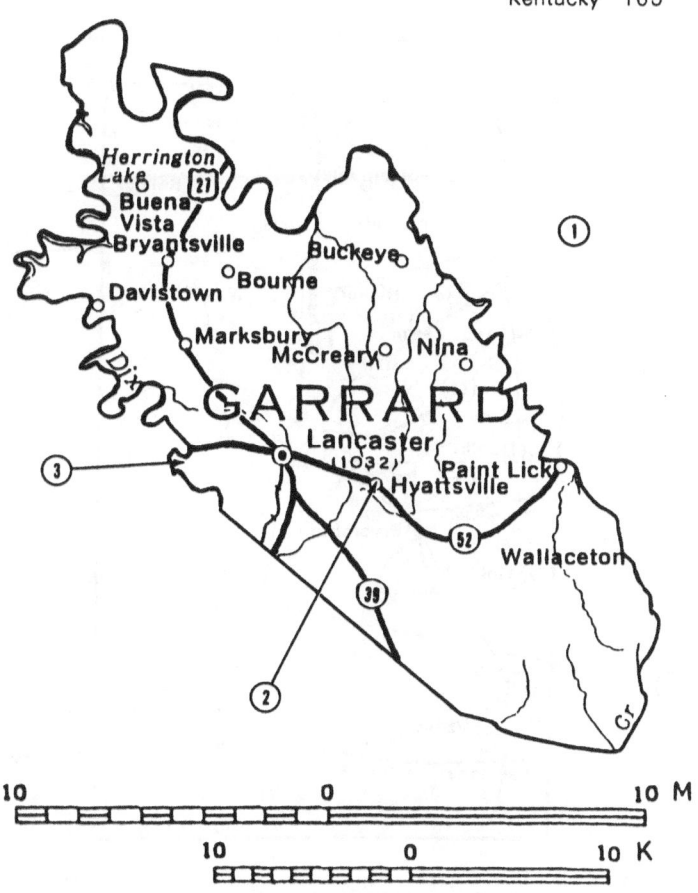

GRAVES COUNTY

1. Farmington area.

    Chalcedony. Chert. Fossils: marine; especially sili-
    cified coral. Opal, wood. Quartz, rock. (35)

2. Hard Money area: 2.5 miles southeast.

    Hematite: small nodules. (3-18-39)

3. Hickory area.

    Chalcedony. Chert. Fossils: marine; especially sili-
    cified coral. Opal, wood. Quartz, rock. (35)

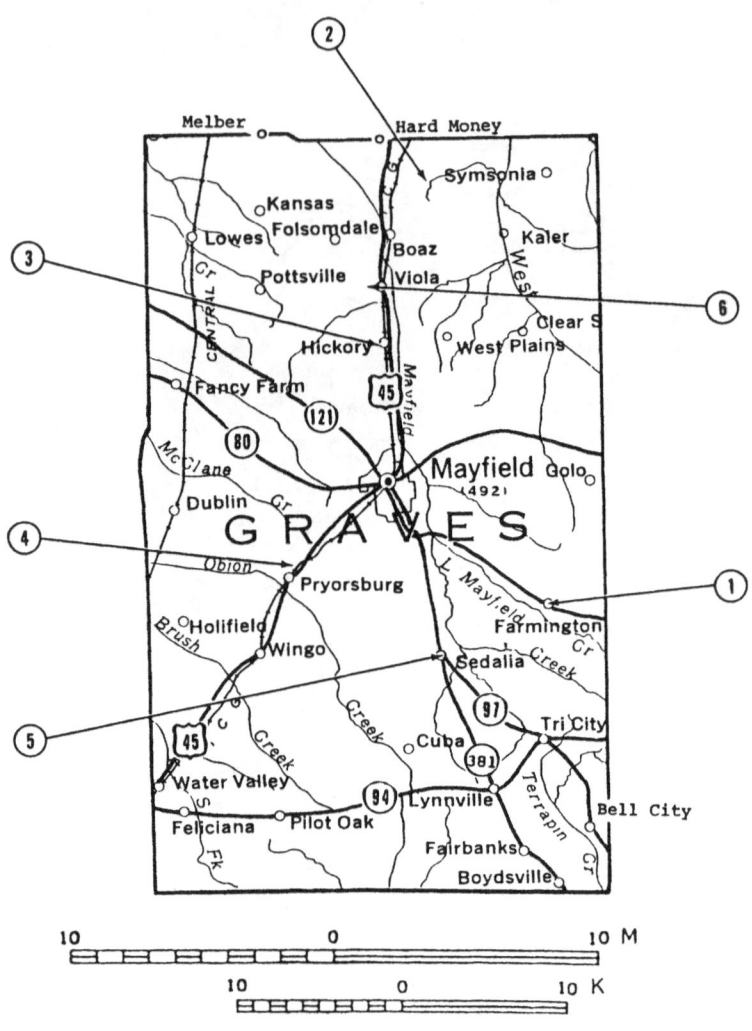

4. Pryorsburg area: just north on US-45.

> Agate. Chalcedony. Chert. Fossils: marine; especially silicified coral. Jasper. Opal, wood. Petrified wood: silicified. Quartz, rock. (35)

5. Sedalia area.

> Chalcedony. Chert. Fossils: marine; especially silicified coral. Opal, wood. Quartz, rock. (35)

6. West Viola area.

> Chalcedony. Chert. Fossils: marine; especially silicified coral. Opal, wood. Quartz, rock. (35)

HARDIN COUNTY

1. Elizabethtown area quarries.

> Quartz: geodes to 14″ diameter variously containing barite crystals, pink calcite crystals, pale blue celes-

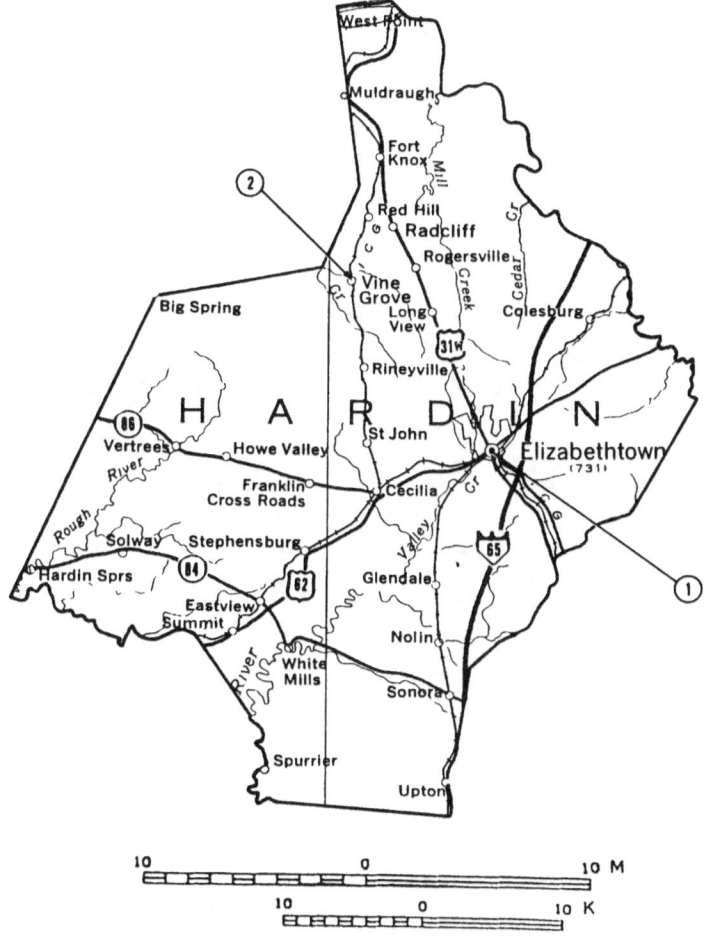

tite crystals, banded chalcedony, goethite, and rock quartz in sparkling drusy and sharply terminated crystals. (12)

2. Vine Grove area quarry.

Quartz: geodes to 14″ diameter variously containing barite crystals, pink calcite crystals, pale blue celestite crystals, banded chalcedony, goethite, and rock quartz in sparkling drusy and sharply terminated crystals. (12)

HARRISON COUNTY

1. Cynthiana area: 3 miles southeast.

Barite. (9-29-limestone)

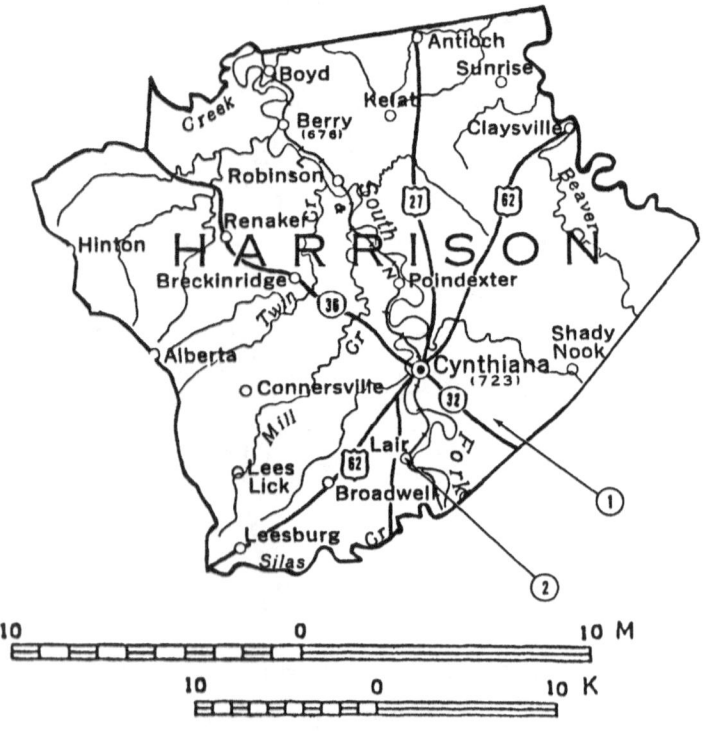

2. Lair area mines and prospects.
  Barite. Galena. (2-21)

HART COUNTY

1. Rowletts area.
  Petrified wood: pieces to 25 lb. (1-13)

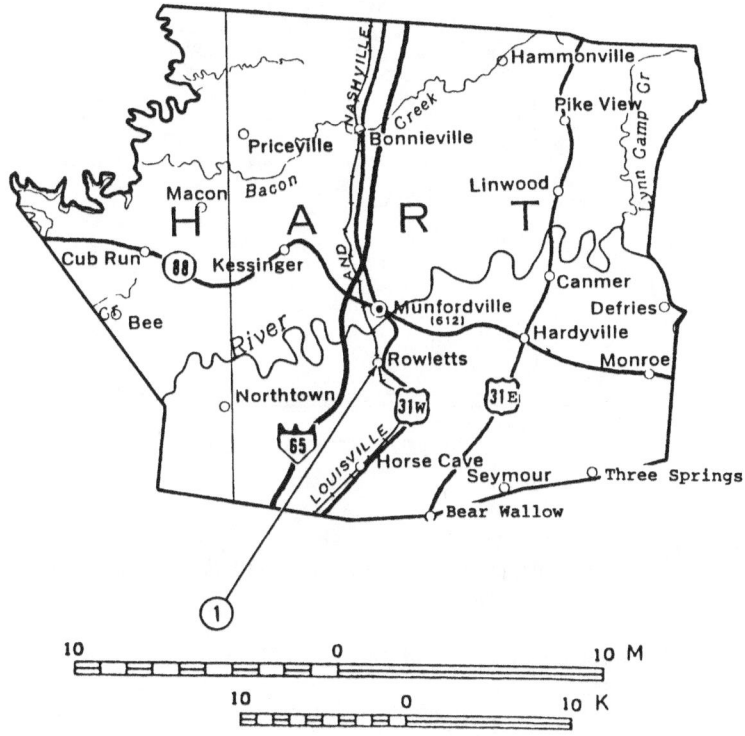

JEFFERSON COUNTY

1. Louisville area: Coral Ridge region; 1.5 miles south and 0.5
  mile east of the old National Turnpike; at Buttermold
  Knob.
  Fossils: coral; silicified. (4)

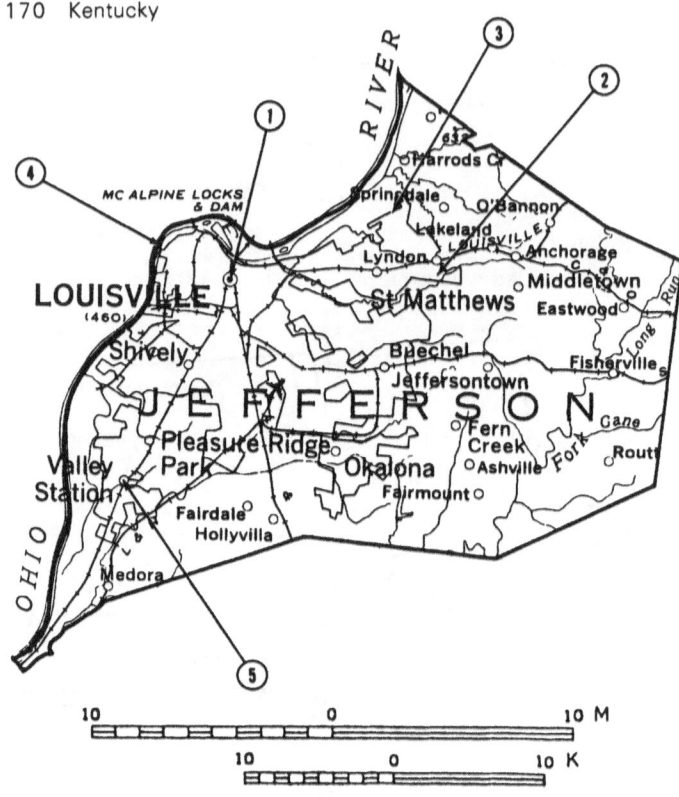

2. Louisville area: east end of city in several stone quarries.
Fossils: marine; brachiopods, cephalopods, corals, etc.
(12-limestone)

3. Louisville area: in and just outside city in road cuts along
US-42.
Fossils, marine: coral; silicified. (4)

4. Louisville area: south shore of Ohio River.
Fossils: marine; brachiopods, cephalopods, corals, etc.
(7-20)

5. Valley Station area: south to Bullitt County border; in area
of Muldraughs Hill; in road cuts.
Quartz geodes to 10″ diameter. (4-14-29)

JESSAMINE COUNTY

1. County-wide.

   Quartz: geodes to 16″ diameter variously containing barite crystals, pink calcite crystals, pale blue celestite crystals, banded chalcedony, goethite, and rock quartz in sparking drusy and sharply terminated crystals. (1-4-7-13-20-29)

2. Nicholasville area: in abandoned mines and mine dumps.
   Barite. Calcite. Fluorite. Galena. Sphalerite. (2)

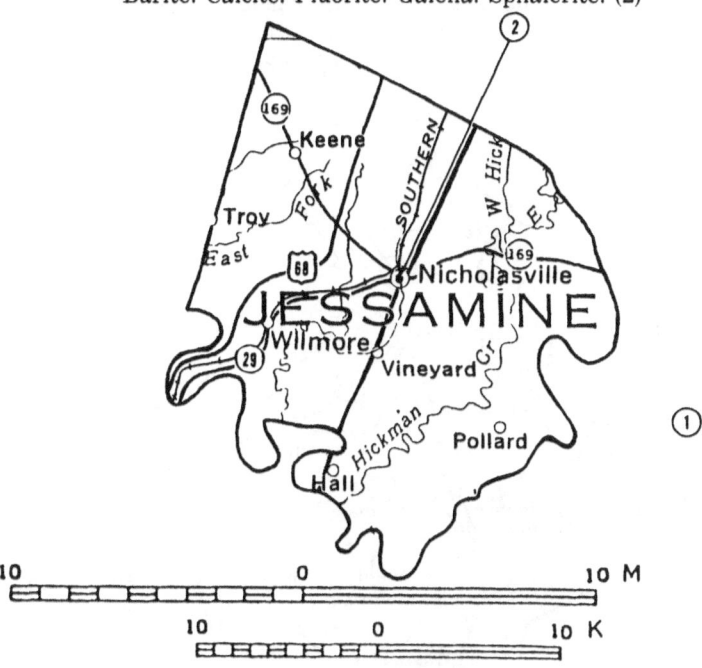

LINCOLN COUNTY

1. County-wide.

   Quartz: geodes, usually 4″ to 16″ diameter (but up to 36″ diameter) variously containing barite crystals, pink calcite crystals, pale blue celestite crystals, banded chalcedony, goethite, and rock quartz in sparkling drusy and sharply terminated crystals. (1-4-7-13-20-29)

2. Crab Orchard area: south on SR-643 (Crab Orchard Road) to the junction of SR-643 and CR-1770; go south across RR tracks and continue 2 miles to unmarked side road at right just before small bridge; turn right and go 0.25 mile to wooden bridge without rails; collect in creekbed and banks; bed of creek liberally studded with geodes (the majority of which have disintegrated interiors, but occasional fine ones are found).

Calcite and Quartz: geodes and nodules to 24" diameter; containing barite crystals, pink calcite crystals, pale blue celestite crystals, banded chalcedony, goethite, and rock quartz in sparkling drusy and sharply terminated crystals. (1-4-7-13-20-29)

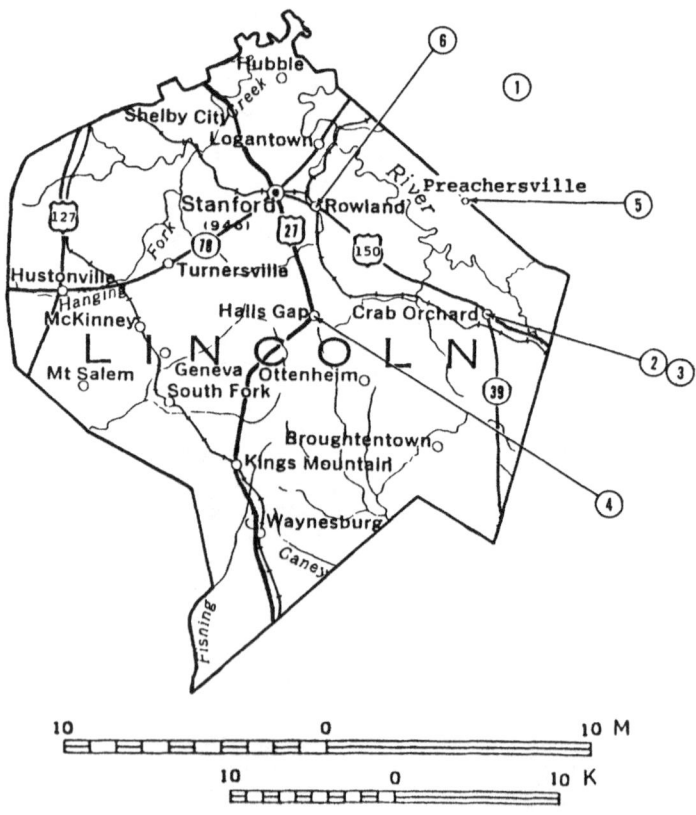

3. Crab Orchard area: south on SR-643 (Crab Orchard Road) to the junction of SR-643 and CR-1770; go south across RR tracks and continue 2 miles to unmarked side road at right just before small bridge; turn right and go 2.3 miles to wooden railless bridge; do not cross bridge—park on pull off to right just before it; collect upstream and down in creekbed and banks; abundant geodes and nodules, many with disintegrated interiors, but occasional fine specimens are found.

> Calcite and Quartz: geodes and nodules to 16″ diameter; variously containing barite crystals, pink calcite crystals, pale blue celestite crystals, banded chalcedony, goethite, and rock quartz in sparkling drusy and sharply terminated crystals. (1-4-7-13-20-29)

4. Halls Gap area: especially good geode collecting downstream from headwaters of the Green River, on both sides, from Miracle to New Bethel Church.

> Quartz: geodes to 16″ diameter variously containing barite crystals, pink calcite crystals, pale blue celestite crystals, banded chalcedony, goethite, and rock quartz in sparkling drusy and sharply terminated crystals. (1-4-7-13-20-29)

5. Preachersville area: abandoned barite pits and prospects.

> Barite: tabular crystals. Calcite. Quartz: geodes to 16″ diameter variously containing barite crystals, pink calcite crystals, pale blue celestite crystals, banded chalcedony, goethite, and rock quartz in sparkling drusy and sharply terminated crystals. (10-21)

6. Rowland area: numerous prospects just south of SR-52.

> Barite. Calcite. Celestite. Goethite. Fluorite. Quartz, rock. (21)

LIVINGSTON COUNTY

1. Birdsville area: on SR-137; at two abandoned quarries.

> Calcite. (12)

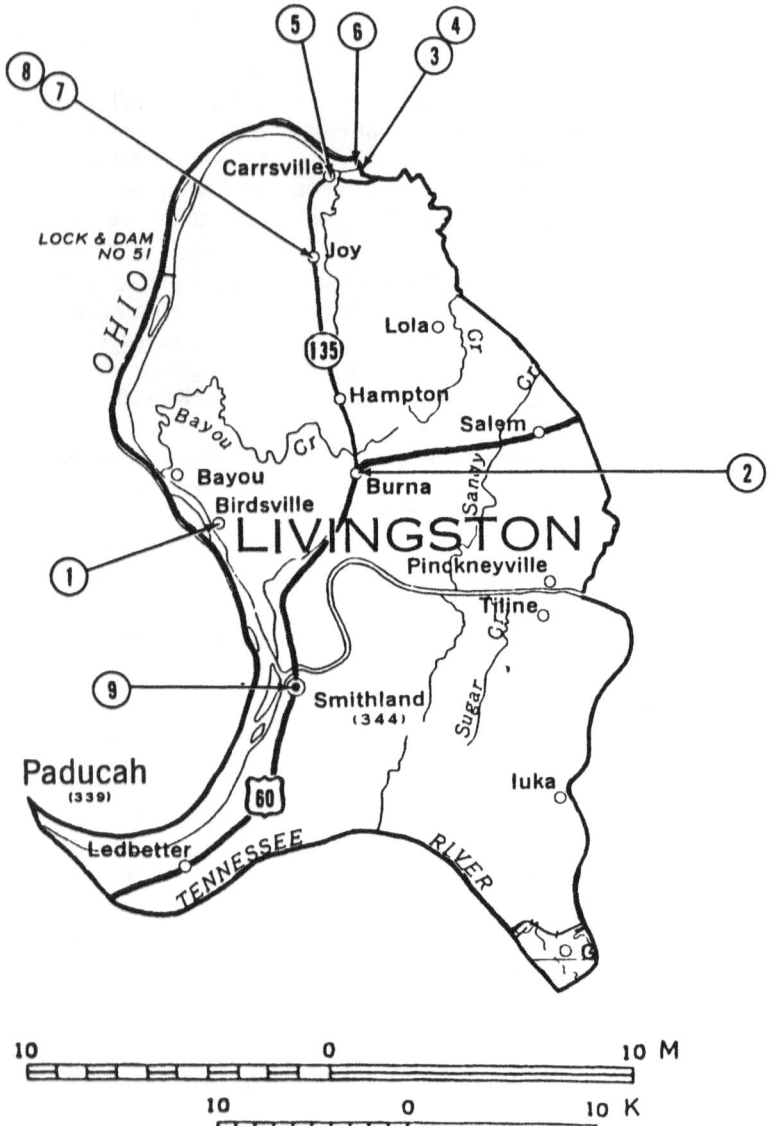

LOCK & DAM NO 51

Carrsville

Joy

I 35

Lola

Hampton

Salem

Bayou

Bayou
Birdsville

Burna

LIVINGSTON

Pinckneyville

Tiline

Smithland
(344)

Paducah
(339)

Iuka

Ledbetter

TENNESSEE   RIVER

OHIO

10   0   10 M

10   0   10 K

2. Burna area quarries.

   Calcite. Fluorite. Quartz, rock. (12-sandstone)

3. Carrsville area: 1 mile east; near the border of Crittenden County; in a fault exposing brecciated sandstone.

   Fluorite: excellent cubic crystals. (1-5-brecciated sandstone-13-14-29)

4. Carrsville area: 1 mile east; south of SR-387; at the Ellis Mine.

   Fluorite. (2)

5. Carrsville area fluorspar mines.

   Barite. Calcite. Dolomite: crystals. Fluorite: crystals to 4″ on edge; variety of colors; some color-zoning; some with etchings or striations. Galena. Marcasite. Pyrite. Smithsonite. Sphalerite. (2)

6. Carrsville area: just east; on north side of CR-3872; in fault close to the Ohio River.

   Fluorite: cubic crystals. (19-brecciated limestone -29)

7. Joy area fluorspar mines.

   Calcite. Fluorite. Galena. Quartz, rock. (2)

8. Joy area: on SR-133; at the Nancy Hanks Mine.

   Calcite. Fluorite. Pyrite. (2)

9. Smithland area: north on US-60; at the Dyer Hill Mine.

   Calcite. Fluorite. Galena. Sphalerite. (2)

LYON COUNTY

1. County-wide: all along the east shore of Kentucky Lake.

   Calcite geodes to 14″ diameter. Jasper. Quartz, rock. (14-20-29)

2. Eddyville area: in gravel dredging operation.

   Agate. Chalcedony. Chert (Fort Payne var.): well patterned and colored. Jasper. (10)

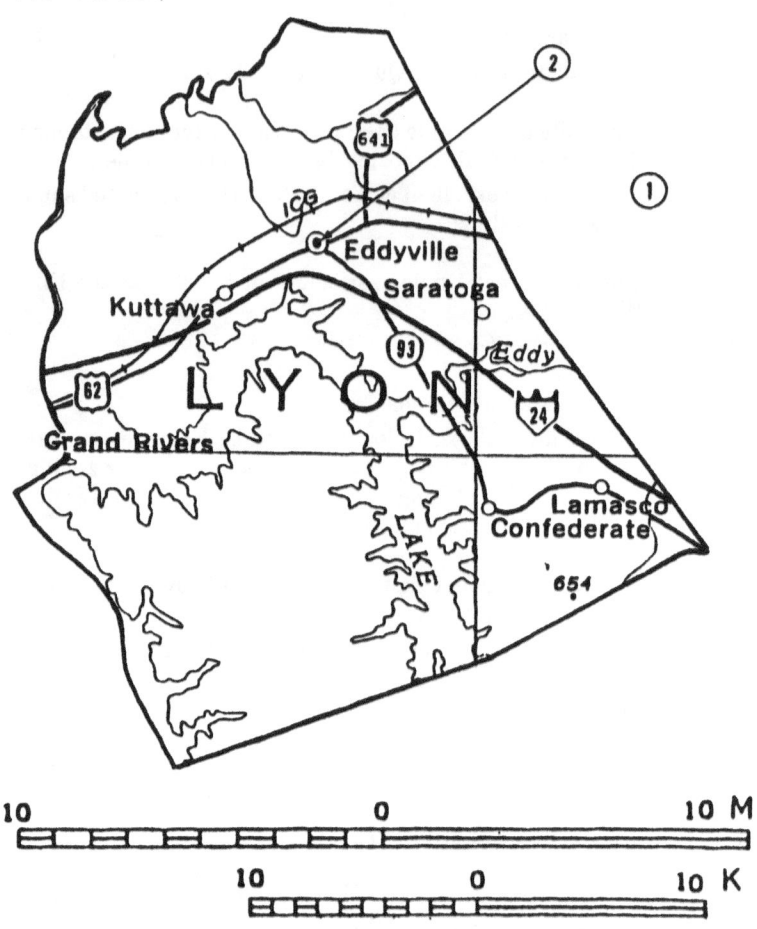

MADISON COUNTY

1. Berea area: southeast on US-25; in road cuts.

   Limestone, oolitic. Quartz: geodes to 9″ diameter variously containing barite crystals, pink calcite crystals, pale blue celestite crystals, banded chalcedony, goethite, and rock quartz in sparkling drusy and sharply terminated crystals. (1-4-13)

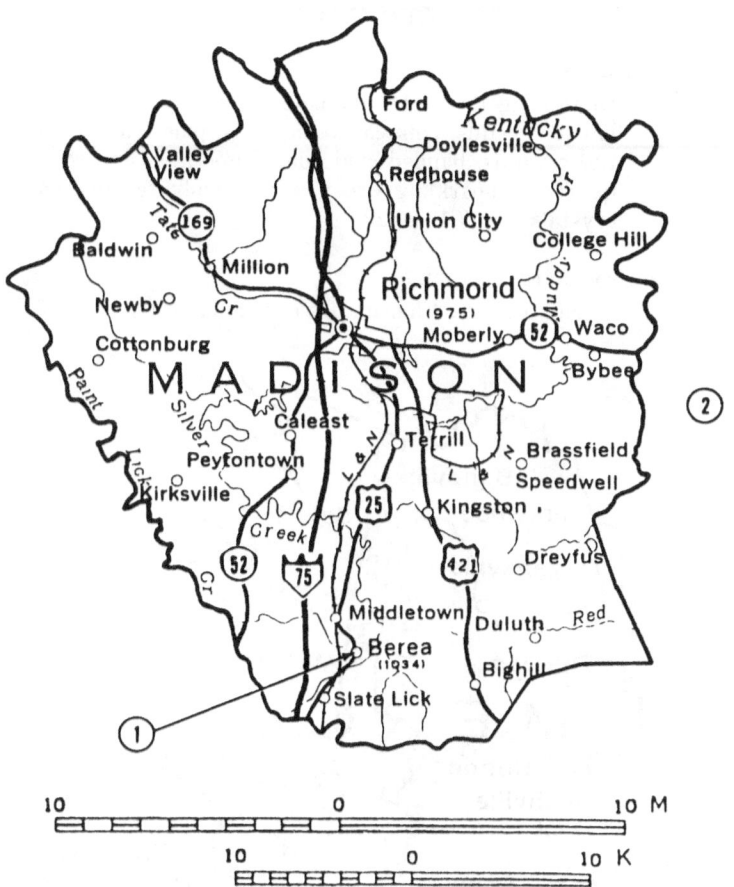

2. County-wide.

> Quartz: geodes to 16″ diameter variously containing
> barite crystals, pink calcite crystals, pale blue celes-
> tite crystals, banded chalcedony, goethite, and rock
> quartz in sparkling drusy and sharply terminated
> crystals. (1-4-7-13-20-29)

MERCER COUNTY

1. County-wide.

   Quartz: geodes to 16″ diameter variously containing barite crystals, pink calcite crystals, pale blue celestite crystals, banded chalcedony, goethite, and rock quartz in sparkling drusy and sharply terminated crystals. (1-4-7-13-20-29)

2. Harrodsburg area mines.

   Barite. Calcite. Fluorite. Galena. Sphalerite. (2)

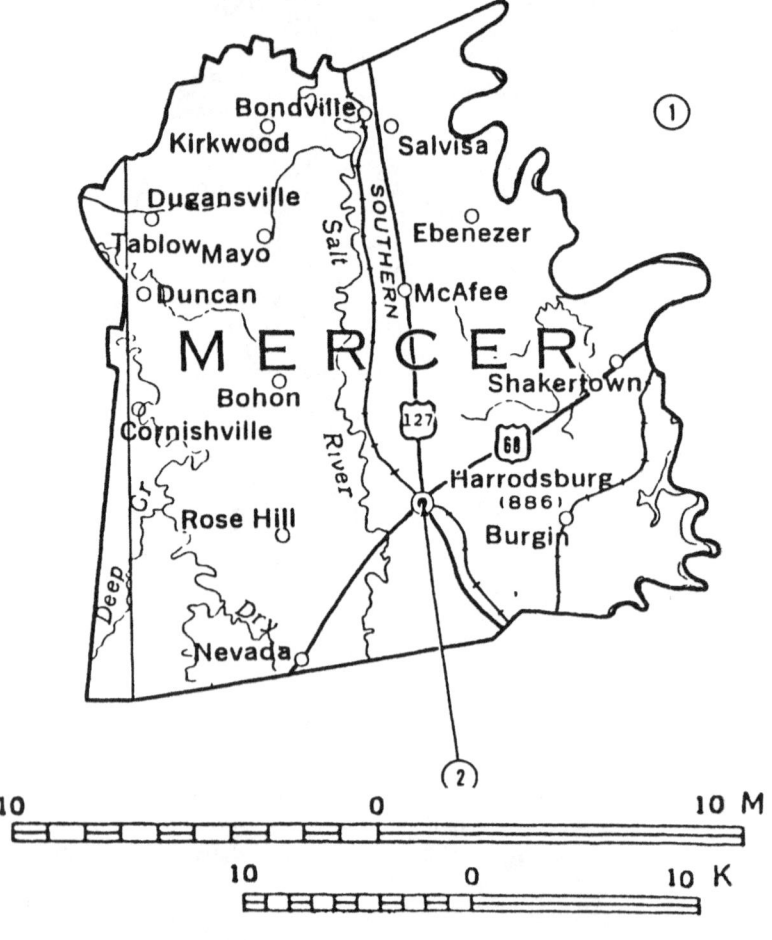

MONROE COUNTY

1. Tompkinsville area quarry.

> Quartz: geodes to 15″ diameter variously containing barite crystals, pink calcite crystals, pale blue celestite crystals, banded chalcedony, goethite, and rock quartz in sparkling drusy and sharply terminated crystals. (12)

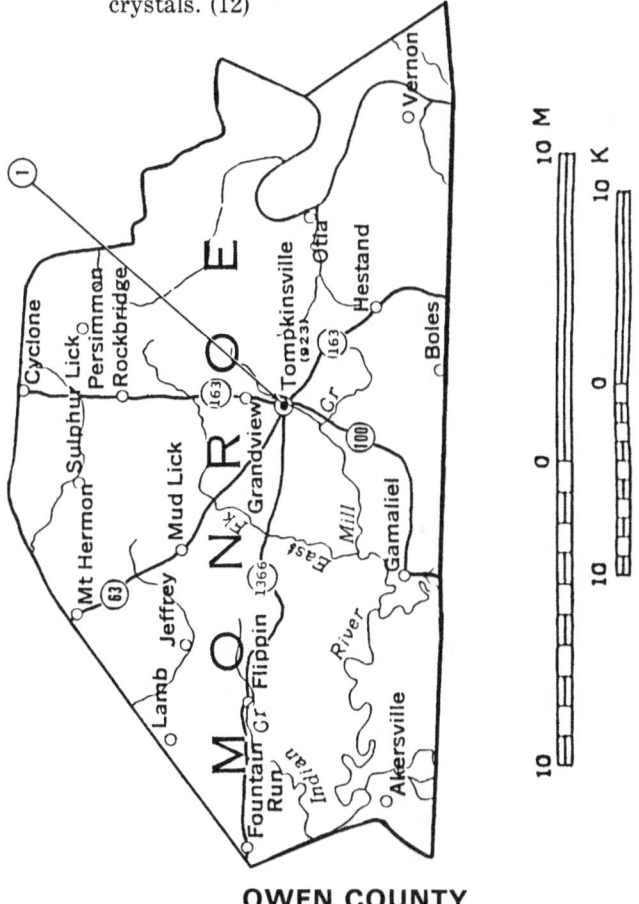

OWEN COUNTY

1. Gratz area: just west of SR-355; along the north side of the Kentucky River; at a series of prospects.

> Barite. Calcite. Fluorite. Galena. Sphalerite. (21)

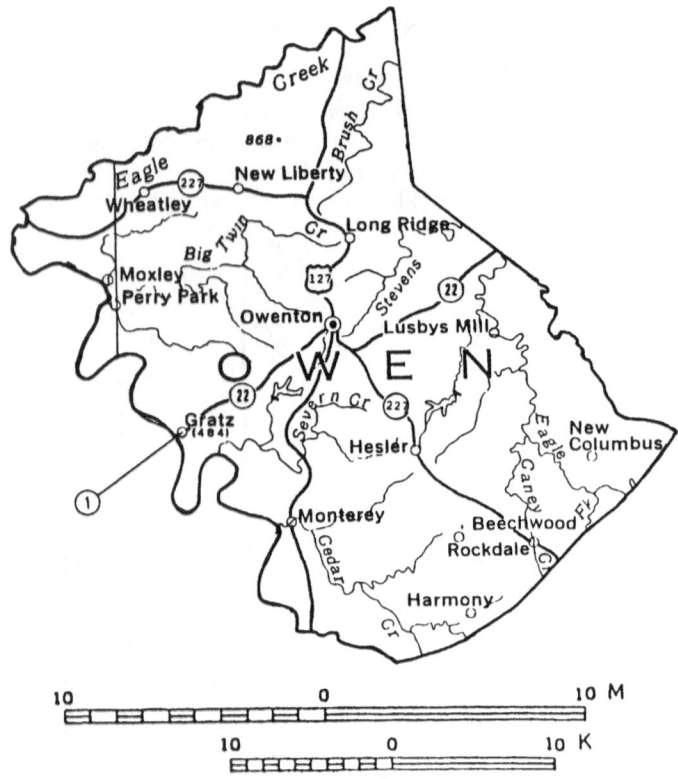

POWELL COUNTY

1. County-wide.

   Quartz: geodes to 16″ diameter variously containing barite crystals, pink calcite crystals, pale blue celestite crystals, banded chalcedony, goethite, and rock quartz in sparkling drusy and sharply terminated crystals. (1-4-7-13-20-29)

ROCKCASTLE COUNTY

1. Boone area: at abandoned limestone quarry.

   Chert. Jasper: nodules to 2″ diameter. Limestone, oo-

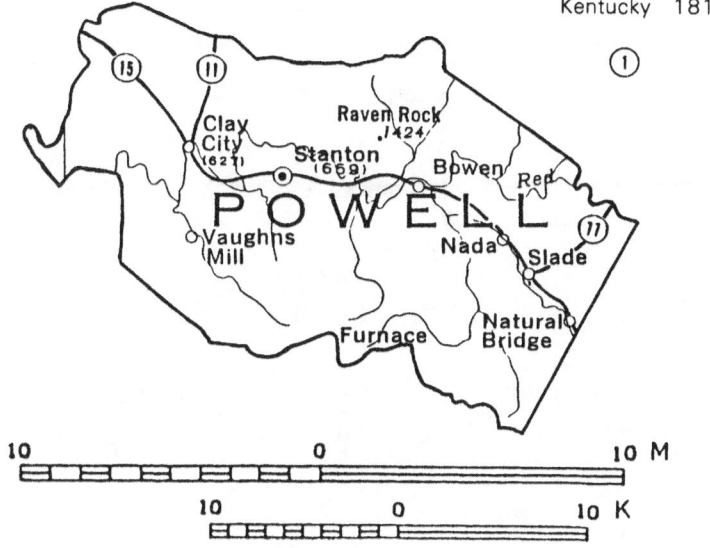

litic. Quartz: geodes to 11″ diameter variously containing barite crystals, pink calcite crystals, pale blue celestite crystals, banded chalcedony, goethite, and rock quartz in sparkling drusy and sharply terminated crystals. (12)

2. Boone area: downstream along both sides of Roundstone Creek to its mouth at Rockcastle River just south of Livingston.

    Chert. Jasper: nodules to 2″ diameter. Limestone, oolitic. Quartz: geodes to 11″ diameter variously containing barite crystals, pink calcite crystals, pale blue celestite crystals, banded chalcedony, goethite, and rock quartz in sparkling drusy and sharply terminated crystals. (1-4-7-13-20)

3. Boone area: in road cut along US-25 south to Mount Vernon.

    Chert. Jasper: nodules to 2″ diameter. Limestone, oolitic. Quartz: geodes to 11″ diameter variously containing barite crystals, pink calcite crystals, pale blue celestite crystals, banded chalcedony, goethite, and rock quartz in sparkling drusy and sharply terminated crystals. (1-4-7-13-20)

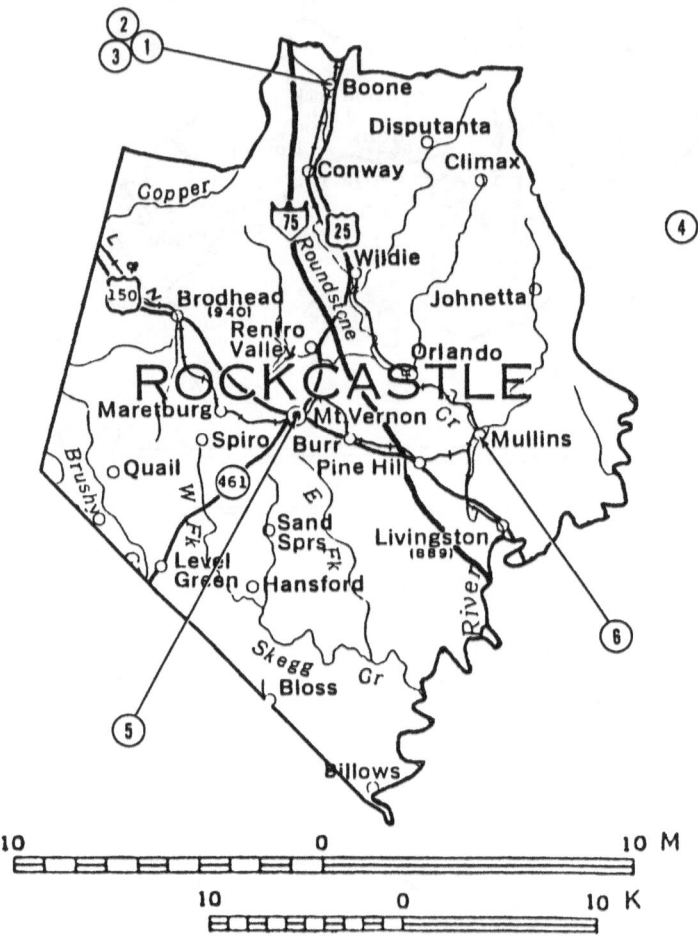

4. County-wide.

Quartz: geodes to 16″ diameter variously containing barite crystals, pink calcite crystals, pale blue celestite crystals, banded chalcedony, goethite, and rock quartz in sparkling drusy and sharply terminated crystals. (1-4-7-13-20-29)

5. Mount Vernon area: just west; at the Kentucky Stone Company Quarry.

> Chert. Jasper. Quartz: geodes to 12" diameter; some of these variously containing barite crystals, pink calcite crystals, pale blue celestite crystals, banded chalcedony, goethite, and rock quartz in sparkling drusy and sharply terminated crystals. (12)

6. Mullins Station area: at the Kentucky Stone Company Quarry.

> Chert. Jasper. Quartz: geodes to 14" diameter variously containing barite crystals, pink calcite crystals, pale blue celestite crystals, banded chalcedony, goethite, and rock quartz in sparkling drusy and sharply terminated crystals. (12)

ROWAN COUNTY

1. Elliottville area stone quarry.
> Fossils: flora and fauna. Jasper: red. (12-limestone)

2. Farmers area stone quarry.
> Fossils: flora and fauna. Jasper: red. (12-limestone)

3. Hays Crossing area stone quarries.
> Fossils: flora and fauna. Jasper: red. (12-limestone)

4. Morehead area stone quarries.
> Fossils: flora and fauna. Jasper: red. (12-limestone)

RUSSELL COUNTY

1. Unidentified stream.
> Diamond: 1 stone; 0.44 ct; specific details unknown; found in 1888. (3)

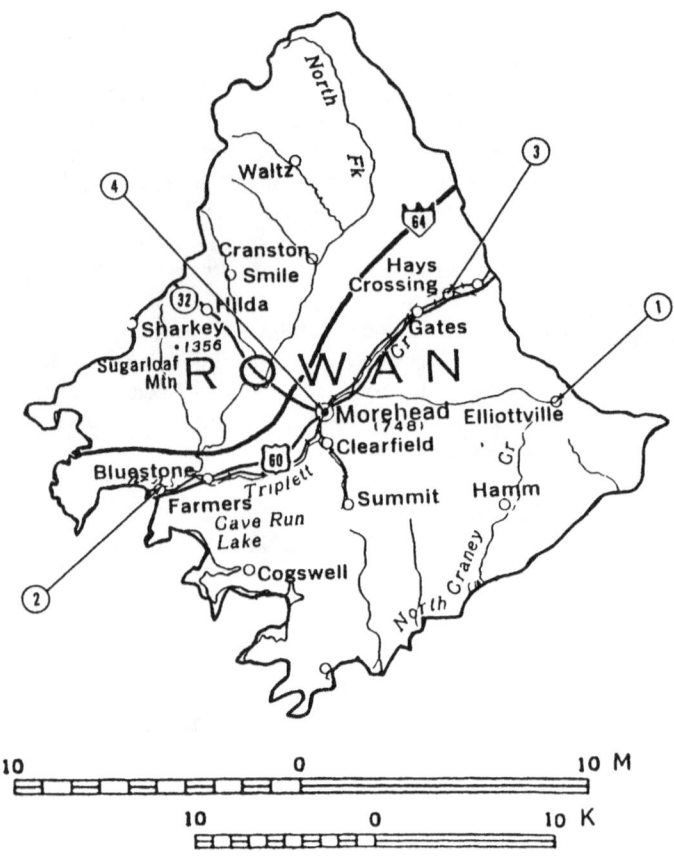

TRIGG COUNTY

1. Fenton area: in the Land Between the Lakes area; along the east shore of Kentucky Lake.

    Chert (Fort Payne var.): well patterned and colored. Jasper. Quartz: geodes to 12″ diameter variously containing barite crystals, pink calcite crystals, pale blue celestite crystals, banded chalcedony, goethite, and rock quartz in sparkling drusy and sharply terminated crystals. (1-13-20)

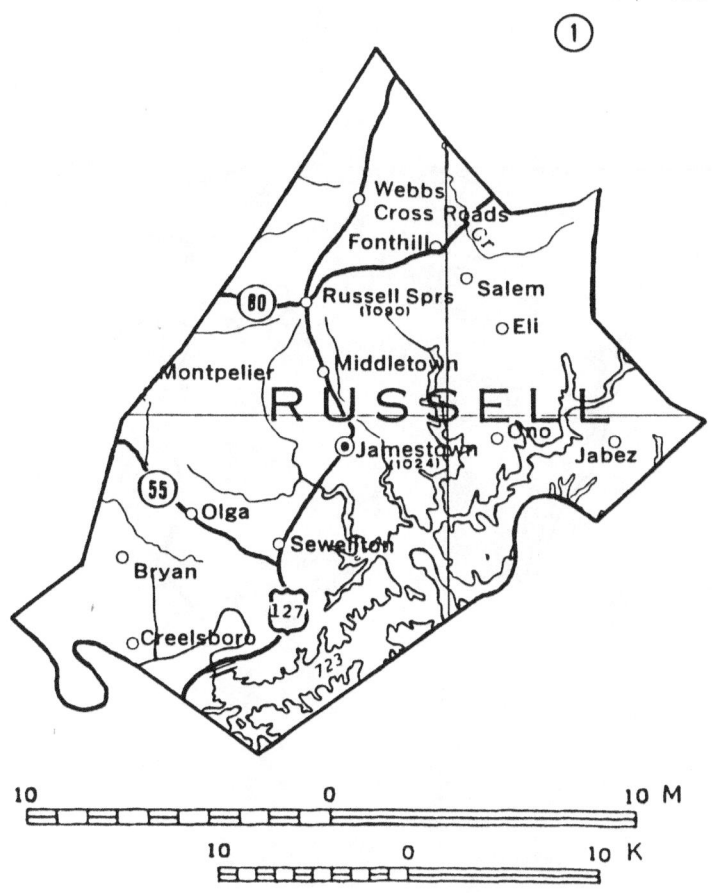

WOODFORD COUNTY

1. County-wide.

   Quartz: geodes to 16″ diameter variously containing barite crystals, pink calcite crystals, pale blue celestite crystals, banded chalcedony, goethite, and rock quartz in sparkling drusy and sharply terminated crystals. (1-4-7-13-20-29)

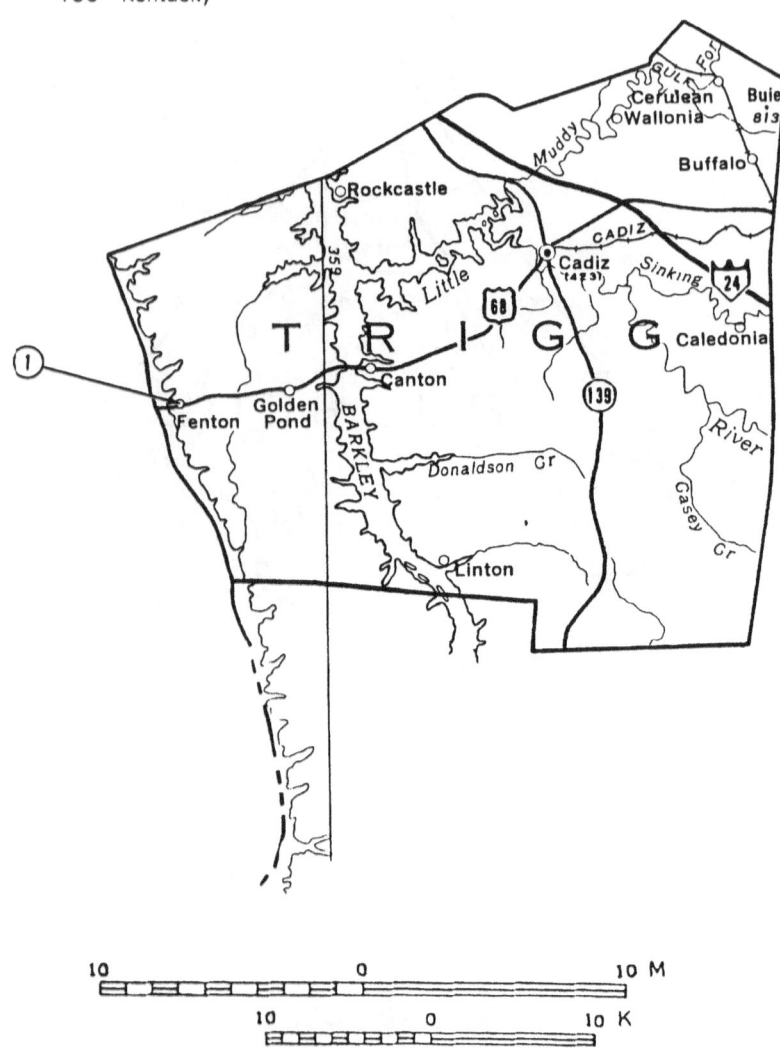

2. Nonesuch area: 2 miles southeast; on Mundys Landing
   Road; directly across from Mundys Landing; at abandoned
   mines.

   Barite. Calcite. Fluorite. Galena. Pyrite. Sphalerite. (2)

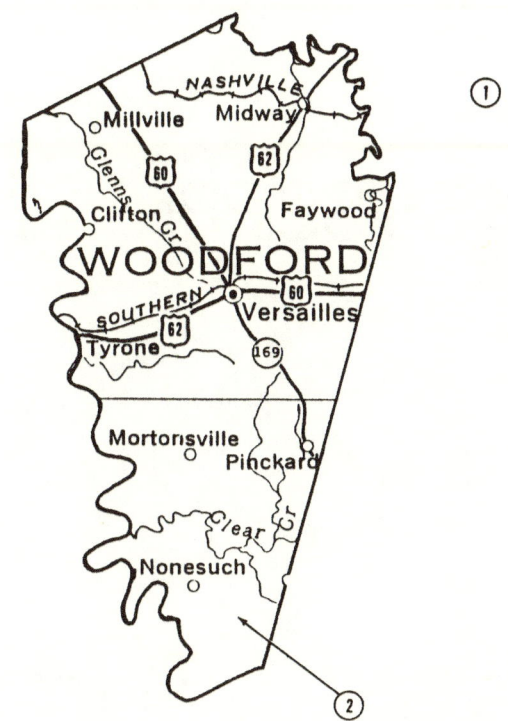

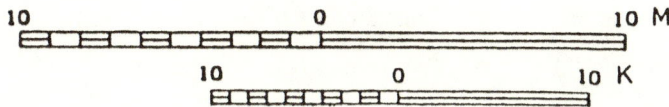

10   0   10 M

10   0   10 K

MISSISSIPPI

Statewide total of 43 locations in the following counties:

Adams (2)	Marion (2)	Washington (2)
Copiah (7)	Stone (2)	Wayne (2)
Franklin (4)	Tallahatchie (2)	Winston (2)
Harrison (2)	Tishomingo (2)	Yazoo (6)
Lawrence (2)	Walthall (2)	
Lee (2)	Warren (2)	

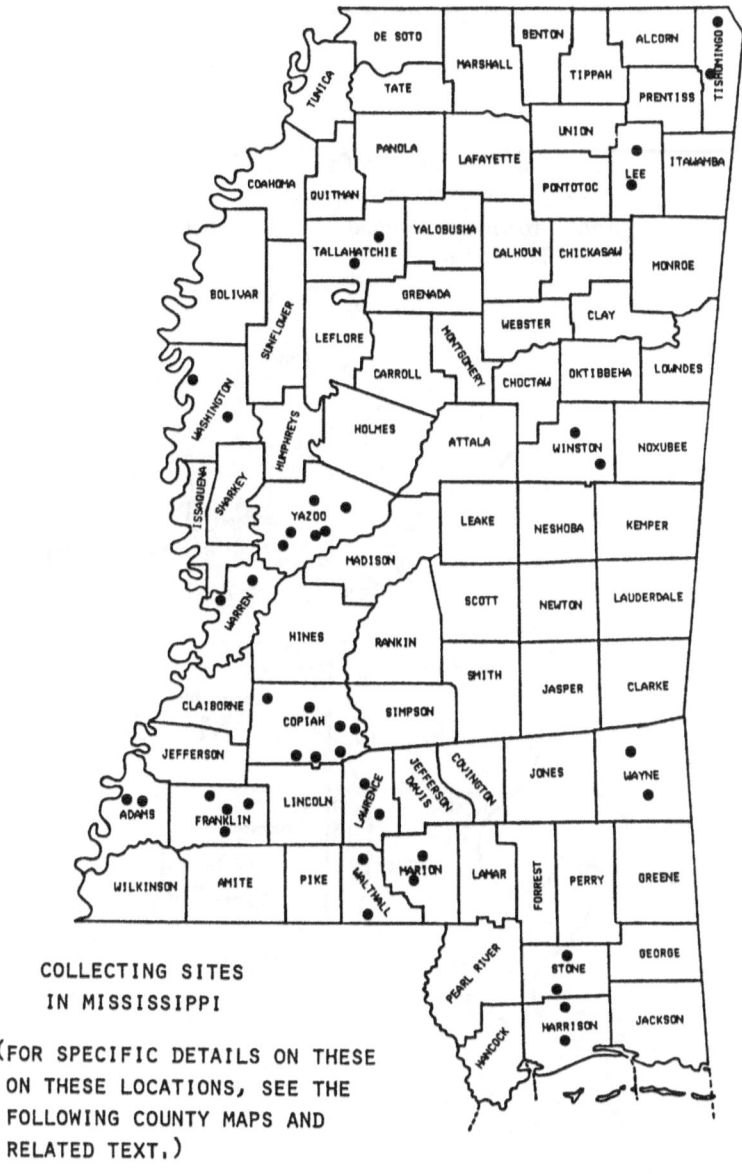

COLLECTING SITES
IN MISSISSIPPI

(FOR SPECIFIC DETAILS ON THESE
ON THESE LOCATIONS, SEE THE
FOLLOWING COUNTY MAPS AND
RELATED TEXT.)

ADAMS COUNTY

1. County-wide in streams and lakes.
    Pearl. (7)

2. Natchez area: southwest to Carthage Point Landing; at
   the St. Catherine Gravel Company.
       Agate, fortification: good. Chert: banded. Jasper. Pe-
       trified bone. (10)

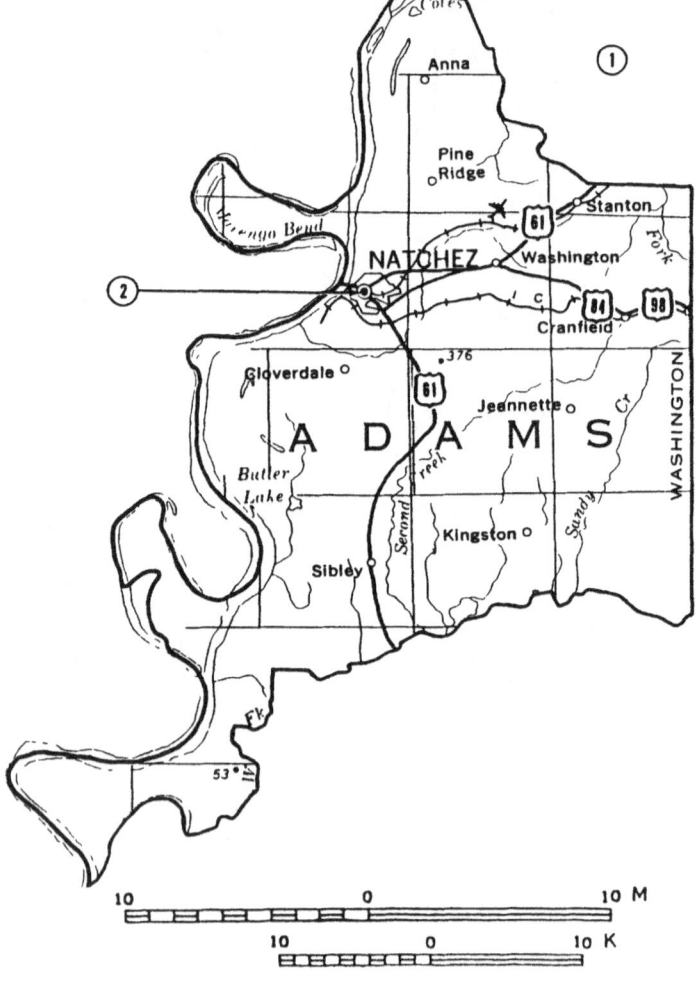

COPIAH COUNTY

1. Carpenter area: 3.5 miles south to Pierre Bayou; at the Traxler Gravel Washing Company.

    Agate. Chert: banded. Petrified wood: palmwood. (10)

2. County-wide in lakes and streams.

    Pearl. (7)

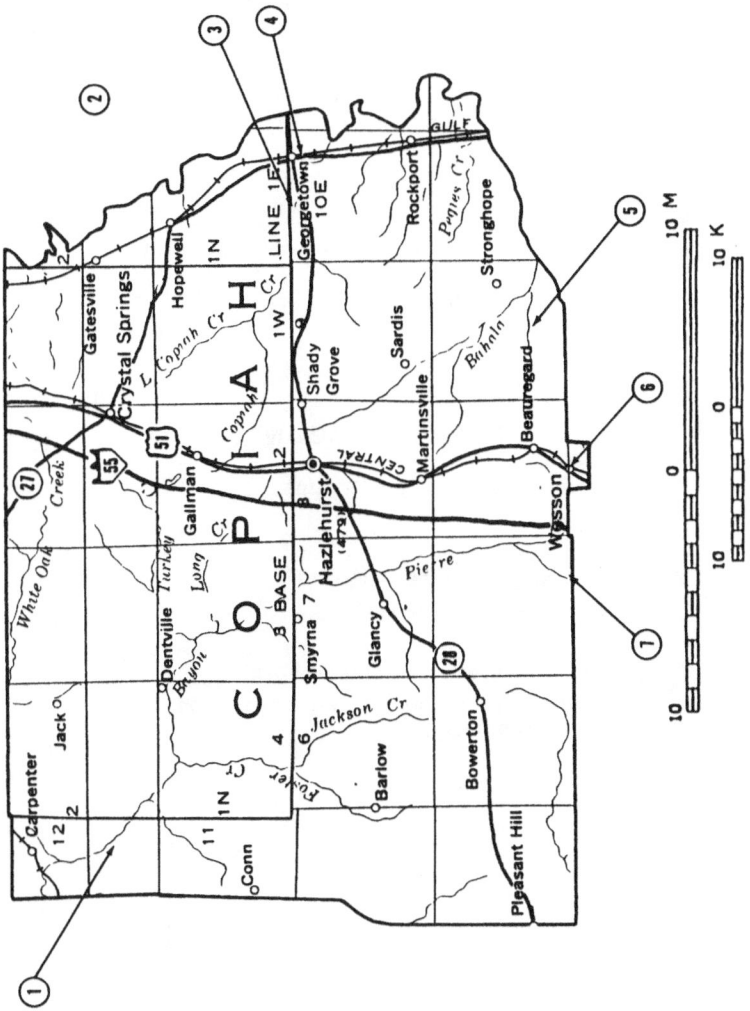

3. Georgetown area: 2 miles west on SR-28; just south of road; on Copiah Creek; at the Green Brothers Gravel Company.

> Agate: good. Fossils: marine. Jaspillite. Petrified wood: some palmwood. (7-10)

4. Georgetown area: south of town on SR-27; west of road; in creek.

> Agate. Petrified wood: some palmwood. (3-7-20-22)

5. Wesson area: 7 miles east-northeast; 6 miles due east of Beauregard; in tributary of Bahala Creek; collect downstream to mouth of creek.

> Agate. (3-7-20)

6. Wesson area.

> Agate, banded: gray; waterworn masses. (3-10)

7. Wesson area: 4 miles west.

> Agate. (10)

FRANKLIN COUNTY

1. County-wide in lakes and streams.

> Pearl. (7)

2. McCall Creek area: 0.5 mile south.

> Agate. Petrified wood: some palmwood. (3-7-20-22)

3. Meadville area: west; in Middle Fork.

> Agate. Chert, banded. Petrified wood: some palmwood. (3-7-20-22)

4. Oldenburg area gravel pit.

> Agate. (10)

1
2
3
4
FRANKLIN
Kennolia
Lucien
McCall Creek
84
Veto
New Hope
Quentin
Eddiceton
Bude
(2232)
Little Springs
98
Dick
Middleton Cr.
Meadville
Kirby
Oldenburg
Morgan
Suffolk
Bunkley
Hamburg
Leesdale
Roxie
White Apple
Knoxville
Garden City
33
10 0 10 M
10 0 10 K

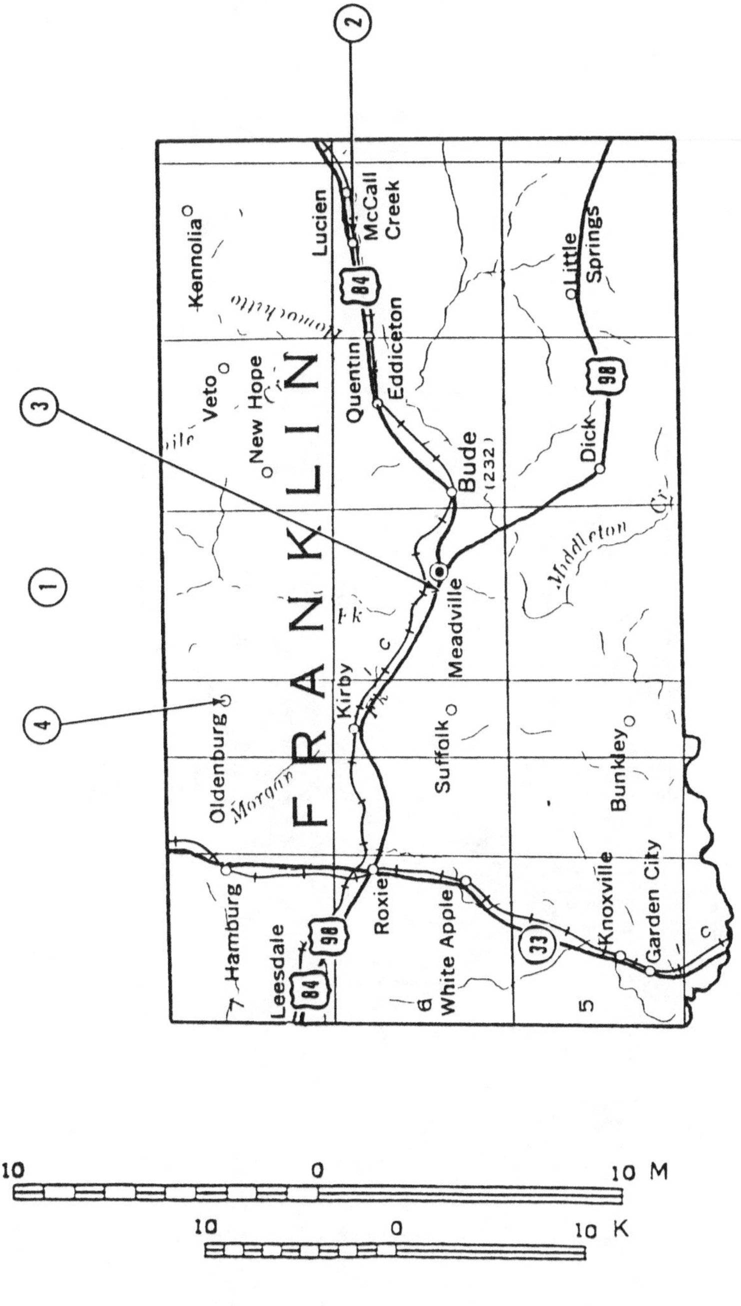

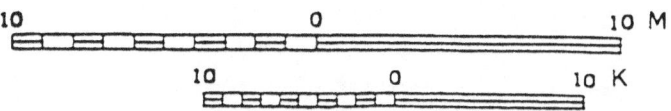

HARRISON COUNTY

1. County-wide in lakes and streams.
     Pearl. (7)

2. Saucier area: in Bell Creek.
     Agate, banded: dark gray. (3-7-20)

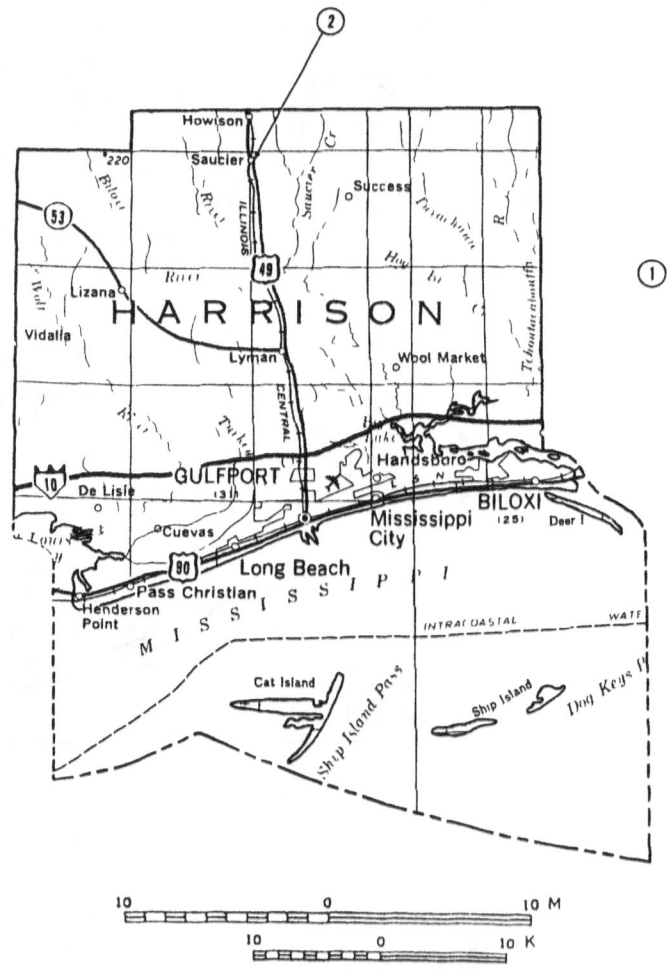

LAWRENCE COUNTY

1. County-wide in lakes and streams.
   Pearl. (7)

2. Wanilla area: south; near Fair River; at the Green Brothers Gravel Washing Company.
   Agate. Chert. (10)

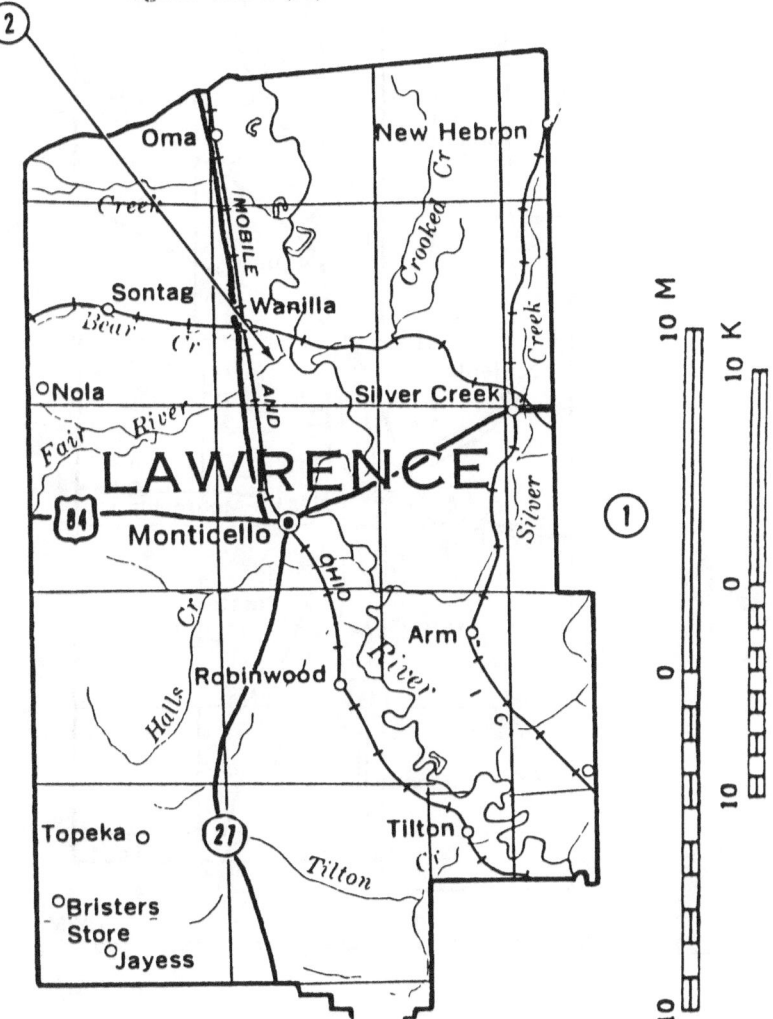

LEE COUNTY

1. County-wide in lakes and streams.
   Pearl. (7)

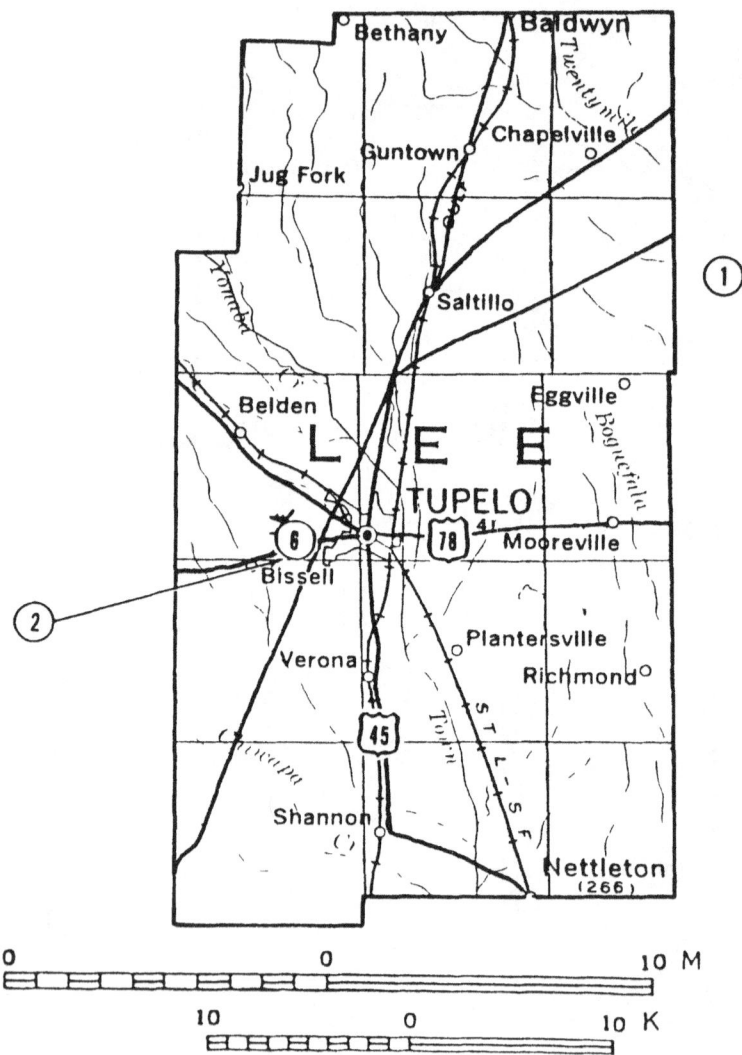

2. Tupelo area: west on SR-6 to 2.4 miles west of the Natchez Trace Parkway; near Bissell.

> Fossils: marine; very good; many are replaced by marcasite. Marcasite. (4-13)

MARION COUNTY

1. County-wide in lakes and streams.

> Pearl. (7)

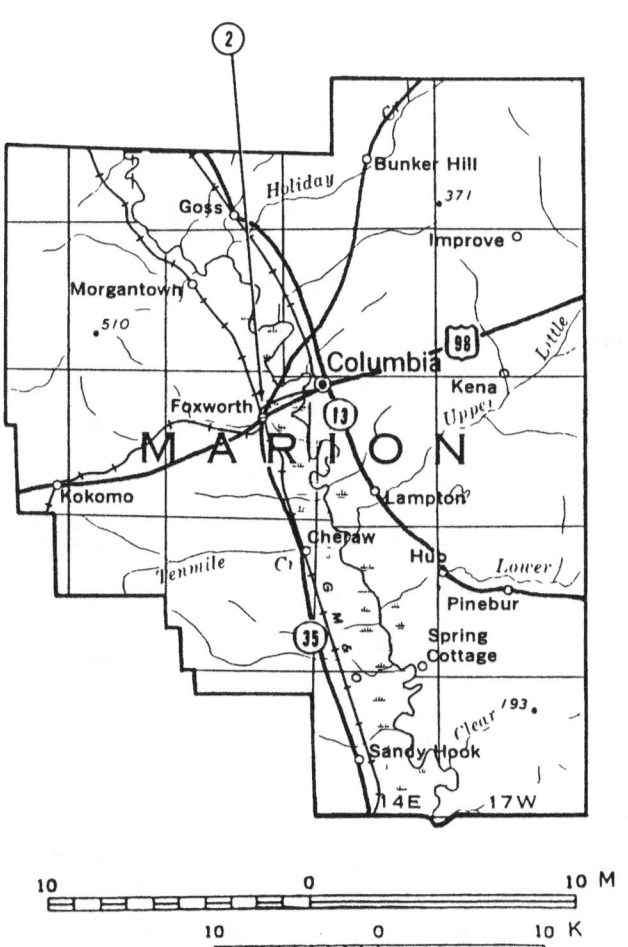

10 0 10 M

10 0 10 K

2. North Foxworth area: at the Pearl River Gravel Washing
   Company.
   Agate. Petrified wood: some palmwood. (10)

STONE COUNTY

1. County-wide in lakes and streams.
   Pearl. (7)

2. McHenry area: 2.5 miles southwest; in Biloxi Creek.
   Jasper. Petrified wood: some palmwood. (3-7-20-22)

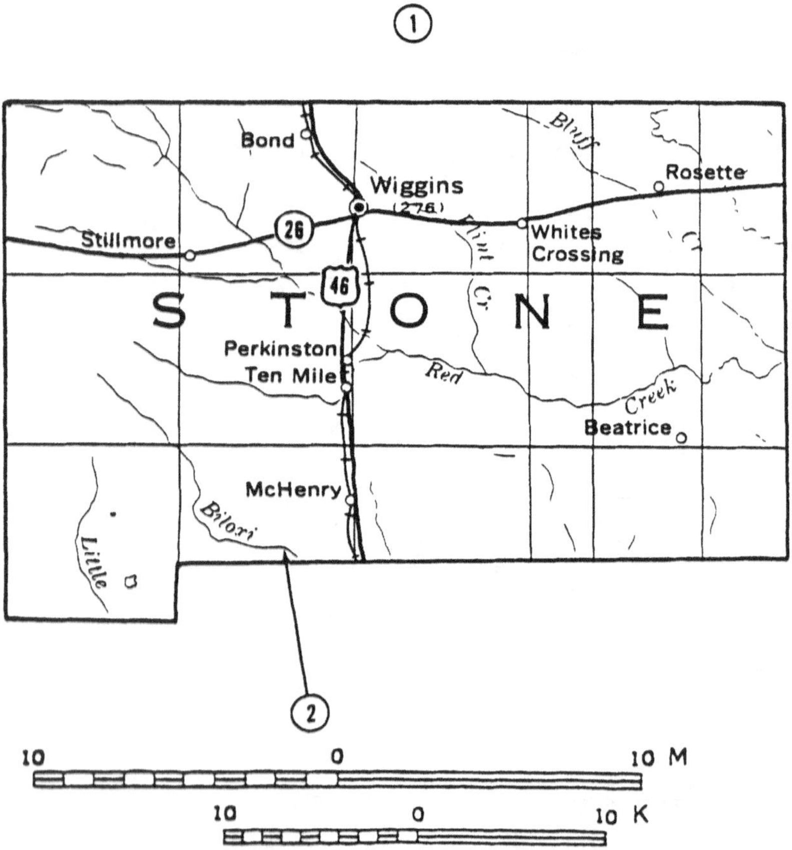

TALLAHATCHIE COUNTY

1. Charleston area.

    Amber (Succinite): yellow to gray; soft; brittle; opaque. (25)

2. County-wide in lakes and streams.

    Pearl. (7)

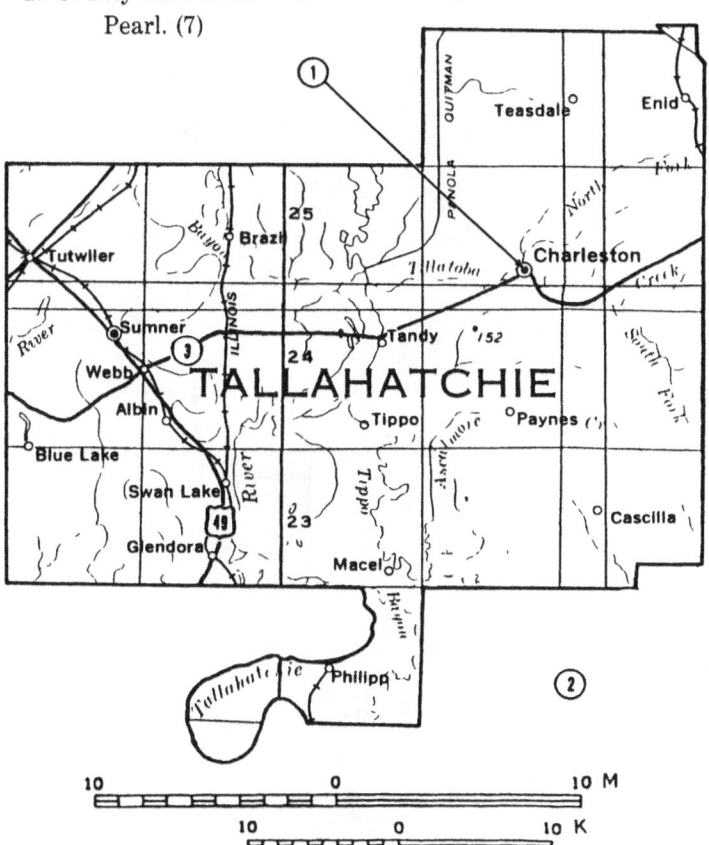

TISHOMINGO COUNTY

1. County-wide in lakes and streams.

    Pearl. (7)

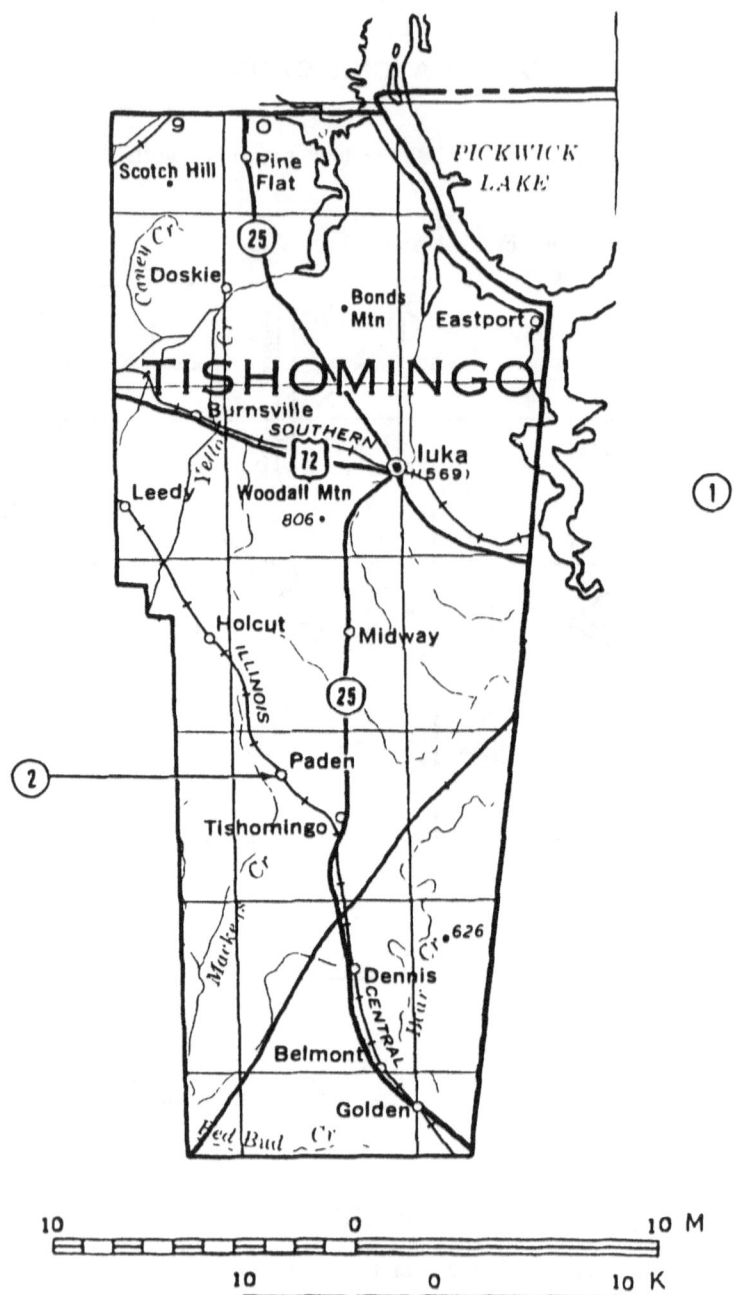

2. Paden area.

> Amber (Succinite): yellow to gray; soft; brittle; opaque. (25)

WALTHALL COUNTY

1. County-wide in lakes and streams.

> Pearl. (7)

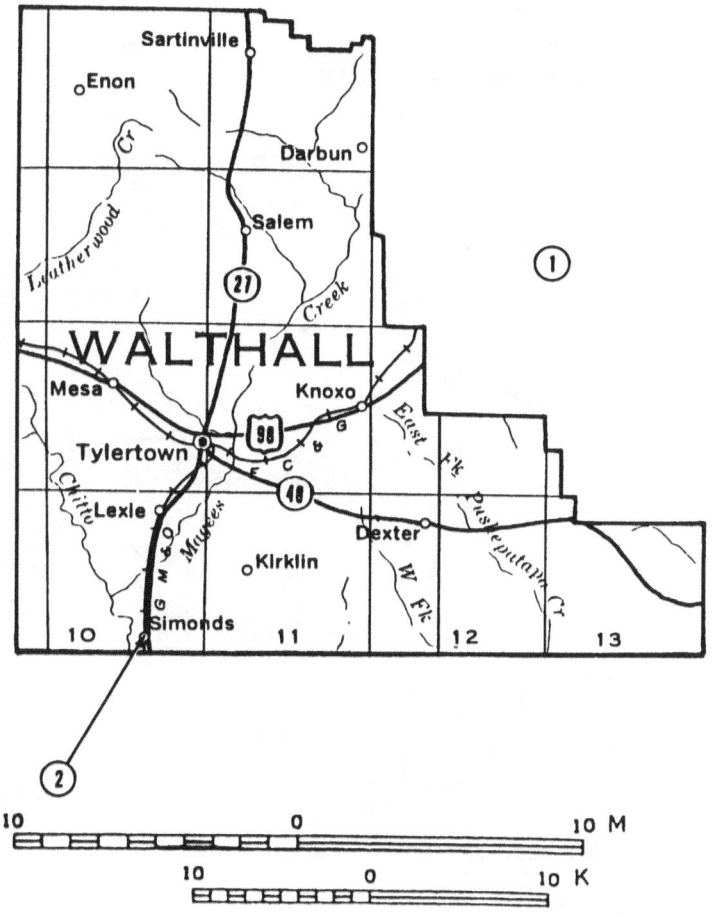

2. Simonds area: along the Bogue Chitto River; just north of the Louisiana border; at the Walthall Gravel Washing Company.

Agate. Chert. Petrified wood: some palmwood. (10)

WARREN COUNTY

1. County-wide in lakes and streams.
   Pearl. (7)

2. Vicksburg area: 6 miles north; at junction of new US-61 and old US-61; in deep road cut.

Fossils: marine; vertebrate and invertebrate; very good; many opalized. (4)

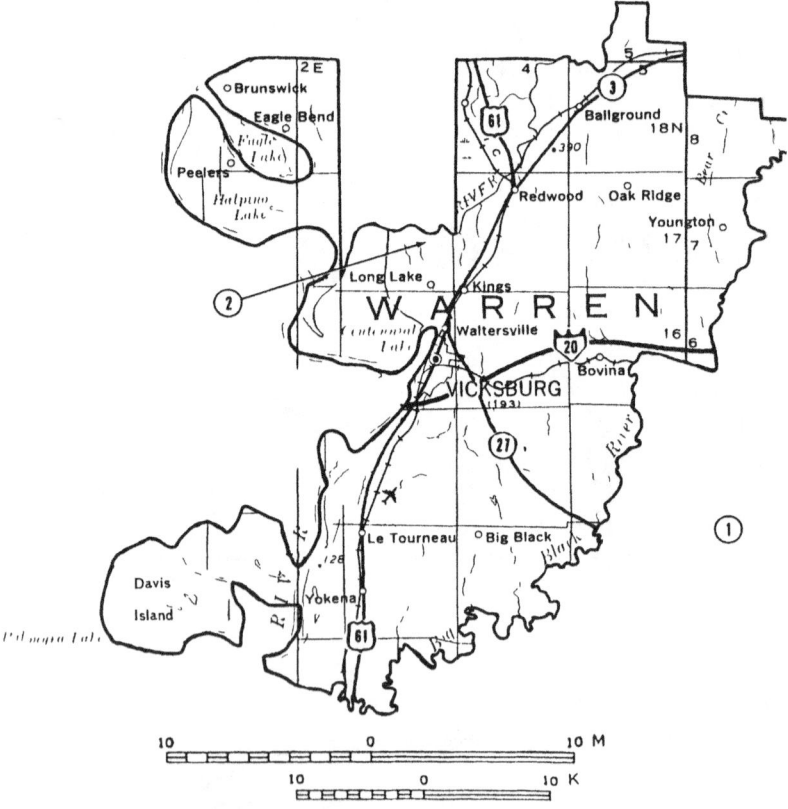

WASHINGTON COUNTY

1. County-wide in lakes and streams.
   Pearl. (7)

2. Greenville area: along shorelines and in the Mississippi
   River.
   Agate. Moonstone. (7-18-30)

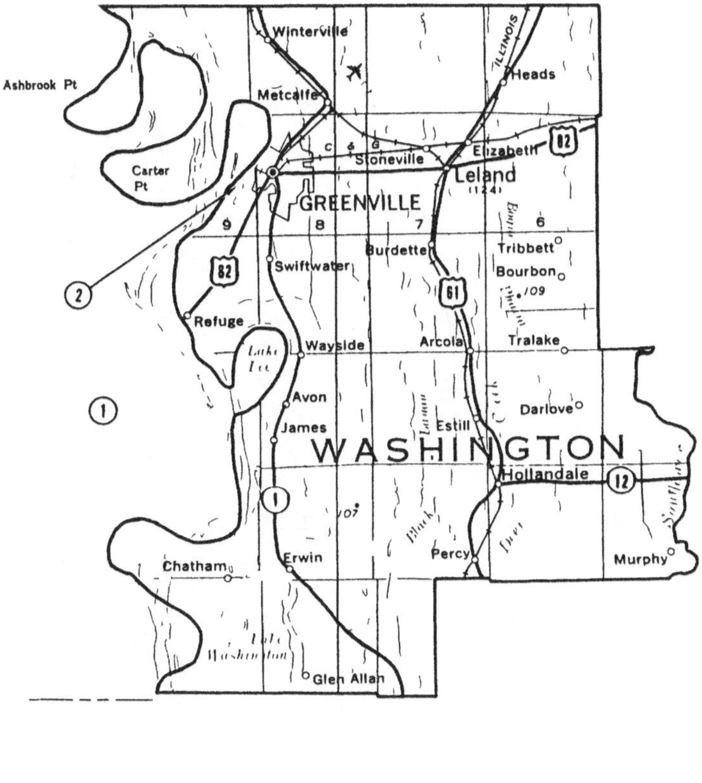

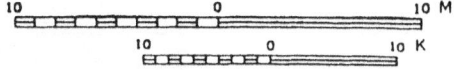

WAYNE COUNTY

1. County-wide in lakes and streams.
     Pearl. (7)

2. Waynesboro area: 7 miles northwest.
     Petrified wood: good palmwood. (1-7-13-20-22)

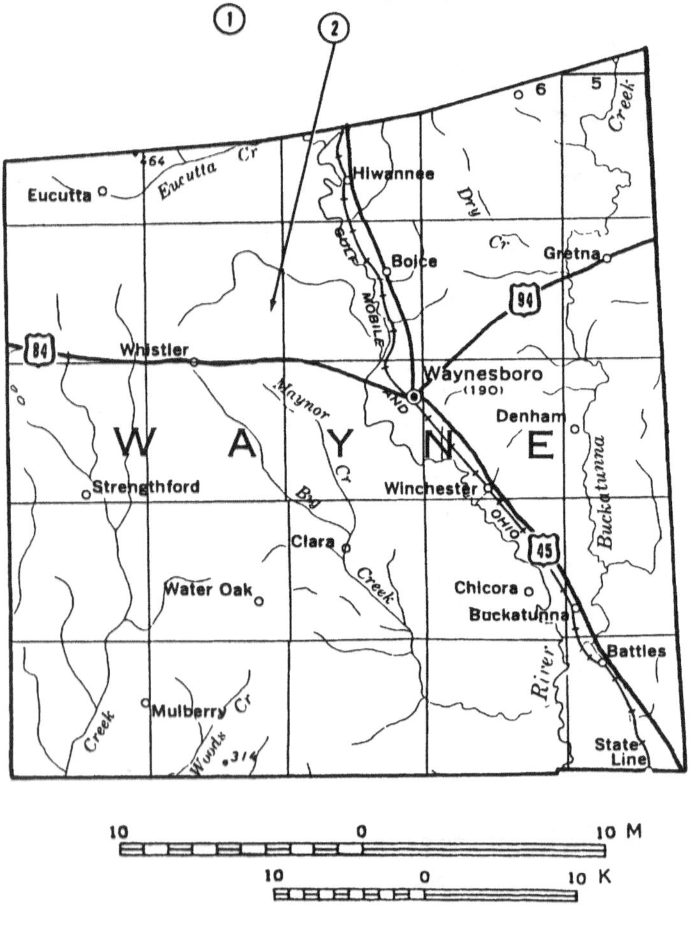

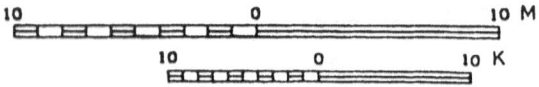

WINSTON COUNTY

1. County-wide in lakes and streams.
   Pearl. (7)

2. Webster area: along SR-25; in Mill Creek.
   Opal, wood. (7)

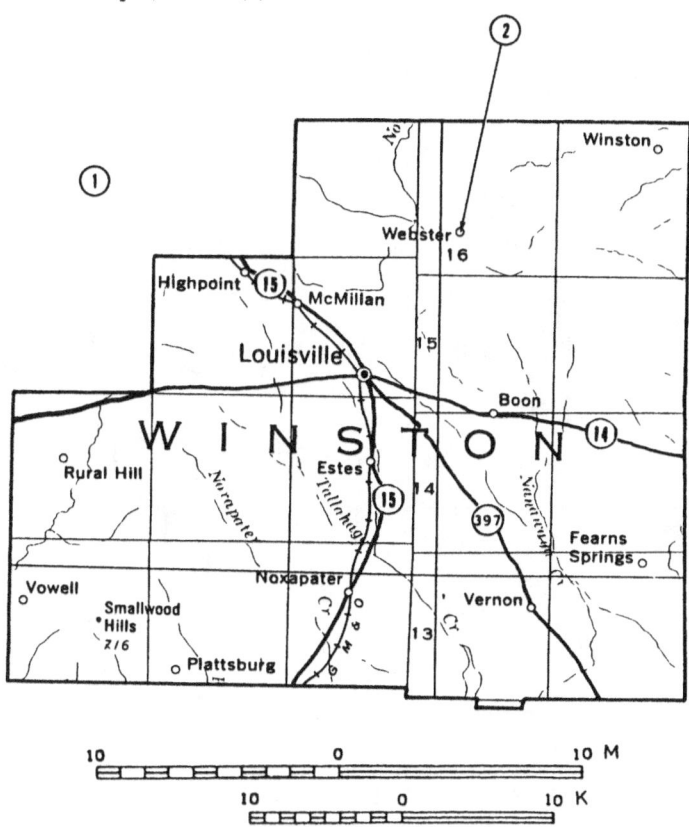

YAZOO COUNTY

1. Anding area: in Perry Creek.
   Agate. Petrified wood: some palmwood. (7-18-20-22)

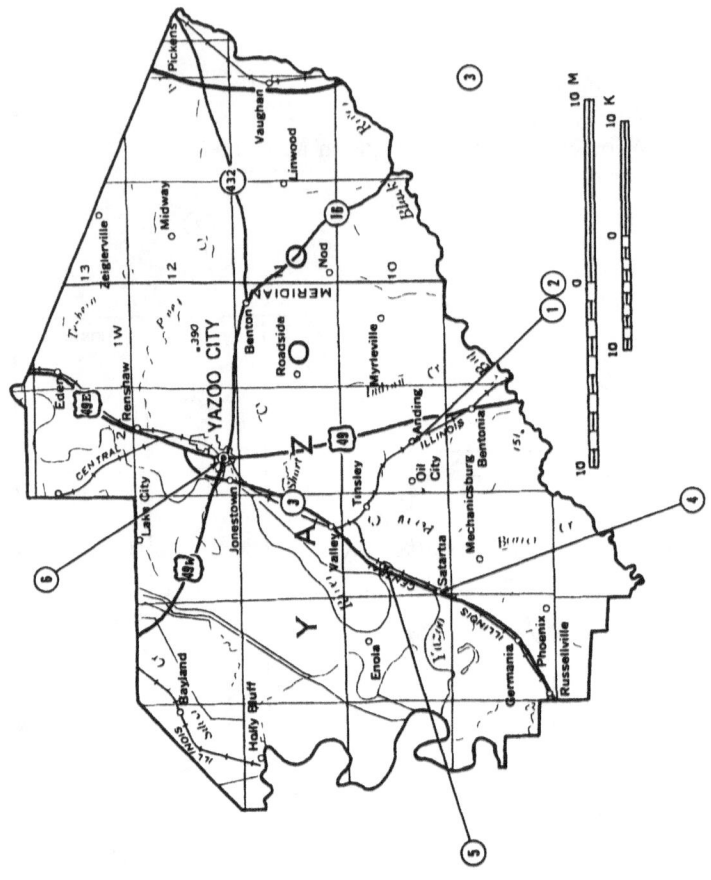

2. Anding area: in Thompson Creek.
      Agate. Petrified wood: some palmwood. (7-18-20-22)

3. County-wide in lakes and streams.
      Pearl. (7)

4. Satartia area: northwest; along the Yazoo River for 3 miles
   upstream.
      Agate. (7-18)

5. Tinsley area: west to the Yazoo River; along both banks.
   Agate. Chert. Fossils: marine; brachiopods, coral. Petrified wood: some palmwood. (3-18-20-22)

6. Yazoo City area: in expansive bed; west to SR-3, east to US-49, south to SR-433.
   Agate. Petrified wood: some palmwood. (1-3-6-13-22)

NORTH CAROLINA

Statewide total of 283 locations in the following counties:

Alamance (2)	Davidson (7)	McDowell (5)
Alexander (23)	Davie (1)	Mecklenburg (2)
Alleghany (4)	Durham (2)	Mitchell (20)
Anson (3)	Forsyth (3)	Montgomery (2)
Ashe (6)	Franklin (2)	Moore (2)
Avery (7)	Gaston (6)	Orange (1)
Buncombe (6)	Granville (1)	Person (1)
Burke (8)	Guilford (1)	Pitt (1)
Cabarrus (2)	Halifax (1)	Polk (1)
Caldwell (2)	Haywood (4)	Randolph (1)
Caswell (6)	Henderson (1)	Richmond (1)
Chatham (4)	Iredell (11)	Rockingham (2)
Cherokee (9)	Jackson (9)	Rowan (1)
Clay (8)	Lincoln (10)	Rutherford (3)
Cleveland (22)	Macon (27)	Stokes (2)
Cumberland (1)	Madison (6)	Surry (4)

Swain (1)   Wake (3)   Wilkes (2)

Union (1)   Warren (4)   Yancey (15)

Vance (2)

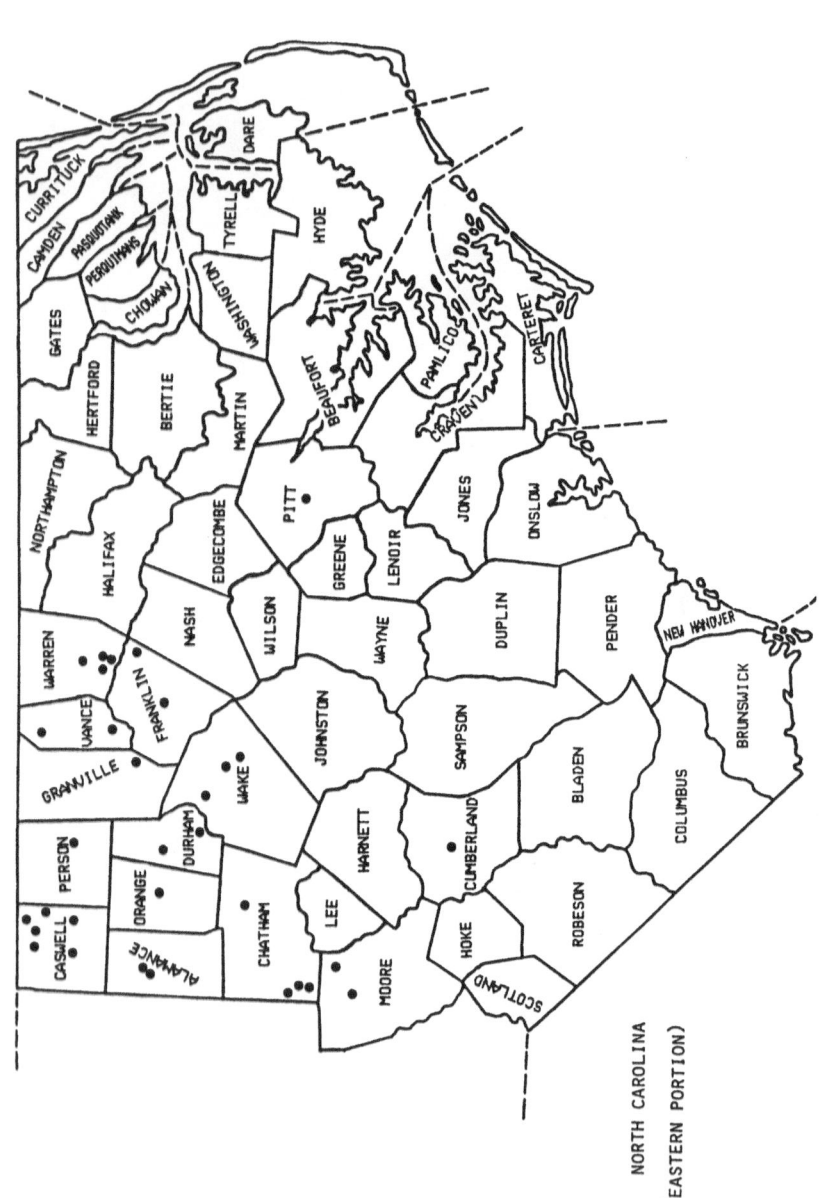

NORTH CAROLINA
(EASTERN PORTION)

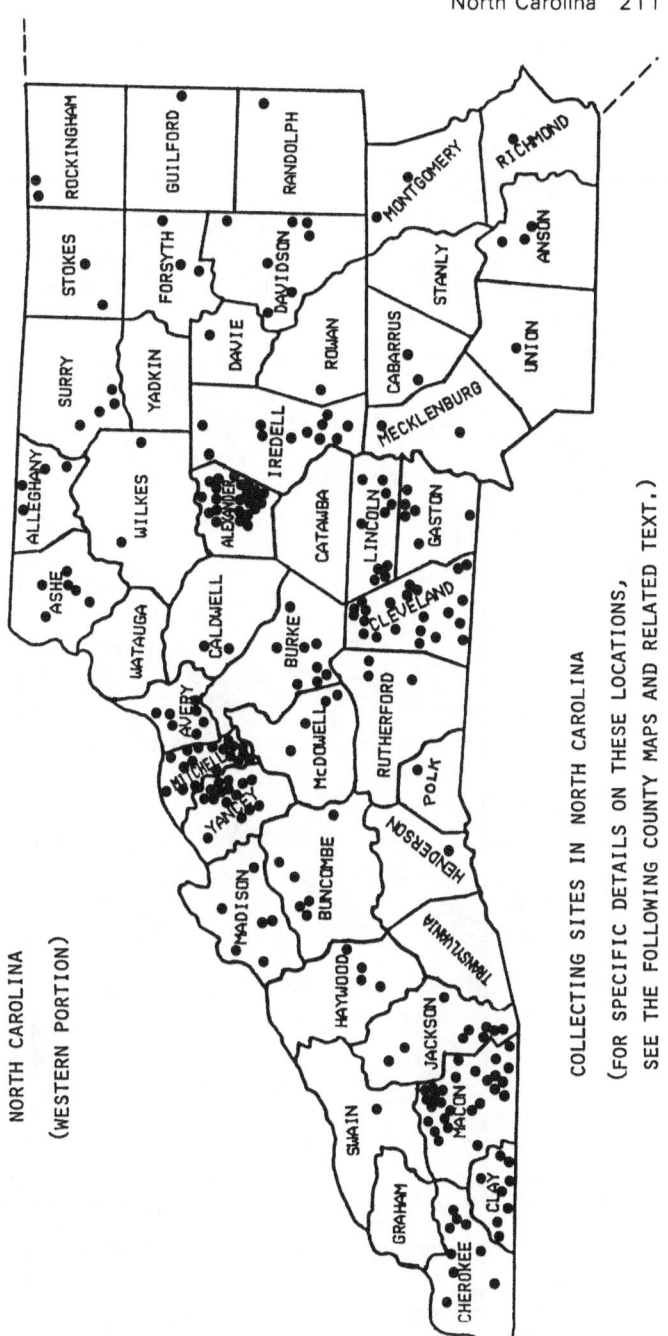

NORTH CAROLINA
(WESTERN PORTION)

COLLECTING SITES IN NORTH CAROLINA
(FOR SPECIFIC DETAILS ON THESE LOCATIONS,
SEE THE FOLLOWING COUNTY MAPS AND RELATED TEXT.)

ALAMANCE COUNTY

1. Burlington area stream valleys.

   Quartz, rock. Quartzite, red: gem quality. Serpentine. (1-3-7-13)

2. Burlington area: at the Superior Stone Quarry.

   Azurite. Chalcosite. Chalcopyrite. Chrysocolla. Copper, native. Cuprite. Fluorite. Galena. Malachite. Pyrite. Sphalerite. (9-12)

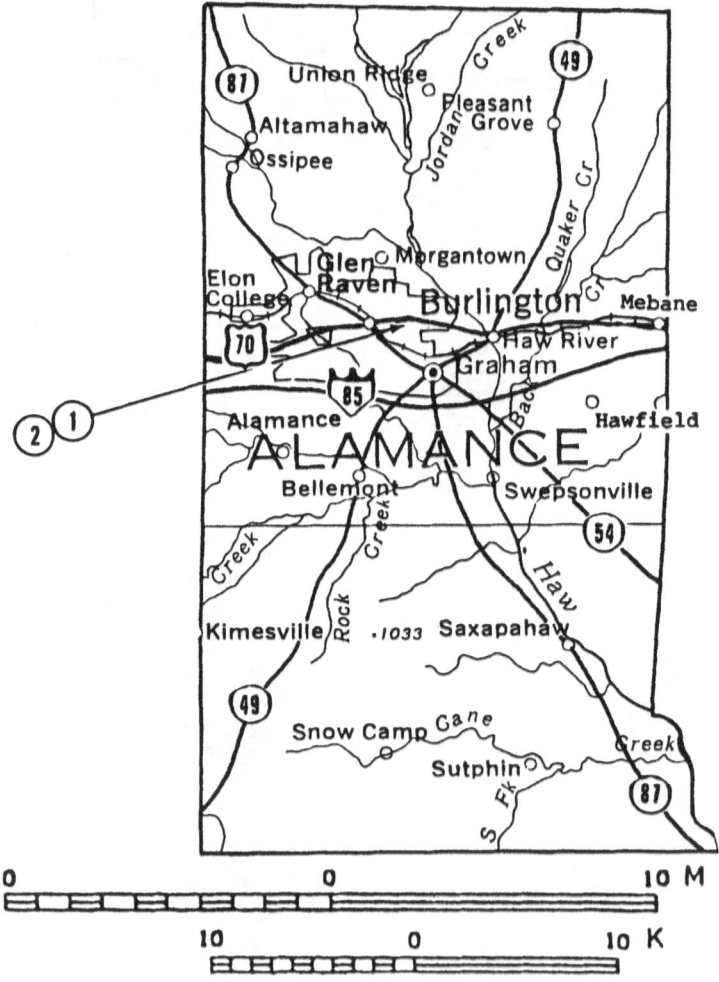

ALEXANDER COUNTY

1. Hiddenite area: 1.5 miles east; on ridge between Davis Creek and the Little Yadkin River.

    Beryl. Quartz, rock. Rutile. (6-21)

2. Hiddenite area: 1.5 miles southeast; near Salem Church; on the Warren property; in 8″-thick vein.

    Aquamarine. Beryl, golden. Calcite. Chlorite. Emerald. Mica. Muscovite: silvery. Spodumene (Hiddenite). (6-16-19)

3. Hiddenite area: 1 mile north; 0.2 mile west of Hiddenite Mine; 300 yards from the Statesville-Taylorsville Road.

    Emerald. Spodumene (Hiddenite). (2-6-16)

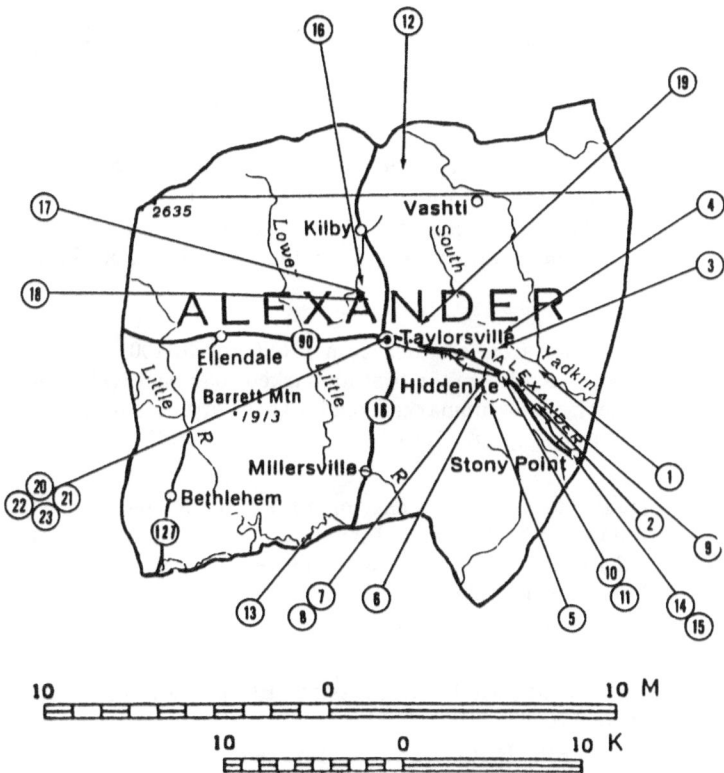

4. Hiddenite area: 1.5 miles north and just northwest; near Salem Church; on the Osborne Lackey property.
   Emerald. (6)

5. Hiddenite area: 1 mile south-southwest to the old Hiddenite Mine; then just beyond to the old Payne Mine and dump.
   Beryl. Quartz, rutilated. (2)

6. Hiddenite area: 0.75 mile southeast; at the O. F. Patterson Mine.
   Beryl. (2)

7. Hiddenite area: 0.5 mile west; in small prospect.
   Emerald. Spodumene (Hiddenite). (21)

8. Hiddenite area: 0.5 mile west; at the American Gem Mining Syndication prospect.
   Emerald. Spodumene (Hiddenite). (21)

9. Hiddenite area: 0.25 mile east; at the Ellis Prospect.
   Aquamarine. Calcite. Chalcopyrite. Dolomite. Emerald. Monzanite. Muscovite. Quartz, rock. Pyrite. Rutile. Spodumene (Hiddenite). Tourmaline, black. (2-21)

10. Hiddenite area: southeast on SR-90 to CR-1001 on left; follow CR-1001 to the first unmarked road past Hiddenite School; follow unmarked road a short distance to mine.
    Emerald. (2)

11. Hiddenite area: southeast on SR-90 to CR 1001 on left; follow CR-1001 to right turn on CR-1498; follow latter to CR-1508; then take CR-1508 and follow signs to the Rist Mine (CF payable at the Rist Mine office).
    Emerald. Quartz, rock. (2)

12. Kilby area: 2.6 miles northeast; on Black Oak Ridge; at the Wike Prospect.
    Beryl. Biotite. Muscovite: pale cinnamon books. Quartz, rock. (21)

13. Millersville area: southwest; a short distance northwest of the Catawba River Dam; 80' west of the Hickory Lake shoreline.

> Apatite: green. Biotite. Muscovite.

14. Stony Point area.

> Citrine: gem quality. (29)

15. Stony Point area.

> Quartz, rutilated: clear; fine gem quality. Rutile: deep red. Spodumene (Hiddenite). (1-5-limonite and quartz -13-14-16-24)

16. Taylorsville area: 2 miles north-northwest of All Healing Springs; on the Thomas Barnes property.

> Beryl: golden, green, yellow. (1-19-quartz -24)

17. Taylorsville area: 1.75 miles north-northeast of All Healing Springs; on the James Chapman property.

> Beryl: golden, yellow. (1-6-13)

18. Taylorsville area: 1.5 miles north-northwest of All Healing Springs; on the Eli Barnes property; west side of a small hill 200 yards northwest of the farmhouse.

> Beryl: golden. (6)

19. Taylorsville area: 1.25 miles east-northeast of All Healing Springs; on the John Webster property.

> Beryl. (6)

20. Taylorsville area.

> Citrine; gem quality. (29)

21. Taylorsville area.

> Rutile: bright red. (5-limonite and quartz -29)

22. Taylorsville area: near Liberty Church.

> Rutile; fine gem quality; crystals to 0.66" diameter and 6" long. (1-3-5-limonite and quartz -13-16)

23. Taylorsville area: to White Plains; near Millholland's Mill.

> Rutile: bright red. (5-limonite and quartz -16-24)

ALLEGHANY COUNTY

1. Cherry Lane area: southwest on the Blue Ridge Parkway
   to the old Brenegar Cabin parking lot; collect on hill above
   the lot. (National Park permit required.)

   Citrine: gem-quality crystals. Quartz, rock. Quartz,
   smoky. (1-13-24)

2. Sparta area: 3 miles northeast on SR-18; turn left and go
   0.5 mile to fork; bear left and go 0.5 mile west to mine
   dumps.

   Rhodonite. (2)

3. Sparta area: 2.9 miles north on SR-18; at the Manganese
   Mine.

   Brucite. Galaxite. Manganite. Pyrolusite. Quartz,
   rock. Quartz, smoky. Rhodochrosite. Spessartite. Te-
   phroite. (2)

4. Twin Oaks area: 1.5 miles north; 1.5 miles south of the
   Virginia border.

   Garnet, almandine (aka "Rhodonite" locally); rare;
   pinkish violet to purple. (5-19-manganese ore -21)

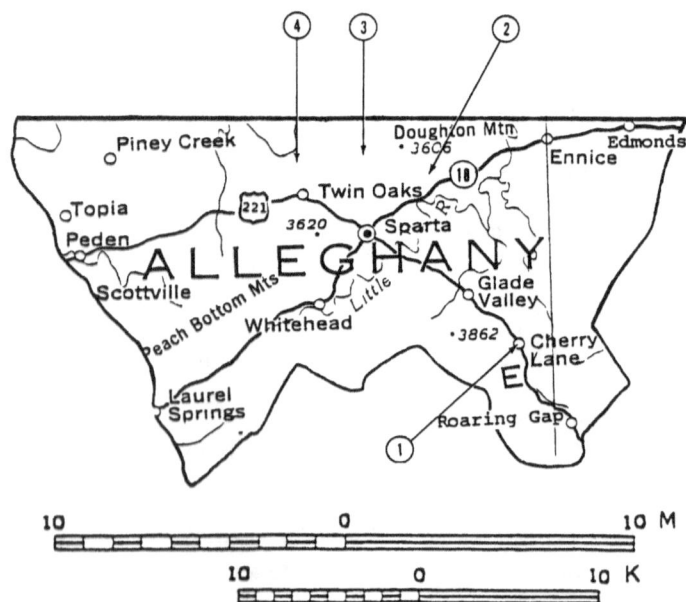

ANSON COUNTY

1. Ansonville area mines.

   Calcite. Galena. Garnet. Gold. Pyrite. Rutile. Siderite. Sphalerite. (2)

2. Wadesboro area: 2 miles southeast; at the Hamilton Mine (aka Bailey Mine).

   Gold. (2-19-quartz)

3. Wadesboro area: at the Jesse Cox Mine.

   Gold. (2)

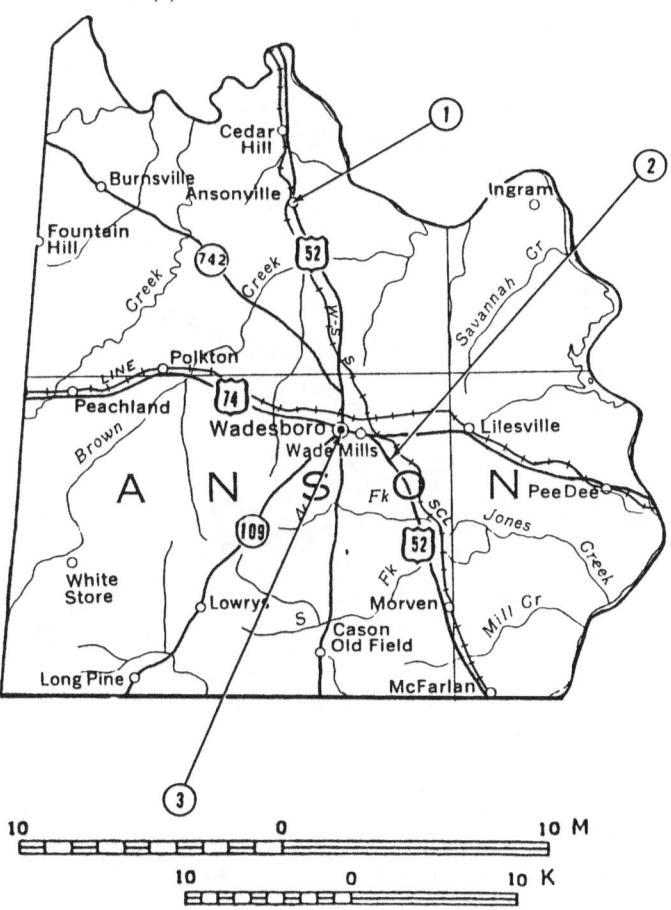

ASHE COUNTY

1. Beaver Creek area: 1.5 miles southwest; near top of small hill; at the South Hardin Mica Mine.

    Aquamarine. Beryl, golden. (2)

2. Fig area: 2 miles northwest of Elk Crossroads; 0.75 mile south of Black Mountain; at the Walnut Knob Mine.

    Aquamarine. (2)

3. Jefferson area: northwest on SR-88 to Ore Knob; just north; at the Ore Knob Copper Mine.

    Arsenopyrite. Bornite (aka "Peacock Ore"). Chalcopyrite. Malachite. Pyrite. (2)

4. Warrensville area: north-northeast; near Long Shoal Creek; on a spur of Phoenix Mountain.

    Quartz, rock: gem quality; individual crystals to 300 lb; to 15" thick; to 29" long. (19)

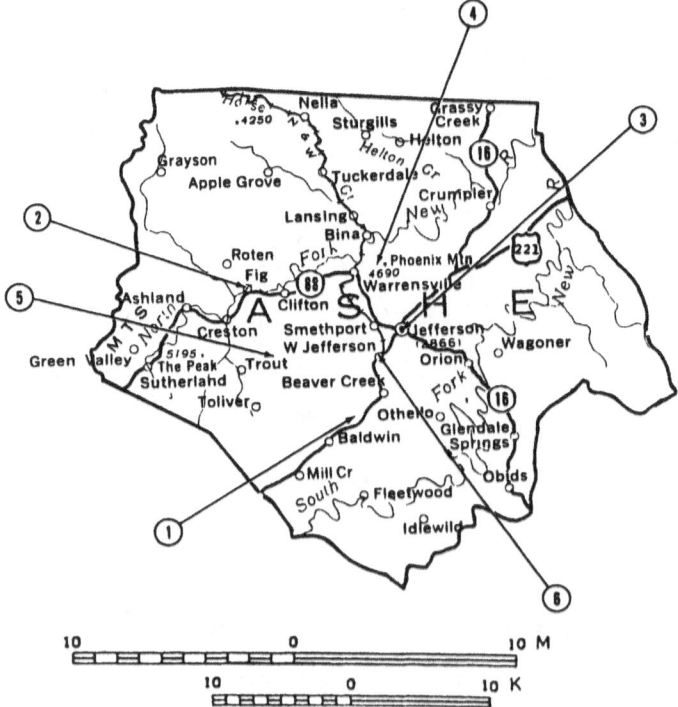

5. West Jefferson area: 5 miles on the road toward Stradford; within 100 yards of Brown's Store; at the Old Copper Mine.
   Arsenopyrite. Azurite. Copper, native. (2)

6. West Jefferson area: west; at the Old Gold Mine.
   Malachite: excellent specimens. (2)

AVERY COUNTY

1. Cranberry area: 1 mile south on US-19E; at the Cranberry Iron Mine.
   Epidote: superb emerald green; fine pink; gem quality. Garnet. Kyanite. Limonite. Magnetite. Olivine. Samarskite. (20)

2. Cranberry area: south on US-19E to road cut before reaching Minneapolis.
   Feldspar: green crystals. Moonstone. Thulite. (4)

3. Elk Park area: at the Rich Gap Mine.
   Garnet, almandine: rich red crystals. (2)

4. Plumtree area: 2 miles east; on Plumtree Creek; at the Elk Mica Mine.
   Garnet. (2)

5. Plumtree area: 2 miles east on Plumtree Creek; at the Slippery Elm Mine.
   Garnet. (2)

6. Plumtree area: 2 miles north on Roaring Creek Road; at the Burleson Mine (CF).
   Moonstone. (2)

7. Plumtree area: 0.8 mile northeast; on Plumtree Creek; at the Plumtree Mine.
   Feldspar: fine gem-quality crystals; pale blue-green. (2)

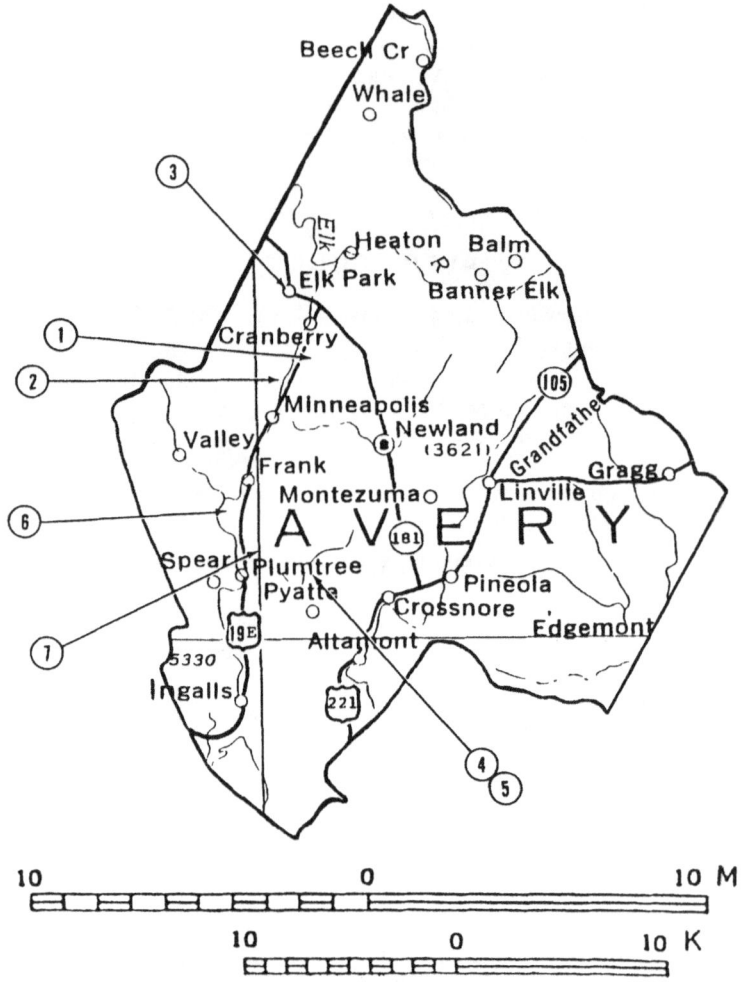

BEECH Cr

Whale

Heaton Balm

Elk Park

Banner Elk

Cranberry

Minneapolis

Newland (3621)

Valley

Frank

Montezuma

Linville

Gragg

A V E R Y

Spear

Plumtree

Pyatte

Pineola

Crossnore

Edgemont

Altamont

Ingalls

| 10 | 0 | 10 M |

| 10 | 0 | 10 K |

BUNCOMBE COUNTY

1. Alexander area: 1.5 miles southeast of Balsam Gap.
   Kyanite. (1-13)

2. Alexander area: at Balsam Gap; on Lookout Mountain.
   Beryl: fine gem quality. (2)

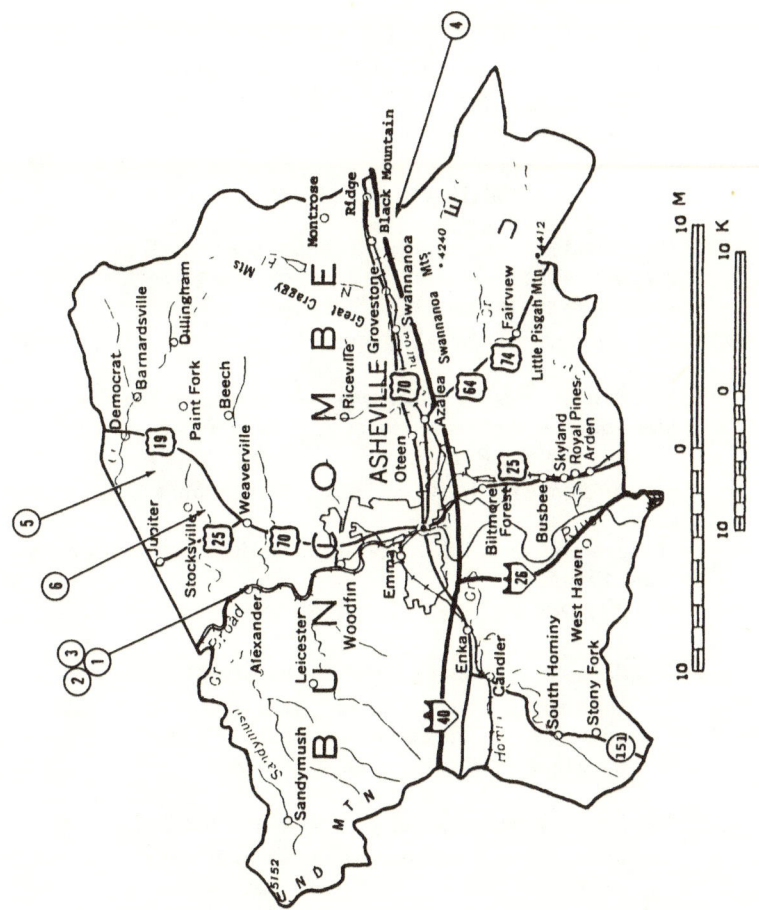

3. Alexander area: at Balsam Gap; at and near the Balsam Gap Mine.

   Kyanite. (1-2-13-14-29)

4. Black Mountain area: 1.5 miles southeast; in an unnamed prospect.

   Beryl: fine gem quality. (21)

5. Jupiter area: 4 miles east on SR-197 to gravel road on left; follow gravel road 0.25 mile north; at the Goldsmith Mine.

   Moonstone. (2)

6. Weaverville area: 1.7 miles north to first road on right past Pleasant Gap Church; 0.5 mile east to mine on creek.
    Serpentine. (2)

BURKE COUNTY

1. Bridgewater area: 8 miles west-southwest; in the South Mountains; 1 mile east of the Joel Walker Prospect, which is 0.5 mile southwest of Walker Knob.
    Aquamarine. Beryl, golden. (16)

2. Bridgewater area: 8 miles west-southwest; in the South Mountains; 0.5 mile southwest of Walker Knob; on hill; at the Joel Walker Prospect.
    Aquamarine. Beryl, golden. Feldspar. Mica. (21)

3. Bridgewater area: 1.25 miles south-southeast; at the Burkmont Prospect.
    Aquamarine. Beryl, golden. (21)

4. Brindletown area: along Brindletown Creek.
    Diamond; 2 stones; found in placer operations in 1843 and 1893; specific details unknown. Quartz, rutilated. Quartz, smoky; gem quality. (3-17-21)

5. Morganton area: series of prospects at town limits.
    Mica: reddish transparent sheets to 4" wide. (21)

6. Salem area: 3.2 miles south and slightly west; 1.25 miles south and slightly west of Burkmont Mountain.
    Beryl: green through chartreuse to yellowish brown; most transparent, some translucent. Mica: clear cinnamon; good quality. (21)

7. Salem area: 2.4 miles south and slightly west; 0.25 mile south of the crest of Burkmont Mountain.
    Beryl. Mica: clear cinnamon; in sheets to 4" diameter. (21)

8. Valdese area: on Laurel Creek; at the Tweedy Mine.
    Garnet. (2)

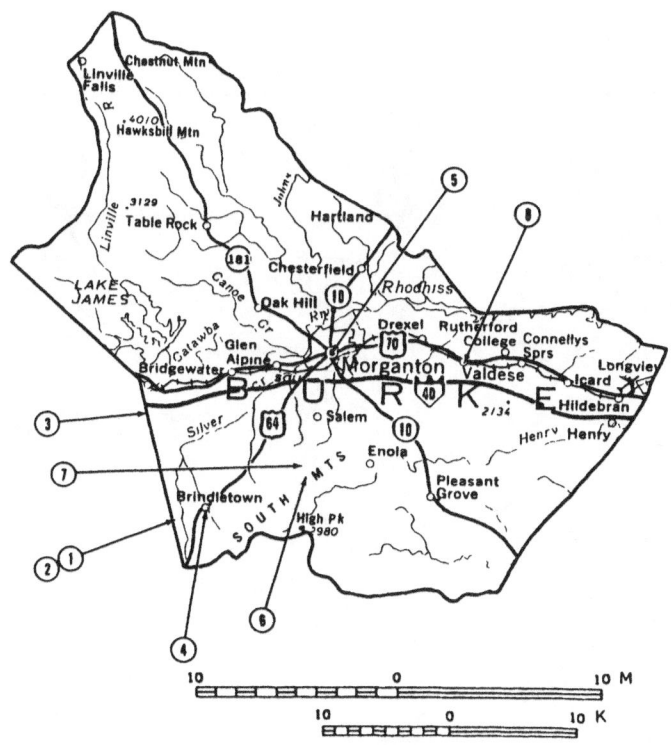

CABARRUS COUNTY

1. Concord area.
    Quartz, rutilated: fine. (1)

2. Harrisburg area.
    Agate: fine. (1-3)

CALDWELL COUNTY

1. Collettsville area: 1.7 miles north.
    Epidote: massive. Pyrite: excellent cubic crystals. (4)

2. Collettsville area: just south; at the Hercules Mine.
    Gold. (2)

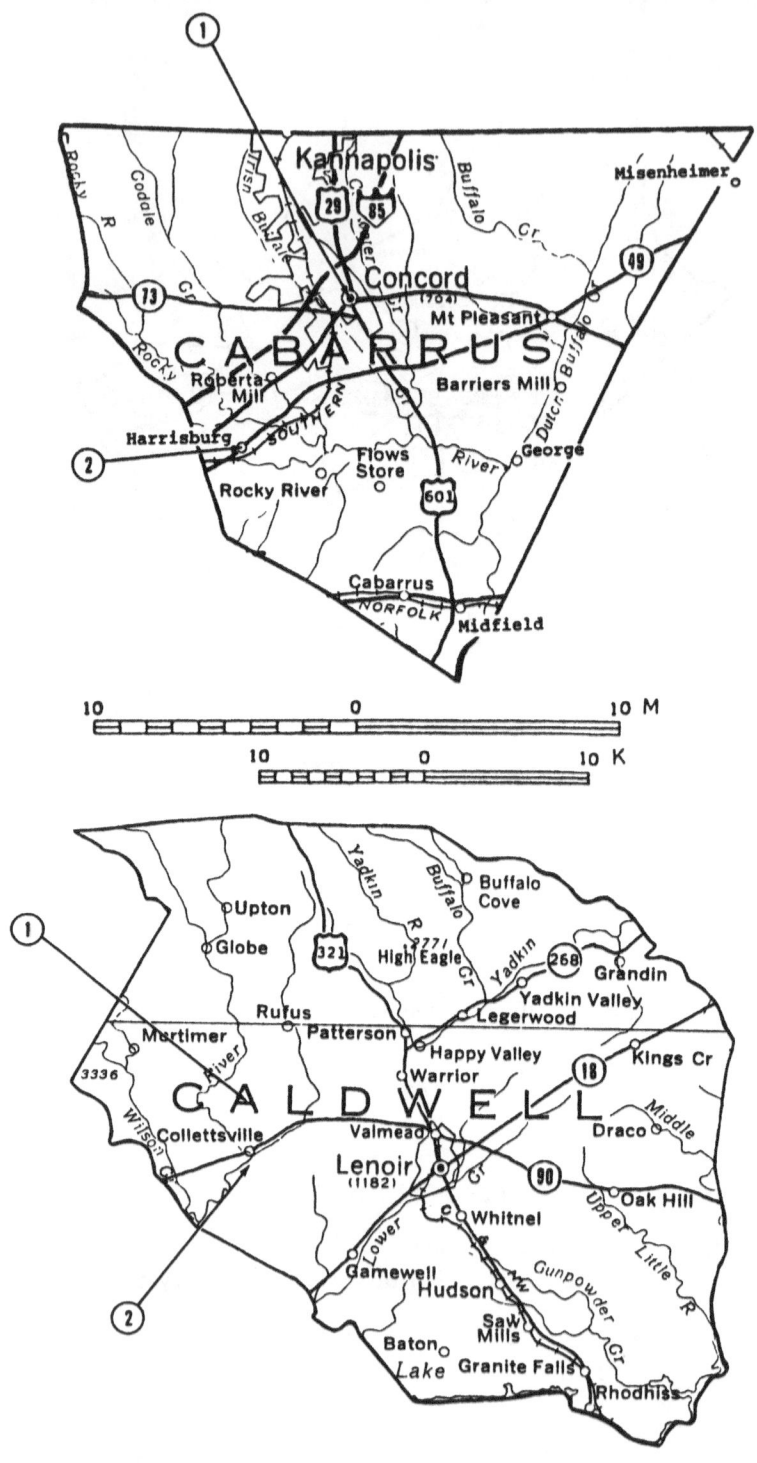

CASWELL COUNTY

1. Blanch area mica mines.
     Allanite. Mica. (2)

2. Leasburg area: 3 miles west; in prospects.
     Chlorite. Epidote. Tourmaline: fibrous crystals. (21)

3. Milton area mica mines.
     Allanite. Mica. (2)

4. Milton area: 3.75 miles southwest; at the Slaughter Prospect.
     Allanite. (21)

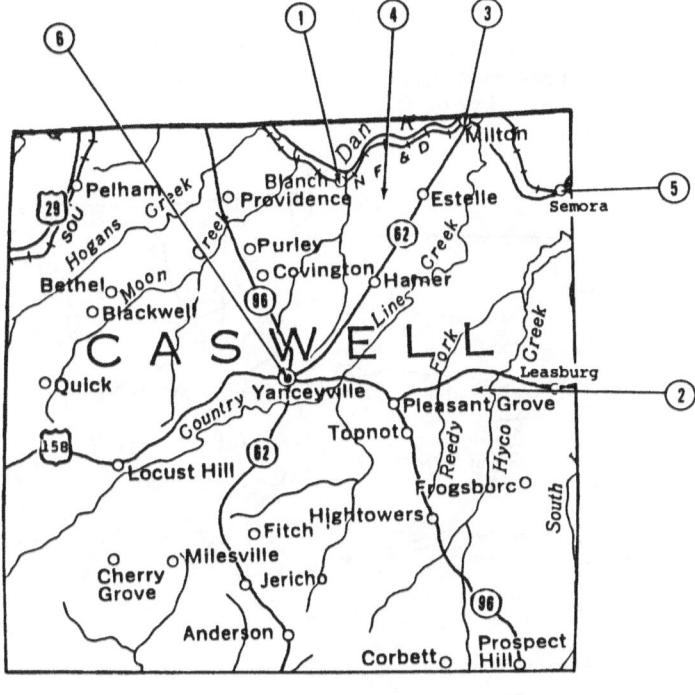

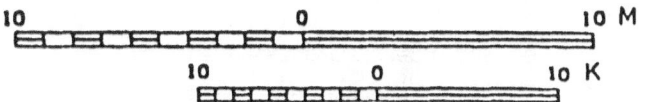

5. Semora area mica mines.
   Allanite. Mica. (2)

6. Yanceyville area mines.
   Albite. Garnet. Quartz, rock. (2)

CHATHAM COUNTY

1. Bennett area mines and prospects.
   Azurite. Chalcedony. Chalcocite. Chalcopyrite. Copper, native. Cuprite. Jasper. Malachite. Pyrite. Pyrolusite. Pyrrhotite. Quartz, rose. (21)

2. Bennett area: 2.6 miles southeast on SR-22 to left turn; 2 miles east to the Bear Prospect.
   Azurite. Bornite. Calcite. Chalcopyrite. Cerussite. Chrysocolla. Cuprite. Galena. Malachite. Pseudomalachite. Pyrite. Pyrrhotite. (21)

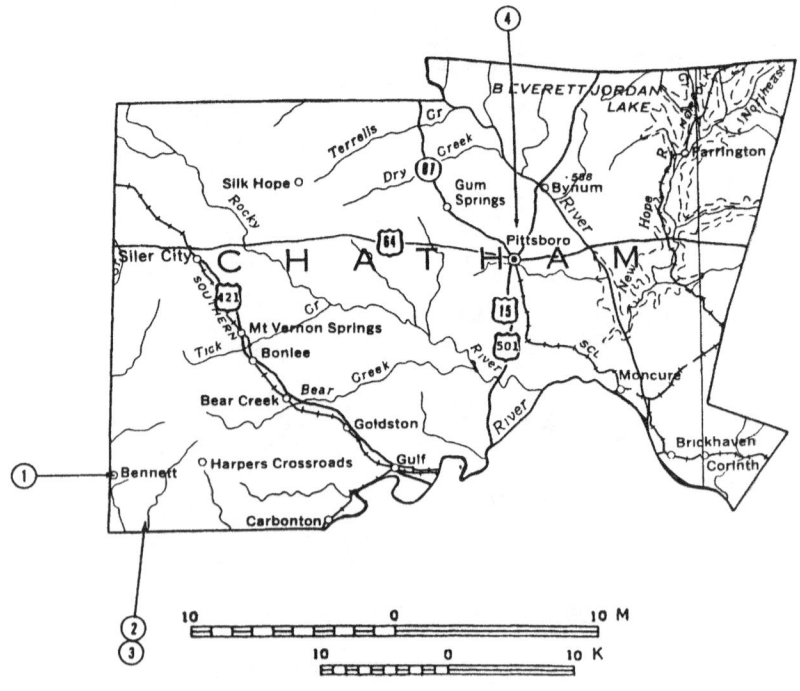

3. Bennett area: 2.6 miles southeast on SR-22 to left turn; then 2 miles east to crossroads; collect to north.

   Azurite. Malachite. Pyrite. Quartz, rock. (1-13-29)

4. Pittsboro area: 1.5 miles north on SR-87; in road cuts.

   Limonite: cubic crystals; pseudomorphs after pyrite. (4)

CHEROKEE COUNTY

1. Boiling Springs area: in Moss Creek.

   Staurolite. (1-3-27)

2. Grandview area: in Hanging Dog Creek.

   Staurolite. (1-3-27)

3. Marble area.

   Staurolite. (1-3-27)

4. Marble area: 1.3 miles north; in Hyatt Creek.

   Calcite (Marble var.). (1-7-20-29)

5. Marble area: 1 mile south; in the Valley River.

   Calcite (Marble var.). (1-7-20-29)

6. Marble area: on the Bettis property.

   Calcite (Marble var.). (1-3-29)

7. Murphy area: in Beaverdam Creek.

   Quartz, smoky: fine. (7)

8. Ranger area: north on Hiwassee Dam Road to Voiles Cabins.

   Quartz, smoky. (1-13-29)

9. Unaka area.

   Staurolite. (1-27)

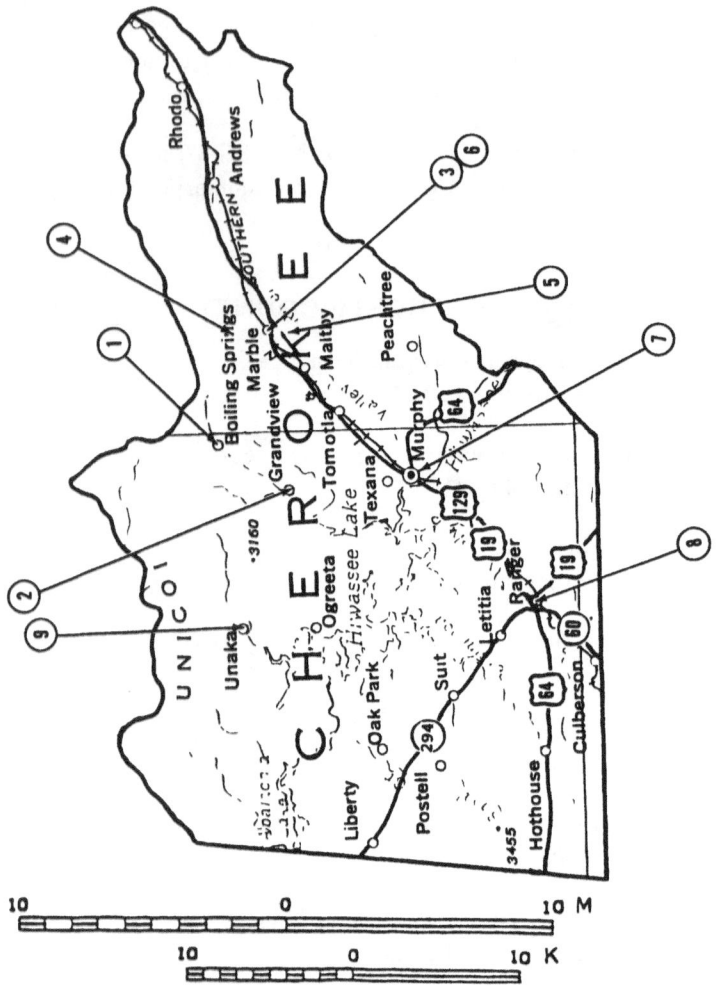

CLAY COUNTY

1. Brasstown area: 1 mile east.
    Staurolite. (1-3-27-28)

2. Brasstown area: southeast from post office on US-64 to the
    Ogden School; take next gravel road on left, which is

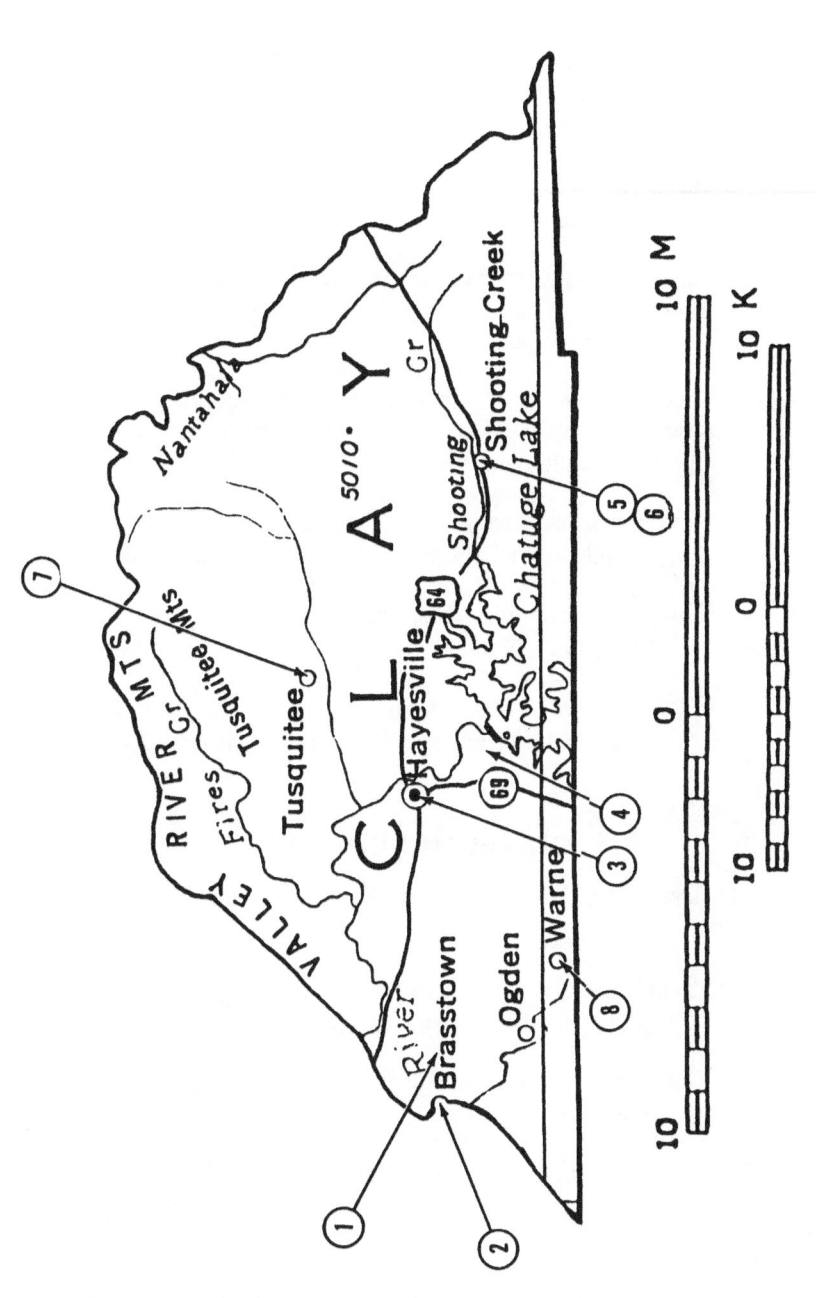

Greasy Creek Road; collect along the left side of the road and in adjacent pastures.

Staurolite. (1)

3. Hayesville area.

Staurolite. (1-3-27-28)

4. Hayesville area: head of Chatuge Lake; at the Bell Creek Mine.

Corundum: pink. (2)

5. Shooting Creek area: 6 miles north of the Georgia border; close to Buck Creek; at the Cullakanee Mine.

Amphibolite: emerald green. Corundum. Ruby. (2)

6. Shooting Creek area: at the Elf Mine.

Amphibolite: bright green. Corundum. Ruby. (2)

7. Tusquitee area: along Tusquitee Creek.

Staurolite. (1-3-27-28)

8. Warne area: at Cat Eye Cut; 600' south of Chestnut Knob.

Amphibolite: bright green. Corundum. Ruby. (6-16)

CLEVELAND COUNTY

1. Casar area: 3.5 miles southwest; at the Elliott Mine.

Beryl. (2)

2. Casar area: 2 miles west.

Quartz, rutilated: clear; colorless; gem quality. (1)

3. Casar area: 2 miles west-southwest.

Quartz, rutilated. (1)

4. Earl area: 1.3 miles east; between SR-198 and Buffalo Creek.

Aquamarine. Garnet. (1-13)

5. Hollybush area: on west side of the Broad River; on the

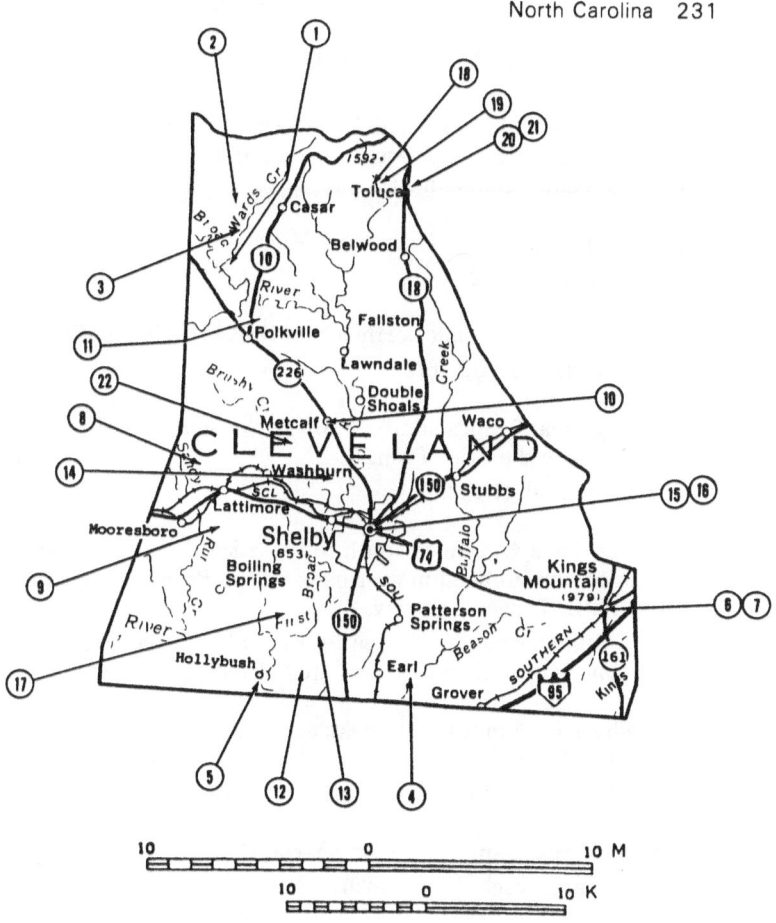

Whisnant property.
   Beryl. (16)

6. Kings Mountain area.
   Diamond: 1 stone; canary; very fine; 0.75 ct; $\frac{5}{16}''$ $\times$ $\frac{3}{16}''$; found in 1893. (3)

7. Kings Mountain area: at the Foote Mine.
   Apatite: fine crystals; honey-colored; transparent. (2)

8. Lattimore area: 1.6 miles northwest on SR-18.
   Mica: good crystals. (1-13-29)

9. Lattimore area: 1.75 miles south; at the C. B. McSwain Mine.

> Biotite. Feldspar: kaolinized. Garnet. Plagioclase. Potash: crystals; colorless; transparent. Quartz, rock. Sillimanite: abundant. (2)

10. Metcalf area: at the Mary Gold Mine.

> Sillimanite. (2)

11. Polkville area: 0.9 mile northeast; at the Gettys No. 1 Mine.

> Marcasite. (2-9-16)

12. Shelby area: 6 miles south on SR-18 to bend of the Broad River; near the Stice Dam.

> Emerald. (1-13-20)

13. Shelby area: 4.75 miles south to the Stice Dam; 0.75 mile west of dam to bend in east bank of the Broad River; on the W. B. Turner property.

> Aquamarine. Beryl. Biotite. Emerald. Muscovite. Olivine. Quartz, smoky: rutilated. Tourmaline. (2)

14. Shelby area: 3 miles northwest; at the Cornwall mine (aka Blanton Mine).

> Microcline: to 3″ diameter. (2)

15. Shelby area: 1 mile south of Sharon Church.

> Quartz, rock. (1-13-14-29)

16. Shelby area: SR-131 to first road on right after the fiber plant; park near waterwheel mill; collect on hillside behind mill and below dam.

> Quartz, smoky. Rutile. (1-13-14-29)

17. Shelby area: south on SR-18 to SR-150; 2 miles west on SR-150 to side road left; 0.5 mile south on side road.

> Quartz, rock. (1-13-14)

18. Toluca area: 1.75 miles west; at the A. F. Hoyle Mine.

> Andesine. Apatite. Biotite: brown. Calcite: white. Epidote. Hornblende: green. Zircon. (2)

19. Toluca area: 1.6 miles west; along a southeast-flowing tributary of Knob Creek; at the A. F. Hoyle No. 2 Mine.

> Apatite: green. Autunite. Calcite. Garnet. Muscovite: green. Perthite: large, well-formed crystals. Quartz, rock. Zeolite. (2)

20. Toluca area: at the G. B. McSwain Mine.

> Garnet: crystals to 1.5″ diameter. (2)

21. Toluca area: west; in road cuts.

> Garnet. (4)

22. Washburn area: 1.75 miles northeast; at the Martin Mine.

> Calcite. Chalcopyrite. Marcasite. Muscovite. Perthite. Plagioclase. Pyrite. Pyrrhotite. Quartz, rock. (2)

CUMBERLAND COUNTY

1. County-wide in streambeds.

> Agate. Chalcedony. Chert. Jasper. Opal: common. Petrified wood: much agatized. (3-7-18)

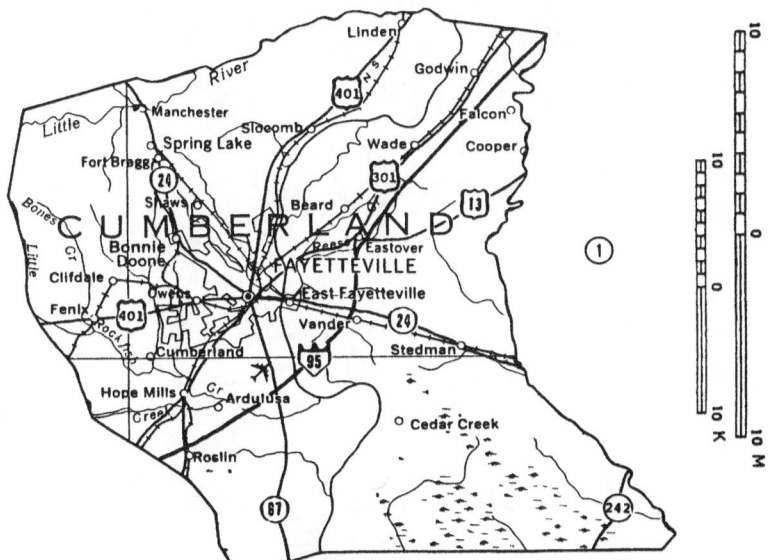

DAVIDSON COUNTY

1. Cid area: 0.25 mile northeast; at the Cid Mine.
   Chalcopyrite. (2)

2. Cid area: 1 mile south; at the Emmons Mine.
   Chalcopyrite. (2)

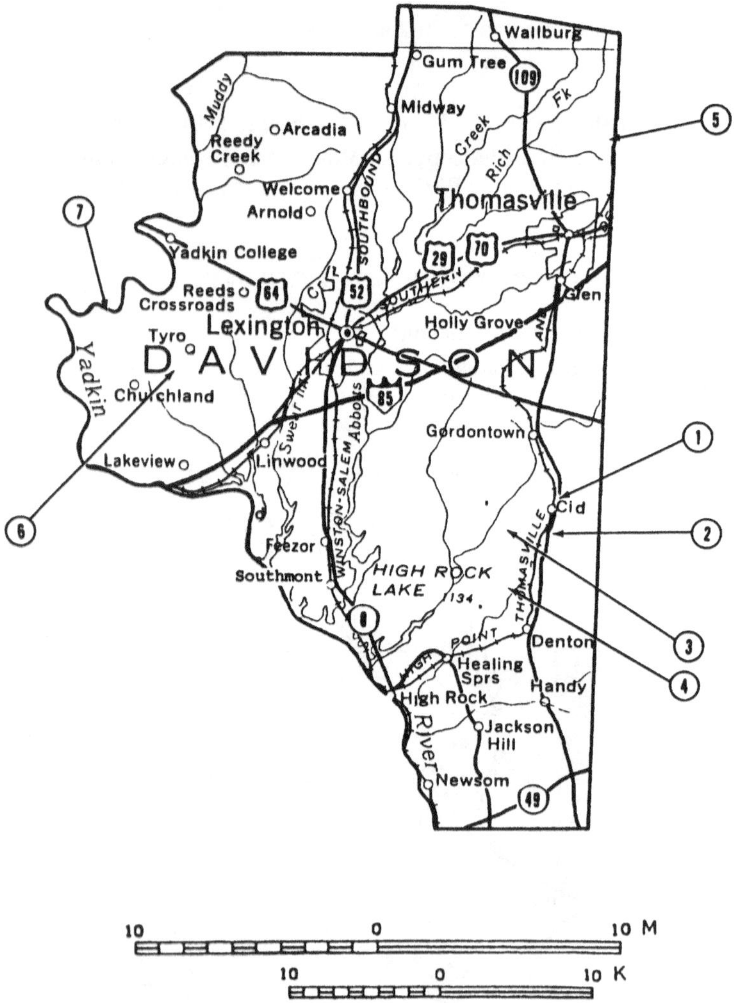

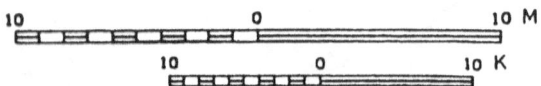

3. Cid area: 2.2 miles west-southwest; headwaters area of Buddle Branch; atop a low round hill; at the Silver Hill Mine.

   Galena. (2)

4. Denton area: 2 miles northwest; on the west side of the Flat Swamp Creek valley; at the Silver Valley Mine.
   Galena (2)

5. Thomasville area: 4.3 miles northeast; a bit east of Silver Hill; atop rounded hill 100′ high; at the Conrad Hill Mine.
   Pyrite. (2)

6. Tyro area: 1 mile southwest; on the Swicegood property; northeast of the house; also 0.2 mile west of the house.
   Amethyst: gem quality. (3)

7. Tyro area: 3.8 miles west-northwest; on the Yadkin River; 1 mile west of Oaks Ferry; on the Hairston property.
   Granite: orbicular. (1-13-29)

DAVIE COUNTY

1. Farmington area: 2 miles east to left turn on SR-801; 1.5 miles north to exposures.
   Autunite. Columbite. (24)

DURHAM COUNTY

1. Bethesda area: in fields west of US-70.
   Petrified wood: silicified. (3-18-20)

2. Weaver area: at the Eno River.
   Petrified wood: silicified. (3-18-20)

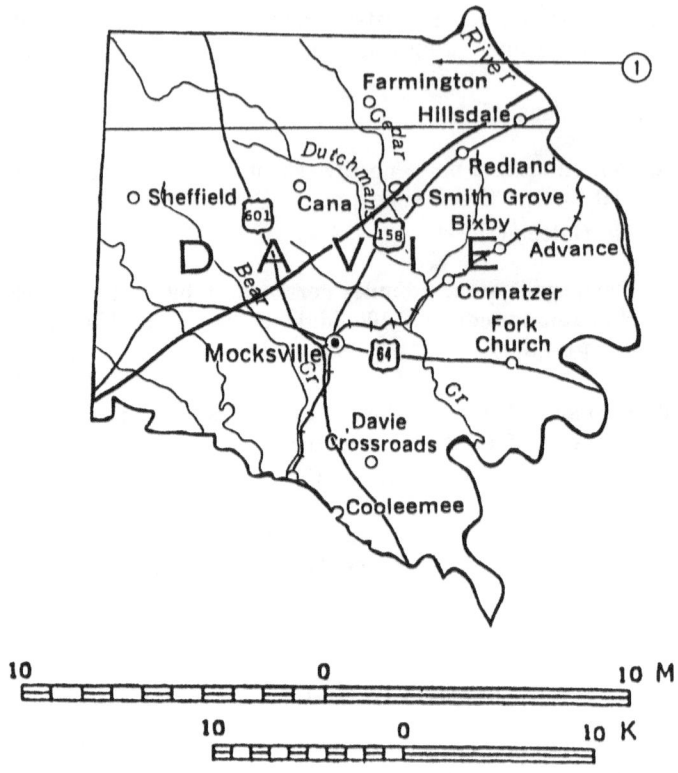

FORSYTH COUNTY

1. Kernersville area.

   Chrysolite. Enstatite (Bronzite var.). Tourmaline. (1-
   3-4-13)

2. Winston-Salem area gravel pits and quarries.

   Garnet: manganese. Halloysite. Hematite. Magnetite.
   (10-12)

3. Winston-Salem area: 4 miles south; at mines.

   Garnet: manganese. Halloysite. Hematite. Magnetite.
   (2)

Rougemont
Quail Roost
Flat
501
WESTERN
Bahama
Little
Willard
Factory
R
Fairntosh
2
River
Weaver
DURHAM
Gorman
85
DURHAM
(414)
Creek
Oak Grove
96
15
Bilboa Lick
Bethesda
(424)
40
SOUTHERN
SOUTHERN
Creek
54
Parkwood
1
10
0
10 M
10
0
10 K

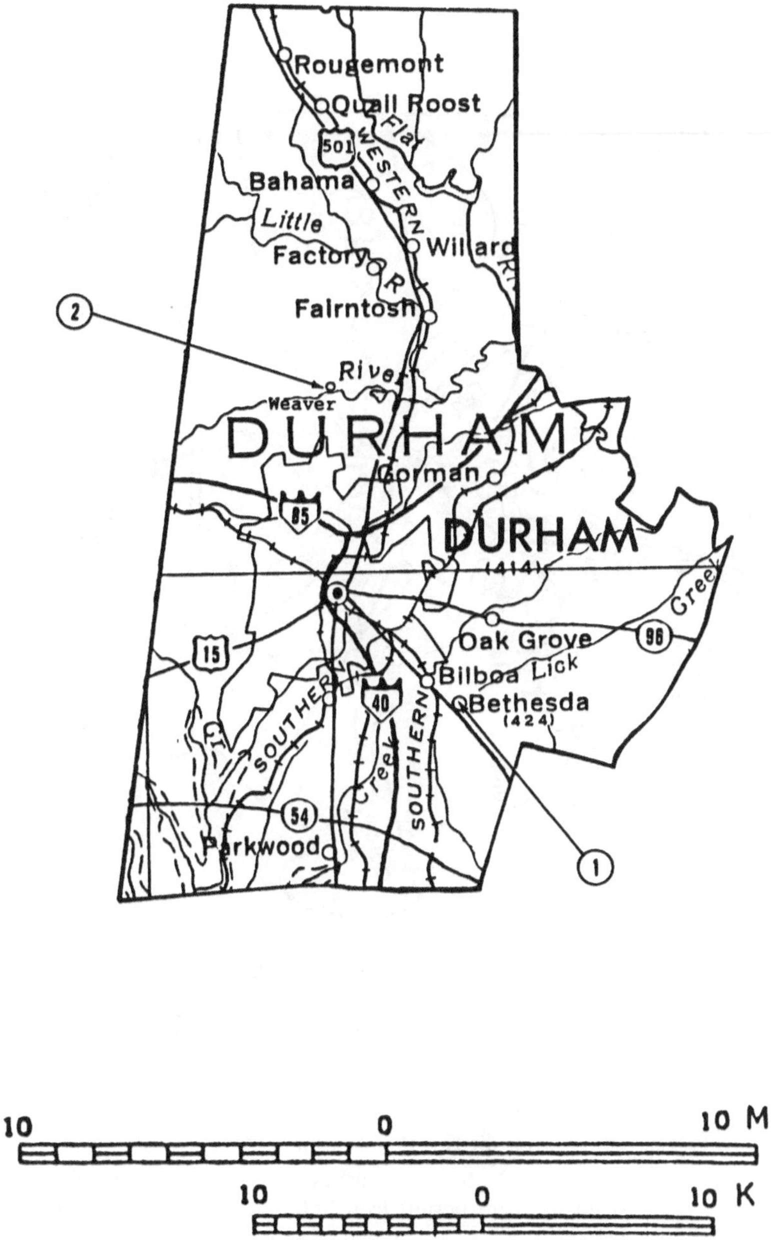

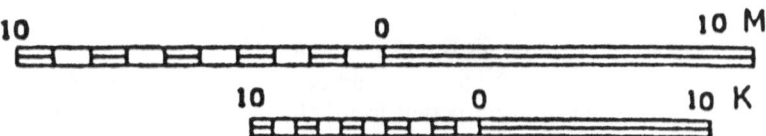

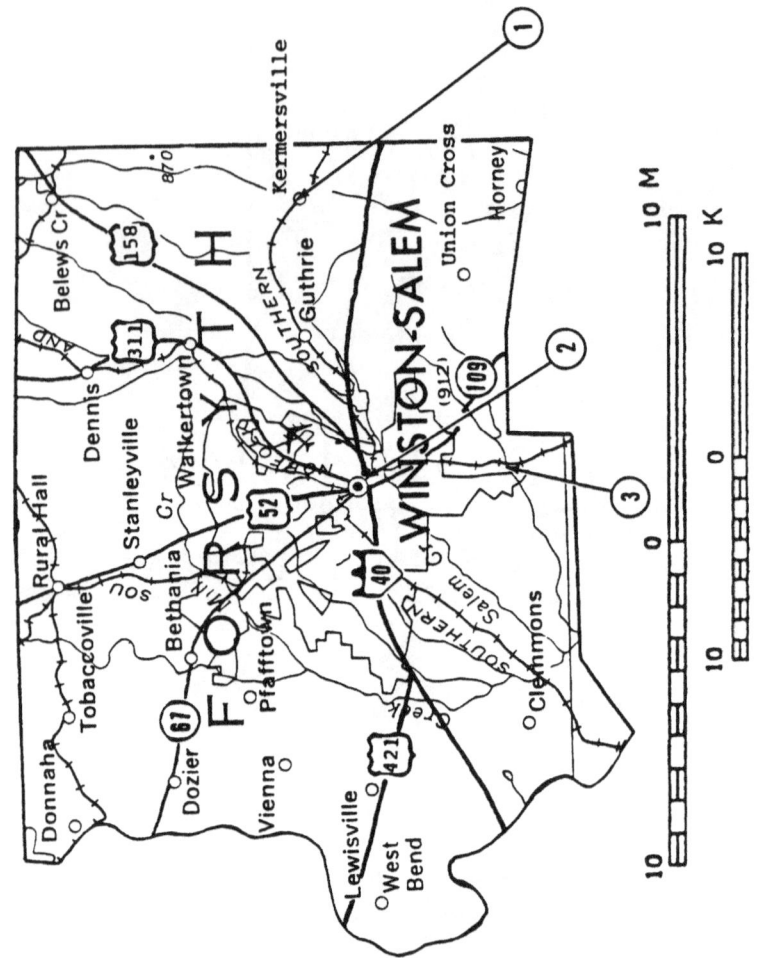

FRANKLIN COUNTY

1. Centerville area: southwest on SR-561; on the Taylor property.

   Amethyst. (1)

2. Louisburg area: at the Portis Mine.

   Diamond: 2 stones; weights unknown, but one described as an octahedron "of fine water"; found in 1852. (2)

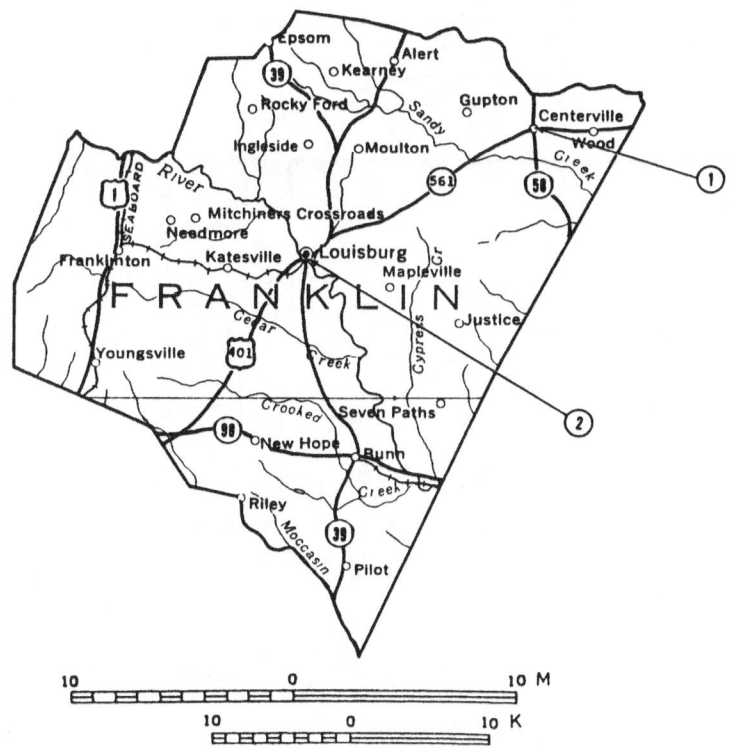

GASTON COUNTY

1. Alexis area: 1.5 miles east; in fields.
    Kyanite. Lazulite. Rutile. (1)

2. Alexis area: on Chubb Mountain.
    Kyanite. Lazulite: rich blue masses. Muscovite. (16)

3. Alexis area: on Chubb Mountain; at the Lowe property.
    Rutile. (1-13-29)

4. Alexis area: on SR-27; near Lincoln County border; at the
    Lowe Commercial Prospect (CF).
    Rutile. (1-10-13-29)

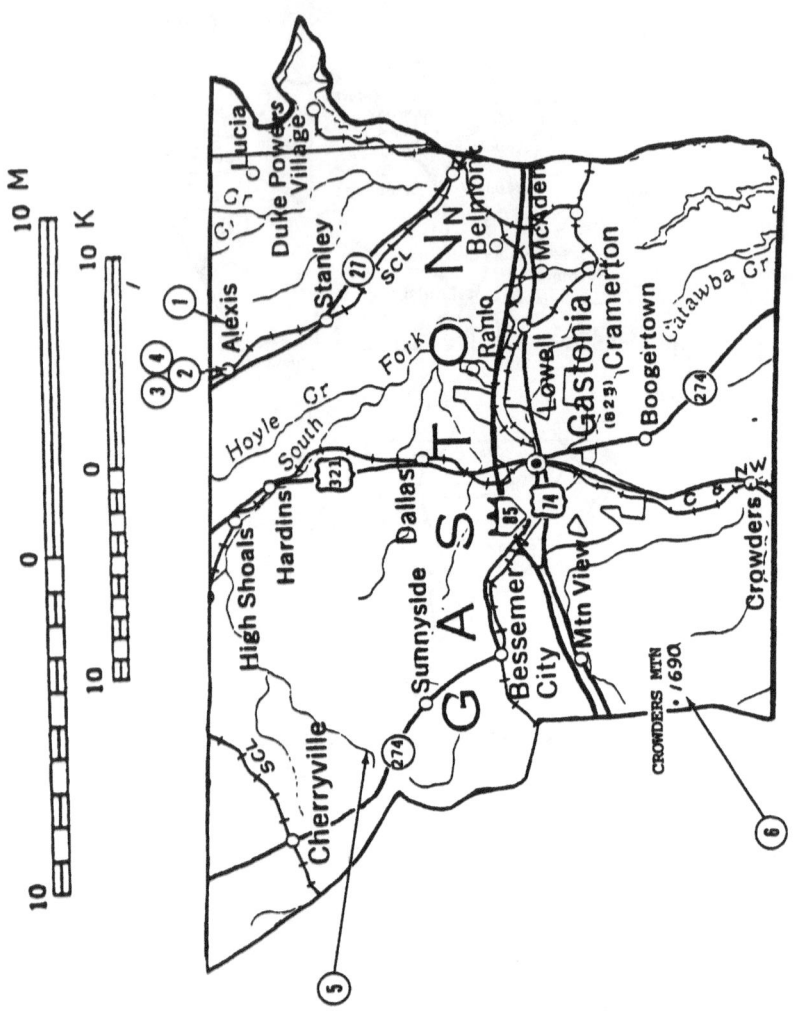

5. Cherryville area: 3.5 miles southeast; on the east side of the Cherrysville–Kings Mountain road; on the M. S. Bess property; at the Big Bess Mine.

> Apatite. Beryl. Biotite. Garnet. Ilmenite. Muscovite. Oligoclase (Sunstone var.). Perthite. Quartz, rock. Sillimanite. Tourmaline. Zircon. (2)

6. Crowders Mountain area.

   Kyanite. Lazulite: blue; crystals and masses. Muscovite. (16)

GRANVILLE COUNTY

1. Wilton area: 2 miles east; at quarry.

   Calcite. Epidote. Feldspar: crystals. Molybdenite. Quartz, rock. (12)

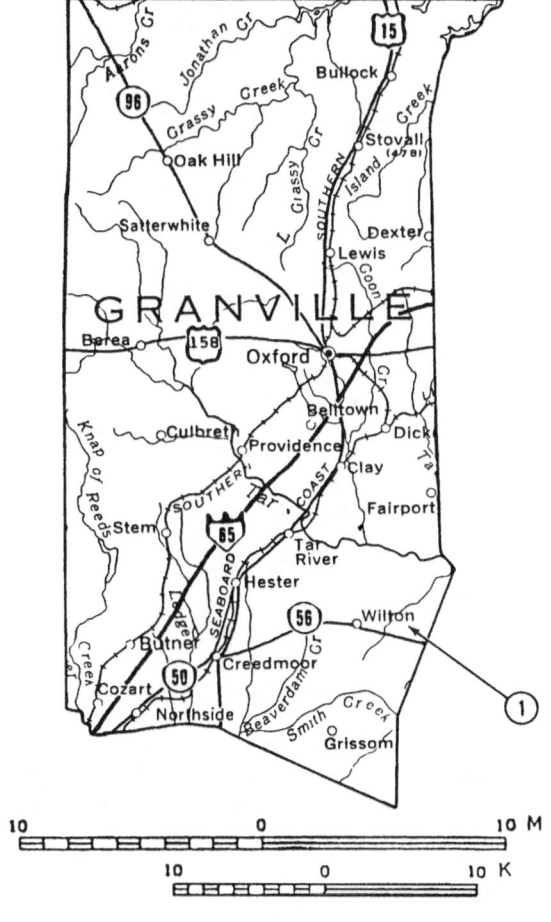

GUILFORD COUNTY

1. Gibsonville area.

> Quartz, green (Aventurine); asbestos inclusions. (1-13-29)

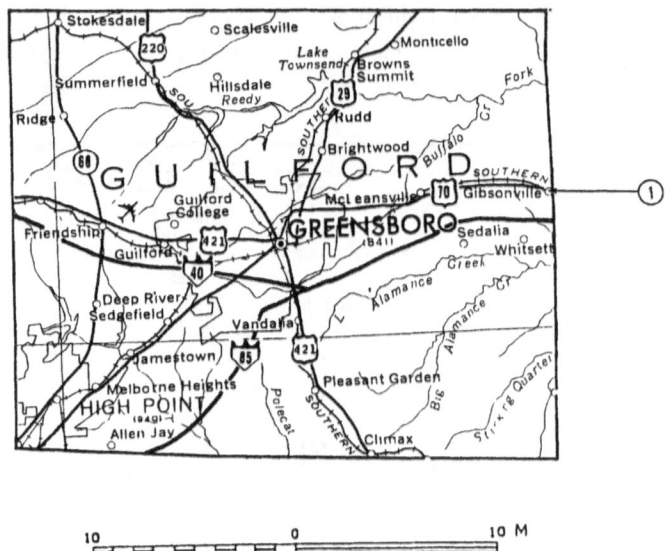

HALIFAX COUNTY

1. Glenview area: 1.7 miles west; at an old gold mine.

> Gold. (2)

HAYWOOD COUNTY

1. Canton area: on Main Street, go across I-40 and then left at the first road past the church; take this to the first gravel road left (CF payable at second house on left); continue to the end of the next road left to the Pressley Corundum Mine.

> Corundum. (2)

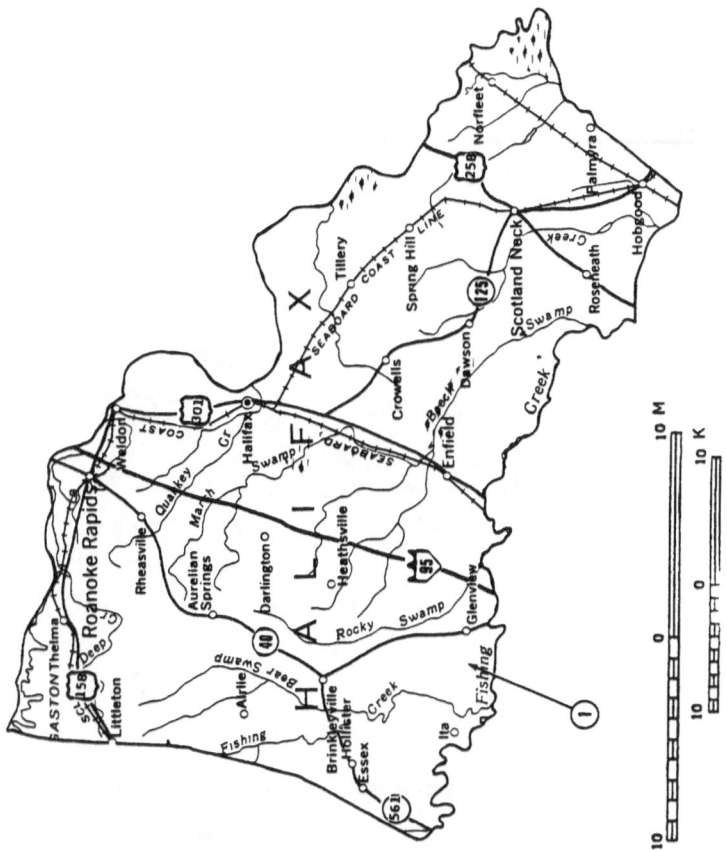

2. Hazelwood area: 7 miles south-southeast to the Sunburst U.S. Forestry Station; get permit here and key to gate; go through gate and follow the road for 10 miles; on slopes of Shining Rock.

    Quartz, rose. (1-13-14-29)

3. Waynesville area: east on SR-276 to the Bethel School; turn right on blacktop road to Shovel Creek community; in diggings and creekbed.

    Sapphire: gray; gem quality. (1-7-13-20-21)

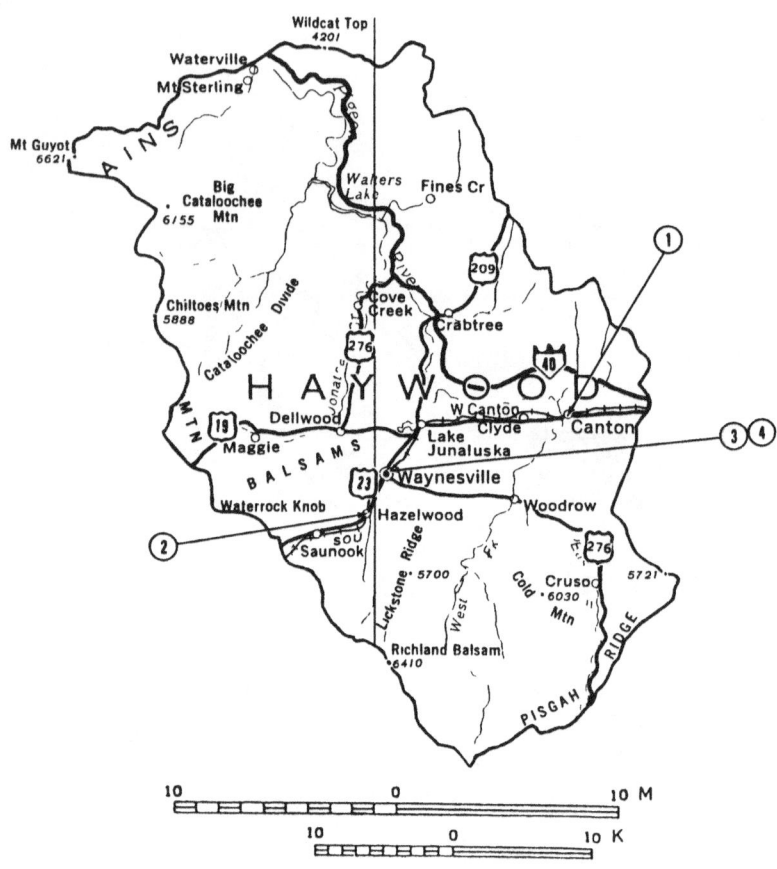

4. Waynesville area: southeast on US-276 to Bethel; at the New Sapphire Mine (CF).

   Sapphire. (2)

HENDERSON COUNTY

1. Tuxedo area: 0.4 mile northeast; turn right on unmarked road and follow along north shore of Lake Summit to the Jones Mine.

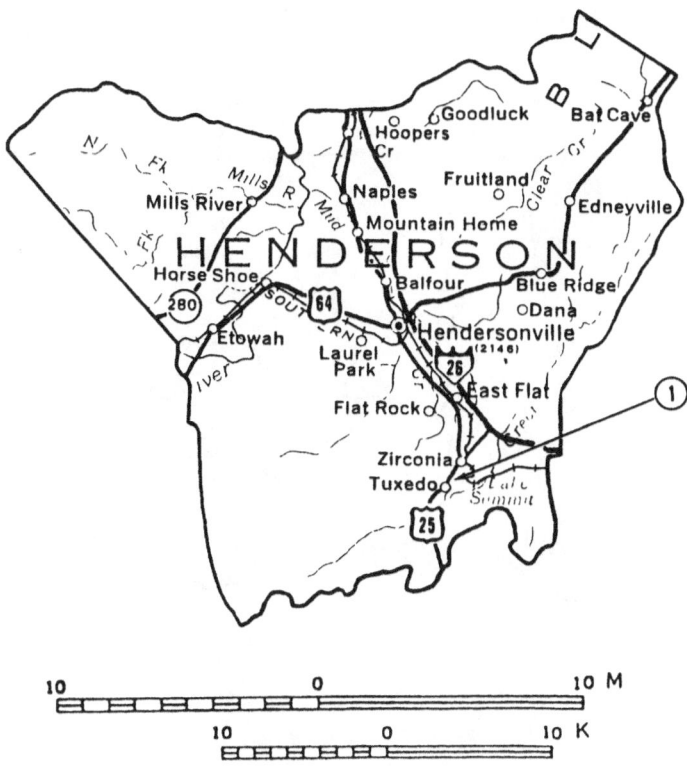

Albite: small colorless crystals. Anatase: tiny needle crystals; pale green to colorless. Auerlite. Calcite. Cryolite. Epidote. Garnet. Microcline. Monzanite. Muscovite. Sphene. Zircon. (2-16)

IREDELL COUNTY

1. Barium Springs area: 0.75 mile northeast; on the Walden property.

   Amethyst: gem quality. (1)

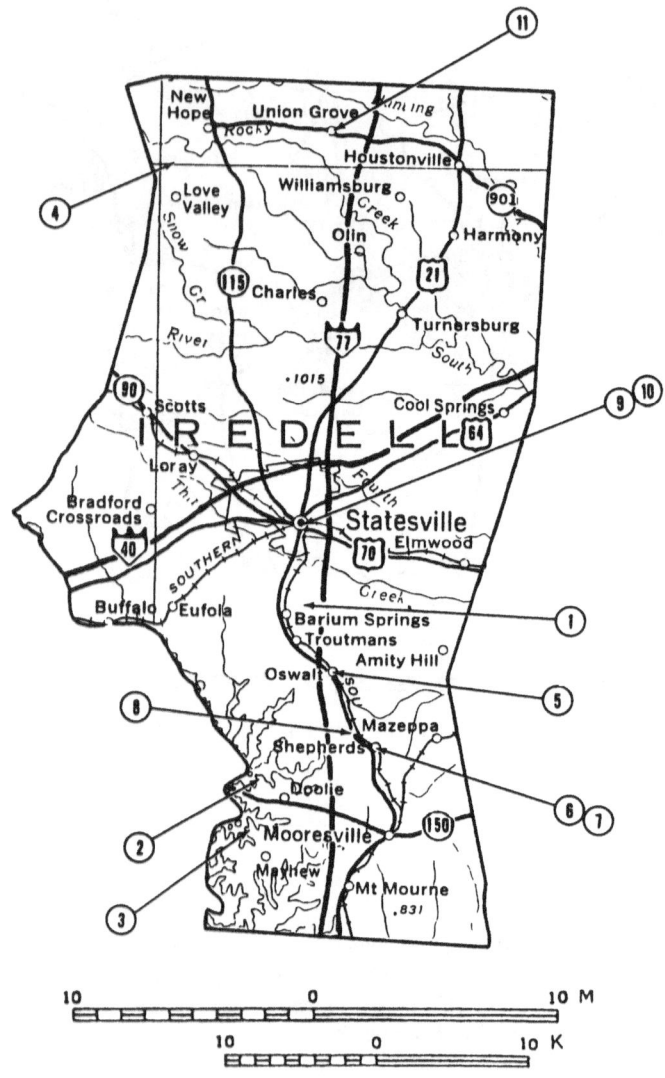

2. Doolie area: 1.3 miles west-northwest; on the Cornelius property.

    Amethyst: gem quality. (1)

3. Mayhew area: 1.2 miles northwest.
   Amethyst: gem quality. (1)

4. New Hope area: 2 miles southwest; at the W. A. Campbell Prospect.
   Beryl. Mica: pale cinnamon books. (21)

5. Oswalt area: in surrounding fields.
   Amethyst. (1)

6. Shephards area: in fields surrounding Shephards School.
   Amethyst. (1)

7. Shephards area: on the Brawley property.
   Amethyst: gem quality. (1)

8. Shephards area: 1 mile northwest; on the Cook property.
   Amethyst: fine. (1-21)

9. Statesville area quarry.
   Oligoclase (Sunstone var.): gray-brown crystals with hematite inclusions. (12)

10. Statesville area: east to Smith Hill; on the slopes.
    Amethyst. (1)

11. Union Grove area: north on SR-901 to Zion Church Road; 4 miles on Zion Church Road to stream; in road bank just after crossing stream.
    Rutile. (4-29)

JACKSON COUNTY

1. Cashiers area: at the Rice Mine.
   Aquamarine. (2)

2. Cashiers area: on the south side of Sapphire Lake; at the Bad Creek Mine.
   Ruby. (2)

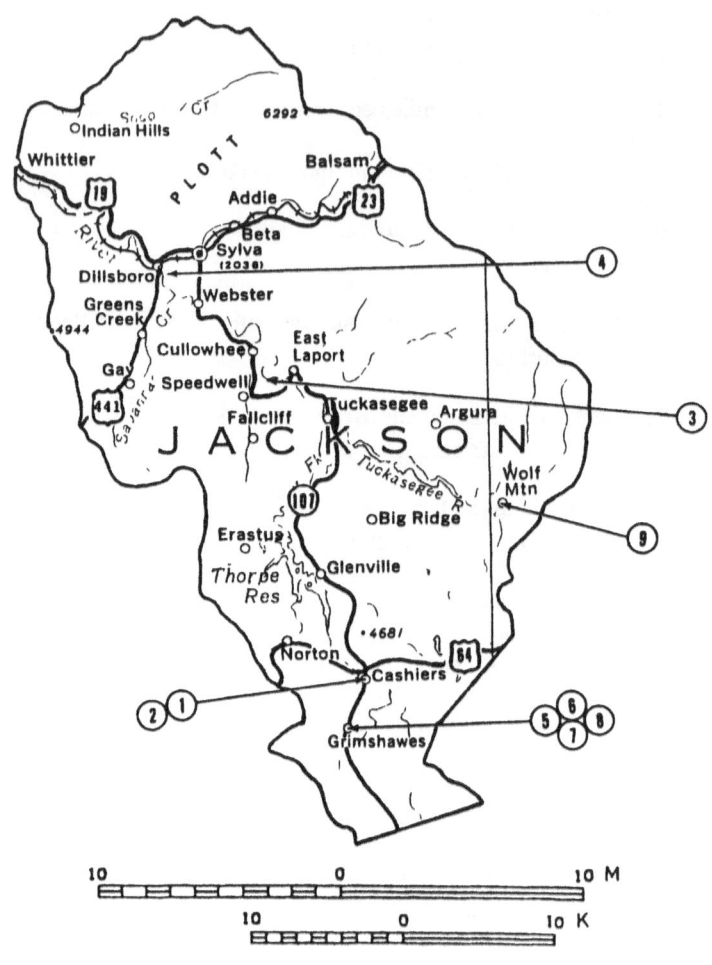

3. Cullowhee area: 1.25 miles south-southeast of the confluence of Caney Fork and Johns Creek; on ridge; at unnamed
   prospect.

   Beryl. (21)

4. Dillsboro area: just southeast; near crest of ridge; at the
   Frady Mine.

   Garnet. Muscovite. Quartz, rock. (2-16)

5. Grimshawes area: 1.75 miles east of the summit of White-side Mountain; 0.5 mile northeast of Whiteside Cove; at the Grimshawes Mine.

Beryl: gem quality. (2)

6. Grimshawes area: on Sugar Loaf Mountain.

Garnet. (1-13)

7. Grimshawes area: on the Whitewater River; at the abandoned Sapphire Mine.

Corundum: blue; small; transparent. Sapphire. (2)

8. Grimshawes area: on the Whitewater River; at the abandoned Whitewater Mine.

Corundum: blue; small; transparent. Sapphire. (2)

9. Wolf Mountain area: in small streams.

Ruby: fine gem quality. Sapphire: blue, yellow. (7-19-narrow, decomposed peridotite)

LINCOLN COUNTY

1. Boger City area.

Diamond: 1 stone; 0.5 ct; transparent; pale green; octahedron; fine; clear; found in placer operation in 1852. (17-21)

2. Denver area: 1.25 miles southwest; on the Goodson property.

Amethyst: gem quality. (1)

3. Flay area: at the Foster Mine (aka Thompson Mine).

Apatite. Biotite. Garnet: brick-red crystals to 0.25" diameter. Muscovite. Pyrite. Quartz, rock. Tourmaline, black. (2-16)

4. Flay area: at the Biggerstaff Mine (aka Deadman Mine).

Garnet: crystals to 2" diameter. (2)

5. Flay area: near where the road crosses Buffalo Creek; at the Carbine Mine.

Beryl. (2)

6. Flay area: near where the road crosses Buffalo Creek; at the Brown Mine.

Beryl. (1)

7. Iron Station area: 1.75 miles northeast; on the J. P. Lynch property (CF).

Amethyst: gem quality. (1)

8. Iron Station area: 2 miles northeast; on the Rendlemann property.

Amethyst: gem quality. (1)

9. Iron Station area: north on CR-1314 to right turn on SR-7S; follow SR-7S to right turn on CR-1509; follow CR-1509 to a right turn on CR-1417; follow CR-1417 to the Reel property; park near trailer (at which pay CF) and walk to digging area.

Amethyst. (1)

10. Iron Station area: on SR-27; on Leepers Creek.

Amethyst: light lavender to deep purple. (2-9)

MACON COUNTY

1. Burningtown area: on trail from there to the Roy Mason Mine (CF).

Ruby. Sapphire. (2)

2. Cullasaja area: 1.4 miles east; 0.5 mile north on SR-64; just west of Crows Branch, which empties into the Cullasaja River; on south slope of Higdon Mountain; on Corundum Hill; at the abandoned Culsagee Mine.

Corundum: fine gem quality; one coarse crystal mass weighed over 300 lb. Enstatite. Ruby. Sapphire: blue, green, yellow; one fine blue crystal found that measured 4″ × 2.5″. (2)

3. Ellijay area: 2.4 miles east; on Sheep Knob Mountain.

Aquamarine. (1-13-29)

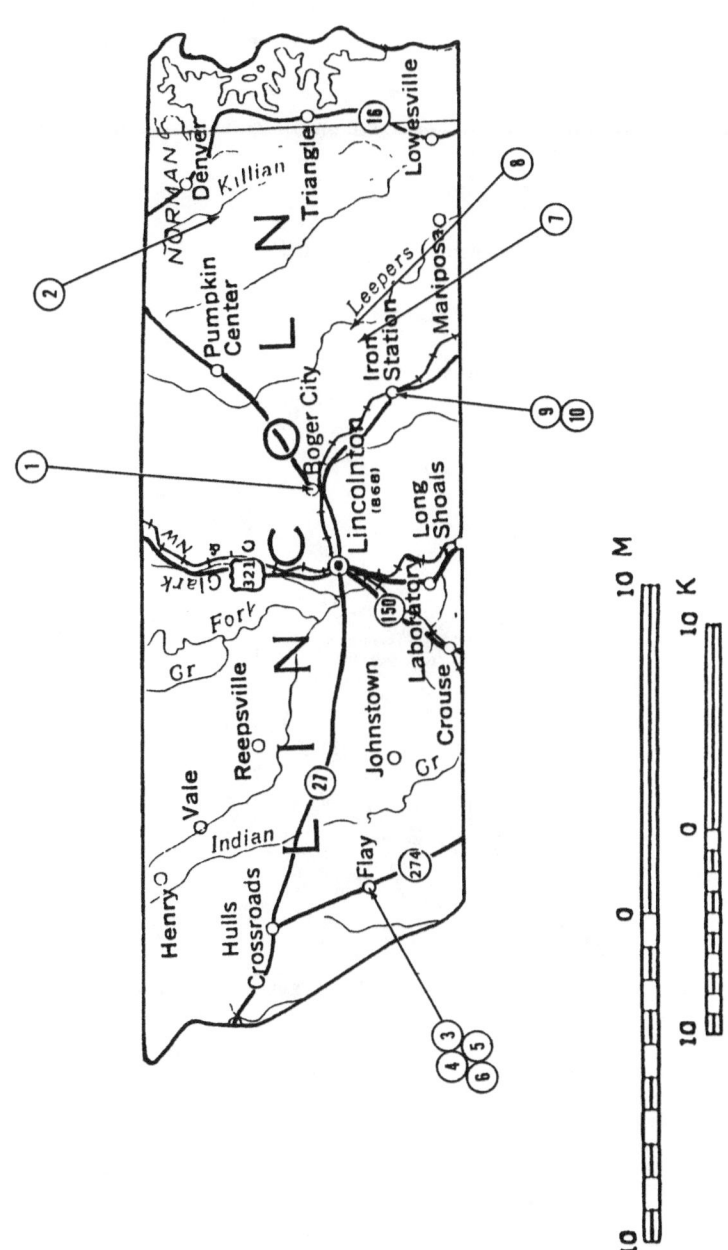

4. Ellijay area: slightly east; then right to the Mincey Mine.
   Corundum: bronze luster; asterism. Ruby: fine gem quality; chatoyant. (2)

5. Franklin area: 2 miles north on SR-28; then east of road at the McCook Mine.
   Garnet, rhodoline. (2)

6. Franklin area: west of SR-28 on Rose Creek Road; at the Four Ks Mine.
   Garnet: asterism. (2)

7. Franklin area: west of SR-28 on Rose Creek Road; just beyond the Four Ks Mine; at the Houston Mine.
   Sapphire. (2)

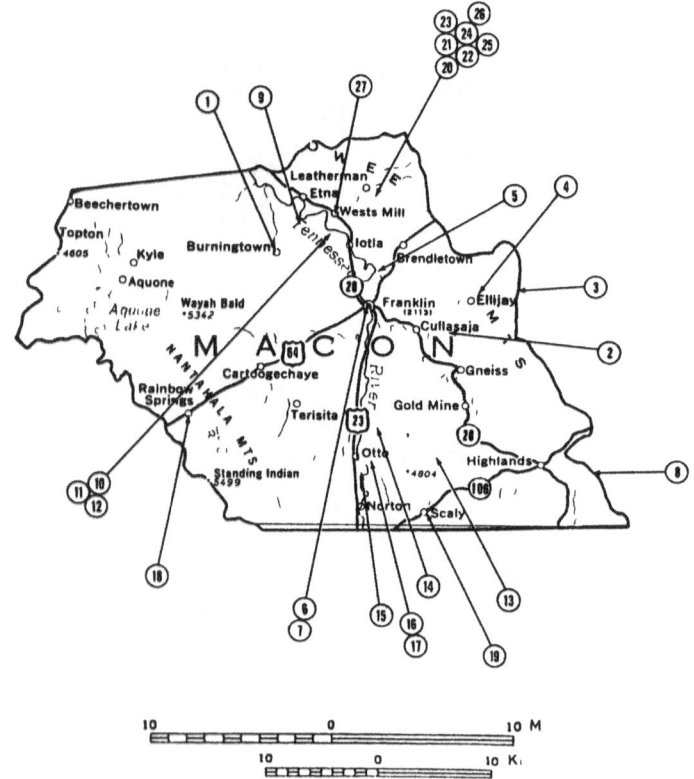

8. Highlands area: 3 miles east on SR-64; take trail up White-side Mountain.

   Garnet, almandine. (1-13-29)

9. Iotla area: 2.8 miles northwest; on southwest slope of Lyle Knob; at the Lyle Knob Mine.

   Biotite. Feldspar. Garnet: pink crystals to 0.5″ diameter. Muscovite: ruby red; transparent. Pyrite. Quartz, rock. (2-16)

10. Iotla area: 1 mile north on SR-28 to Mason Branch Mine sign; follow sign directions 0.25 mile to the Mason Branch Mine.

    Garnet, rhodoline. Sapphire. (2)

11. Iotla area: 1 mile north on SR-28; turn right to head of valley; on Mason Mountain.

    Garnet, rhodoline. Sapphire. (2)

12. Iotla area: in Mason Branch from south slope of Mason Mountain downstream to its mouth at the Little Tennessee River.

    Moonstone. (1-13-24-29)

13. Otto area: 4.5 miles east; near headwaters of Tessentee Creek; at the William Long Prospect.

    Amethyst: gem-quality crystals to 1.5″ diameter; deep colors; frequent banding. (19-21)

14. Otto area: 2 miles northeast of the mouth of Tessentee Creek; at the American Gem and Pearl Company workings.

    Amethyst: gem-quality crystals to 2″ diameter; deep color; frequent banding. (2-21)

15. Otto area: 1 mile south; in an outcrop on headwaters of Tessentee Creek; at the Littfield Beryl Mine.

    Beryl: blue, golden, yellow. (2)

16. Otto area: 0.6 mile east; on Tessentee Creek; at the Long Mine.

    Amethyst. Aquamarine. (2)

17. Otto area: 0.6 mile east on Tessentee Creek; at the Connolly Mine.

    Amethyst. Aquamarine. (2)

18. Rainbow Springs area: southwest on US-64 to bridge over Buck Creek; turn right at bridge and go 1 mile to another bridge; park and walk up mountain to outcrops.

    Garnet, pyrope. Ruby (in Smaragdite). (1-8-9-13-24-29)

19. Scaly area: southwest on SR-106 to mine dumps on Scaly Mountain.

    Corundum. (2)

20. Wests Mill area: 2.5 miles northeast; at the Holbrook Mine (CF).

    Ruby. Sapphire. (2)

21. Wests Mill area: 2.5 miles northeast; at the Jacobs Mine (CF).

    Ruby. Sapphire. (2)

22. West Mill area: 2.4 miles northeast; at the Sheffield Mine (CF).

    Ruby. Sapphire. (2)

23. Wests Mill area: 2.4 miles northeast; at the Schuler Mine (CF).

    Ruby. Sapphire. (2)

24. Wests Mill area: 2.3 miles northeast; at the Gregory Mine (CF).

    Ruby. Sapphire. (2)

25. Wests Mill area: 2 miles northeast; at the Caler Mine (CF).

    Ruby. Sapphire. (2)

26. Wests Mill area: 2 miles northeast; at the Gibson Ruby Mine (CF).

    Ruby. (2-6)

27. Wests Mill area: in Cowee Creek from its mouth on the

Little Tennessee River upstream for 3 miles; in streambed and banks and on In Situ Hill, which is the original source.

Chromite. Gahnite (Zinc spinel). Garnet, almandine: pink to deep red. Gold. Iolite: colorless. Kyanite. Monzanite. Pyrite. Ruby. Rutile. Spinel (Pleonaste): black. Staurolite: transparent. Tremolite. Zircon. (1-6-7-13-29)

MADISON COUNTY

1. Belva area.

   Staurolite. (1-3-27-28)

2. Bluff area: 2 miles southwest; on Roaring Fork Creek; 0.5 mile west of its junction with Meadow Fork.

   Unakite. (1-13-20)

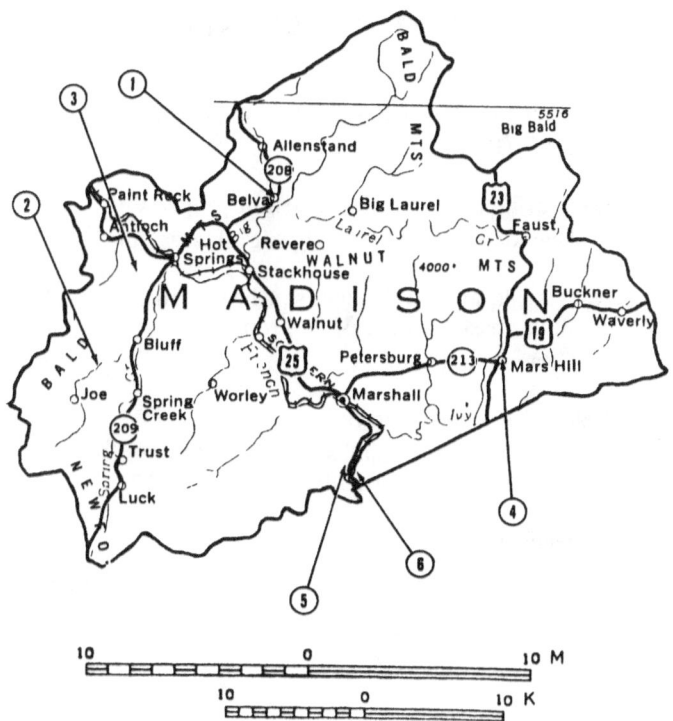

3. Hot Springs area: 1.75 miles west-southwest.
   Unakite. (1-13-29)

4. Mars Hill area.
   Monzanite: fine crystals to 11″ thick. (16)

5. Redmon area: 2 miles southwest to Little Pine Creek; follow road along creek to Roberts Branch; then follow mine road up Roberts Branch to the Little Pine Mine.
   Garnet, almandine. (2)

6. Redmon area: 3 miles south-southwest along the Souther RR on the road to Redmon Dam; turn east on first road north of dam; bear left at fork to the Freeman property; at the Lone Pine Mine.
   Garnet, almandine. (2)

MCDOWELL COUNTY

1. Dysartsville area.
   Diamond: 1 stone; 4.33 ct; fine; pale green; found in 1866. (17)

2. Dysartsville area: on the A. Bright property.
   Diamond: 1 stone; 4.88 ct; high luster. (6)

3. Dysartsville area: 1.2 miles southeast; on north side of SR-26; in stream on the Mills property.
   Corundum. (7-20)

4. Muddy Creek headwaters.
   Diamond: 2 or 3 small stones; weight not given; found during panning; specifics not known. (17)

5. Unspecified location.
   Diamond: 1 stone; 2.375 ct; flawed; white; found in 1877. (17)

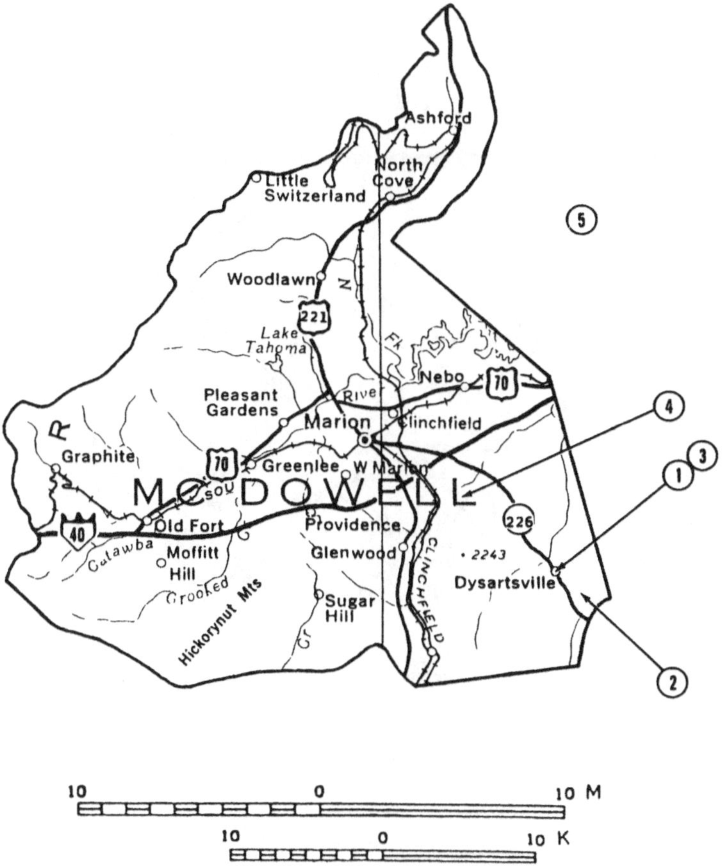

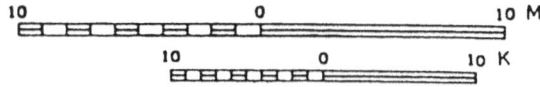

10   0   10 M

10   0   10 K

MECKLENBURG COUNTY

1. Caldwell area.

   Agate: fine. (1-3)

2. Todds Branch.

   Diamond: 2 stones; one weighed 1 ct; weight of other
   not given; fine; in placer; found in 1852. (17-21)

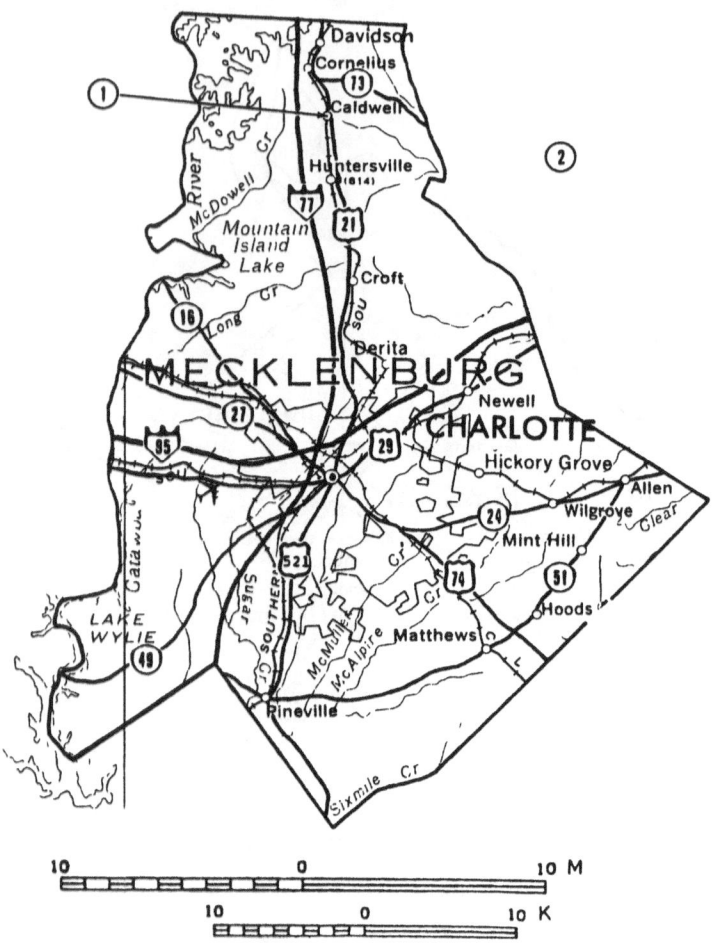

MITCHELL COUNTY

1. Bakersville area: 1 mile north; at the base of Medlock Mountain; on the W. G. Bowman property.

   Oligoclase (Sunstone var.). (6)

2. Bakersville area: 1 mile north; on Medlock Mountain.

   Oligoclase (Sunstone var.). (1-13)

3. Buladean area: 3 miles northeast; at the Sink Hole Mine.
   Albite. Apatite. Beryl. Kyanite. (2)

4. Busick area (Yancey County): follow Forest Service road
   north-northeast into Mitchell County; on the divide be-
   tween Rush Creek and Crabtree Creek; on Big Crabtree
   Mountain at elevation 5,000'.
   Emerald. Feldspar (Albite var.). Garnet. Gummite.
   Quartz, milky: coarse-grained. Tourmaline. (16)

5. Chalk Mountain area: at the Chalk Mountain Mine.
   Torberite. (2)

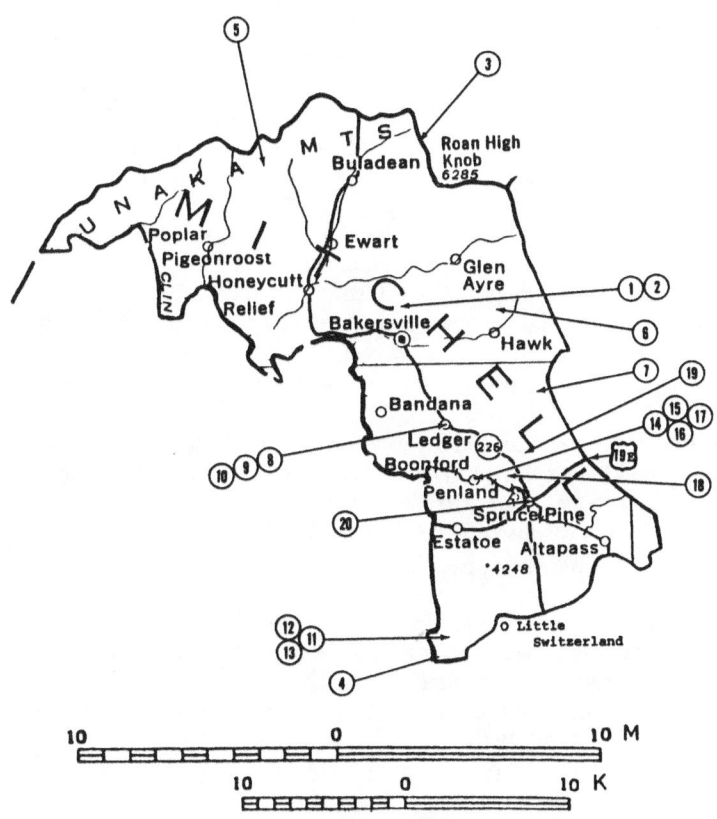

6. Hawk area: 1 mile north on SR-261; at the confluence of Soapstone Branch and the left fork of Cane Creek; at the Hawk Mica Mine.

> Albite: fine gem quality; pale blue-green crystals. Bytownite. Moonstone. Oligoclase (Sunstone var.): pink; thulite rutilations. Steatite (Soapstone). Thulite. (2)

7. Ingalls area (Avery County): southwest to North Branch Toe River; on west side of river along eastern border of Mitchell County.

> Autunite. Moonstone. (1-6-13)

8. Ledger area: east to the Chestnut Flat Mine.

> Beryl. Columbite. Danburite. Garnet. Mica. (2)

9. Ledger area: east to the Old Sol Mine.

> Autunite. Columbite. Garnet, almandine. (2)

10. Ledger area: near the summit of Yellow Mountain.

> Kyanite: deep blue crystals to 2″ long. (5-milky quartz -13-14-29)

11. Little Switzerland area (border of McDowell County): follow Crabtree Road south and remain on it as it curves back northwest; go past CR-1100 on right and continue downhill to sharp right curve; take side road on left to the No. 20 Mine.

> Thulite. (2)

12. Little Switzerland area (border of McDowell County): northwest through underpass beneath the Blue Ridge Parkway to left turn at Crabtree Church onto CR-1100; southwest on CR-1100 to the McKinney Mine.

> Amazonite. Beryl. Thulite. (2)

13. Little Switzerland area (border of McDowell County): northwest through underpass beneath the Blue Ridge Parkway to left turn at Crabtree Church onto CR-1100; bear right onto CR-1104; turn left on CR-1105 and drive to end of road; walk to the Crabtree Mine dump (CF).

> Emerald. (2)

14. Penland area: 1.8 miles on Penland School Road to the Penland RR tracks; 0.5 mile afoot down RR bed to mine dump.

>   Garnet, almandine: deep red; gem-quality crystals to 3" diameter. (92)

15. Penland area: at the Deer Flat Mine.

>   Thulite: masses to 1" thick. (2)

16. Penland area: at the Pine Mountain Mine.

>   Thulite: masses to 1" thick. (2)

17. Penland area: at the Putnam Mine.

>   Thulite: masses to 1" thick.

18. Spruce Pine area: 1.3 miles northwest; at the McChone Mine.

>   Amazonite. Spodumene (Hiddenite): pale yellow; fine gem quality. (2)

19. Spruce Pine area: 1.8 miles north and slightly east; 0.25 mile southwest of English Knob; on the Wiseman property.

>   Aquamarine. Beryl, golden. (6)

20. Spruce Pine area: at the Eric Thomas Prospect.

>   Amazonite. Aquamarine: gem quality. Beryl, golden: gem-quality. (21)

MONTGOMERY COUNTY

1. Eldorado area mines.

>   Azurite. Calcite. Gold. Hydrozincite: very fluorescent. Malachite. Pyrite. Silver. Smithsonite. Sphalerite. (2)

2. Troy area abandoned mine.

>   Leopardite. (2)

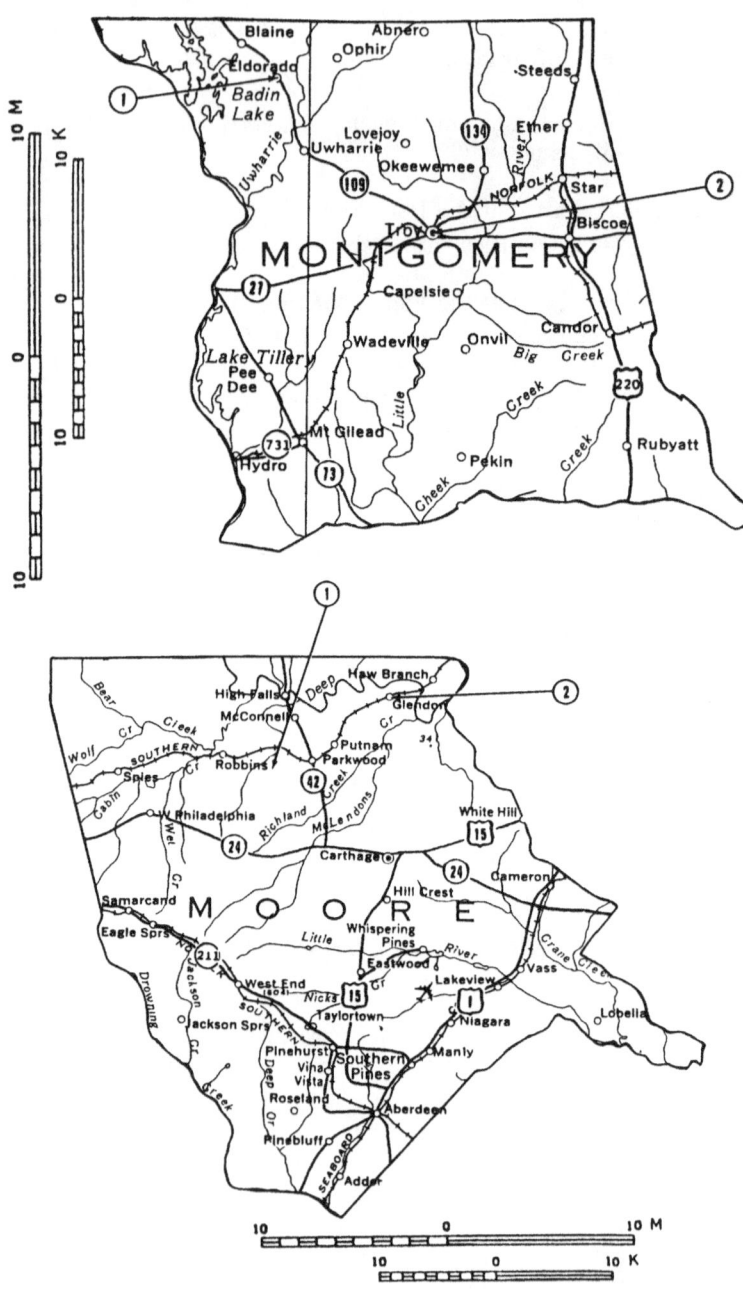

MOORE COUNTY

1. Carthage area: 8 miles northwest; at the Bell Mine.
   Gold. (2)

2. Glendon area mines.
   Fluorite. Hematite. Lazulite. Pyrite. Pyrophyllite. (2)

ORANGE COUNTY

1. Hillsborough area: at the Piedmont Minerals Mine.
   Andalusite. Lazulite. Topaz. (2)

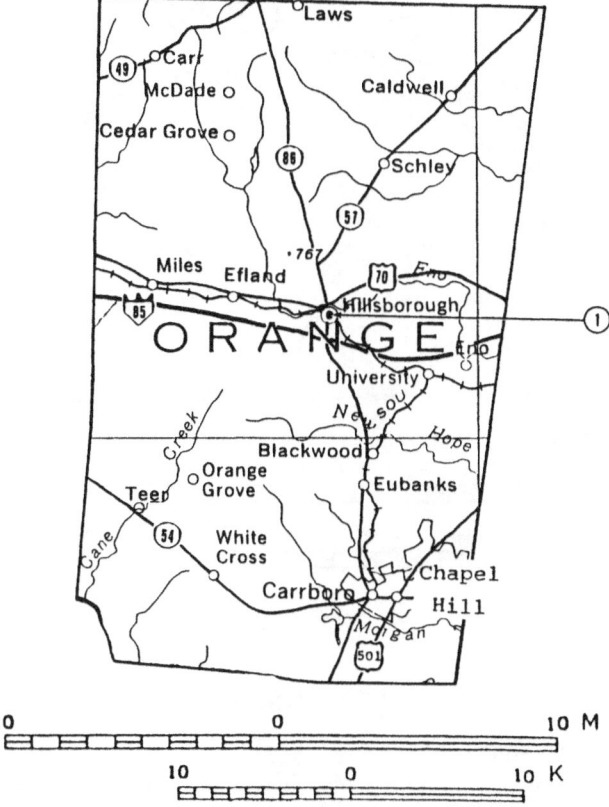

PERSON COUNTY

1. Allensville area: 0.25 mile northeast to side road on left; 1
   mile north to the Durgy Copper Mine.
   Malachite. (2)

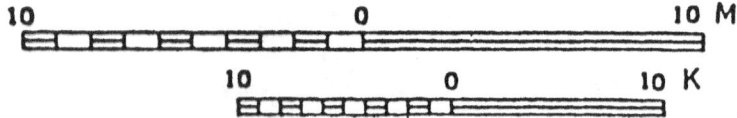

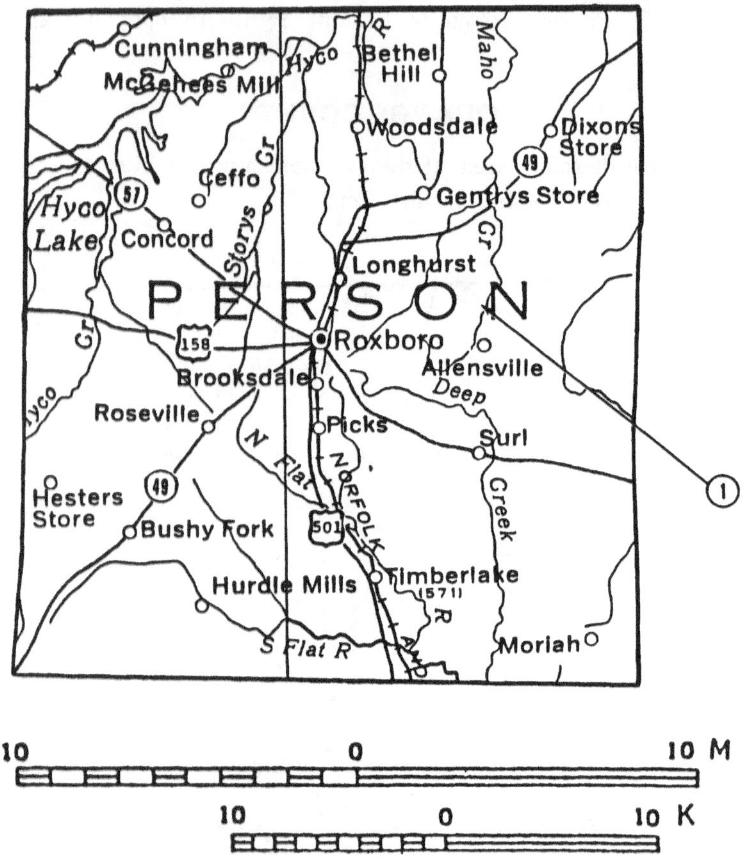

PITT COUNTY

1. County-wide.
   Amber (Succinite): to several ounces. (4-10-11-12-33)

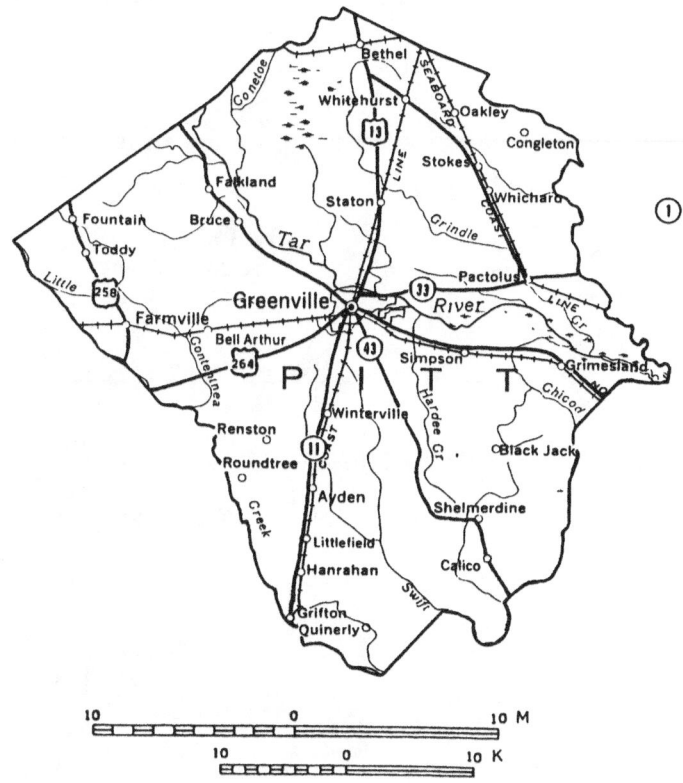

POLK COUNTY

1. Pea Ridge area: 0.5 mile south; at the North Star Mine.
   Feldspar: crystals; gem quality. Garnet. Quartz, blue:
   gem quality. Tourmaline. (2)

RANDOLPH COUNTY

1. Staley area.
   Prophyllite: silver (sometimes greenish) bladed crystals in stellate clusters; to 1″ diameter. (2-16)

Sunny View
Lake Adgers
Pea Ridge
1
Green
River
108
Mill Spring
River
P O L K
Whiteoak
Cr
Sandy Plains
26
Columbus
Saluda
Valhalla
Lynn
9
Green Creek
176
SOU
Tryon
10
0
10 M
10
0
10 K

Archdale
Trinity
Muddy
65
NW
Glenola
Cr
Level Cross
Julian
Redcross
Liberty
311
421
Salem
New Market
Sophia
Randleman
Grays Chapel
1
Flint Hill
Deep
Millboro
Staley
Worthville
49
Central Falls
Cedar Falls
Franklinville
Moueta
North Asheboro
Ramseur
R A N D O L P H
64
Asheboro
(853)
Jackson Creek
22
South Asheboro
49
Farmer
Richland
Coleridge
Brush
Ulah
Creek
Martha
220
New Hope
River
981
Pisgah
Seagrove
Fork
Erect
Creek
10
0
10 M
10
0
10 K

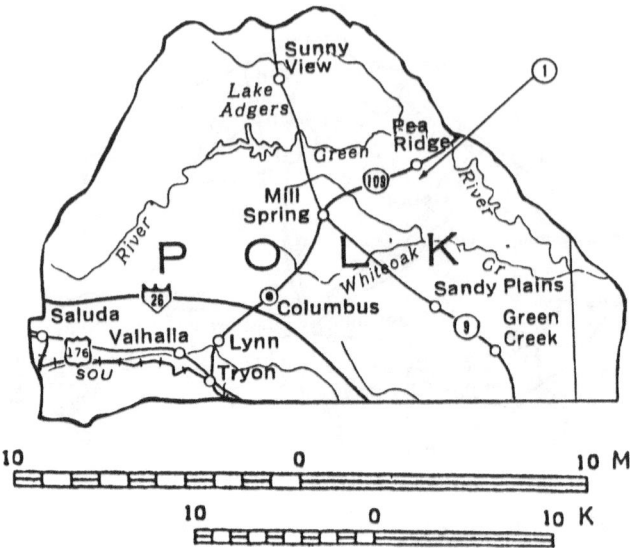

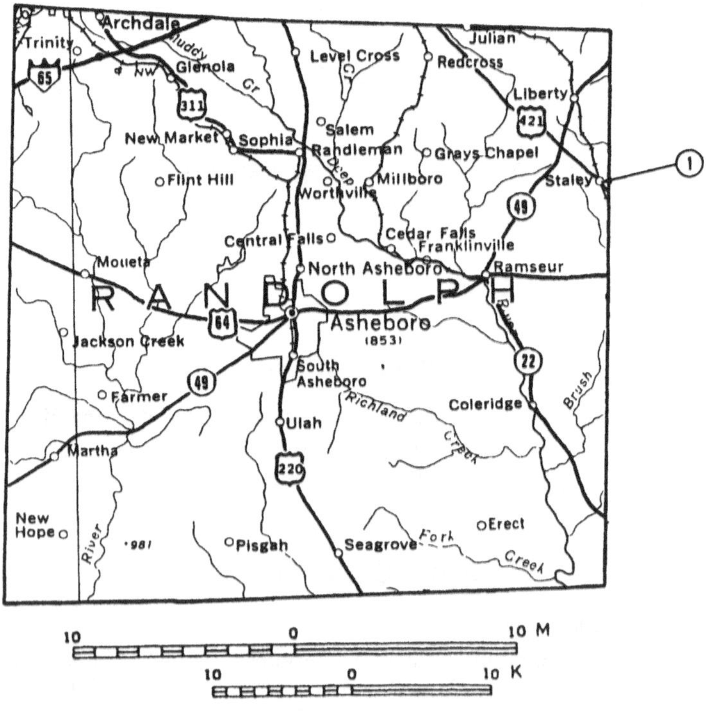

RICHMOND COUNTY

1. Ellerbe area streambeds.
   Petrified wood. (3-7-20)

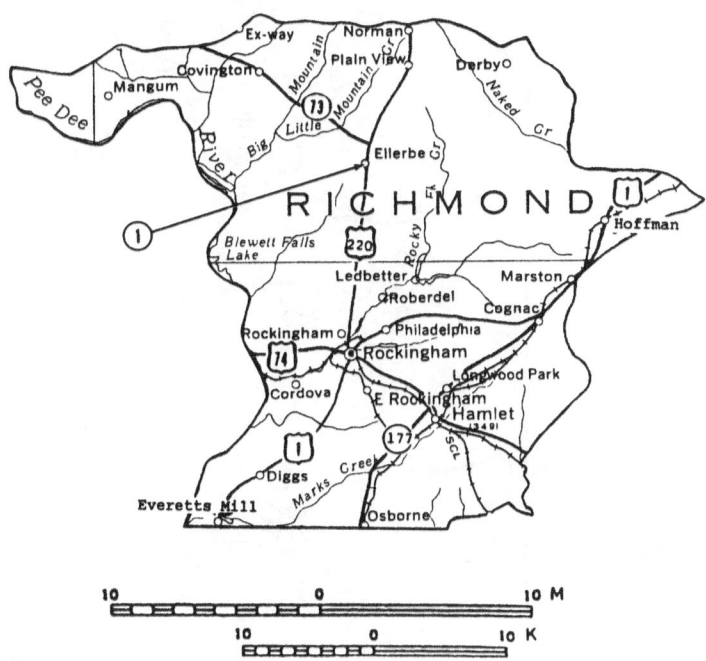

ROCKINGHAM COUNTY

1. Price area: 3.25 miles west-southwest; south side of Sandy Ridge Road; at the Short Tom Smith Mine (aka Ben Smith Mine).

    Garnet. Mica: green. Perthite. Plagioclase: kaolinized. Quartz, rock. (2)

2. Price area: 1 mile west; near the Virginia border; at the Clifton Mine.

    Garnet. Quartz, rock. (2)

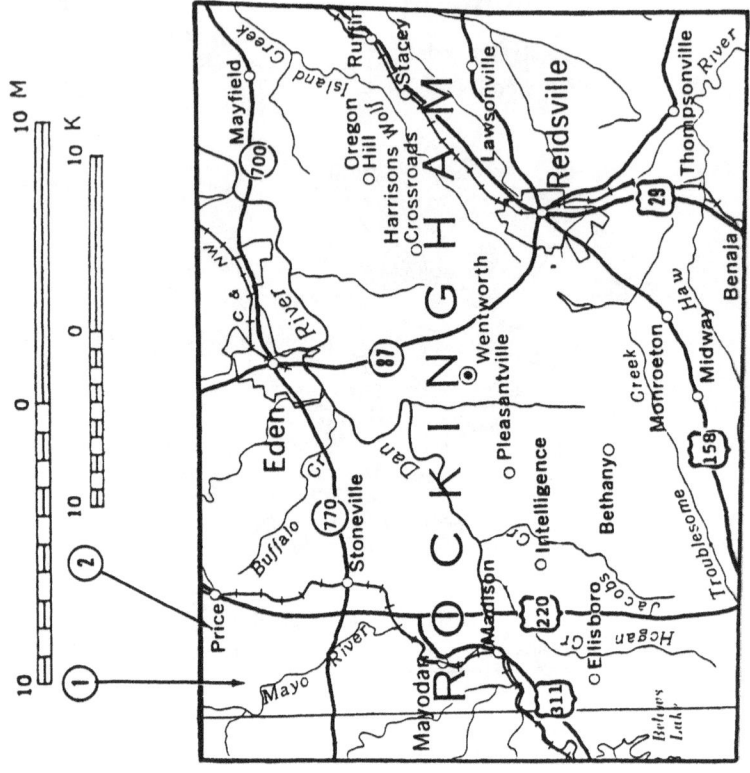

ROWAN COUNTY

1. Mount Ulla area: on the J. T. Eudy property.
   Amethyst: gem quality. (1)

RUTHERFORD COUNTY

1. Ellenboro area: west of US-74 to just outside town; then right on paved road for 1 mile; then right again to the Dycus Mine.
   Beryl. Quartz, rose. (2)

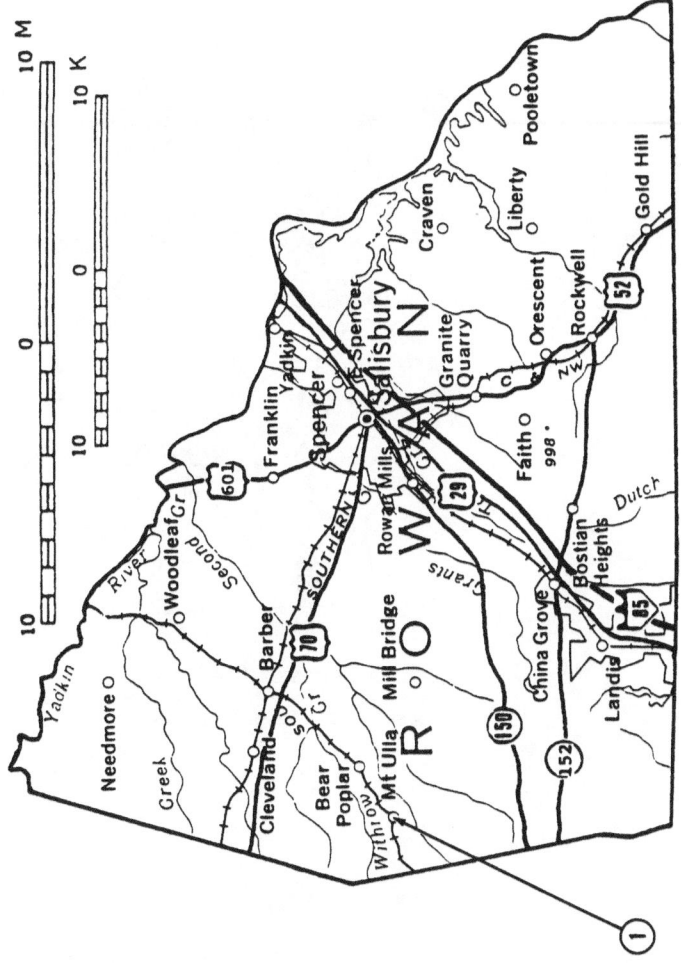

2. Hollis area: at the J. D. Twitty Placer Gold Mine.

    Diamond: 1 stone; 1.33 ct; flawless; pale yellow; found in 1845. (17-21)

3. Hollis area: on the Levinthorpe property.

    Diamond: 1 stone; 0.84 ct; fine; found in placer; date not given. (17)

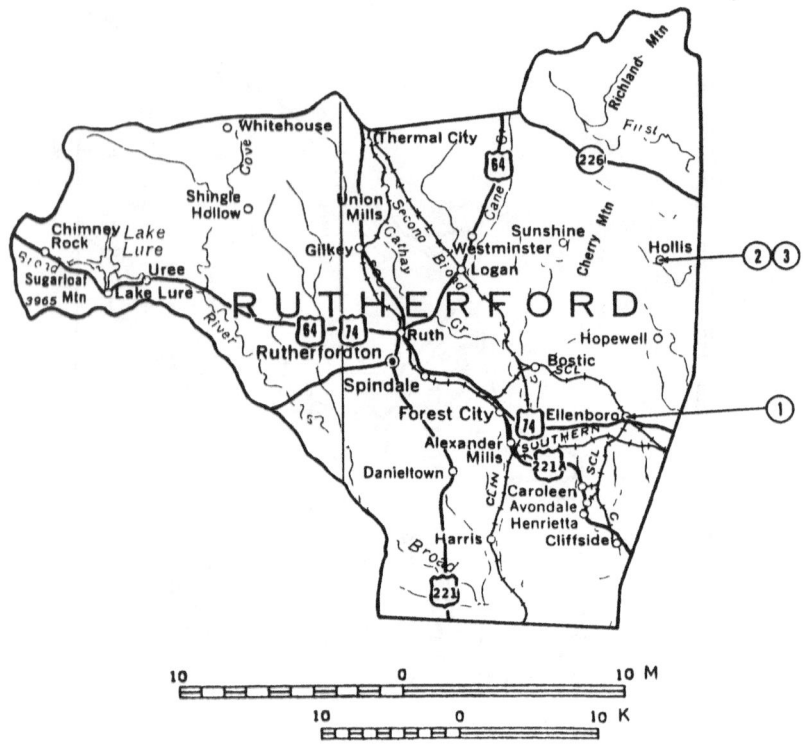

STATEWIDE

1. In such streams as the Cape Fear, Catawba, Irwins, Livingston, Long, Neuse, Roanoke, and Yadkin; also in many lakes.
     Pearl. (7)

STOKES COUNTY

1. Danbury area: upstream and down along the Dan River.
     Carnelian. (3-7-18-20)

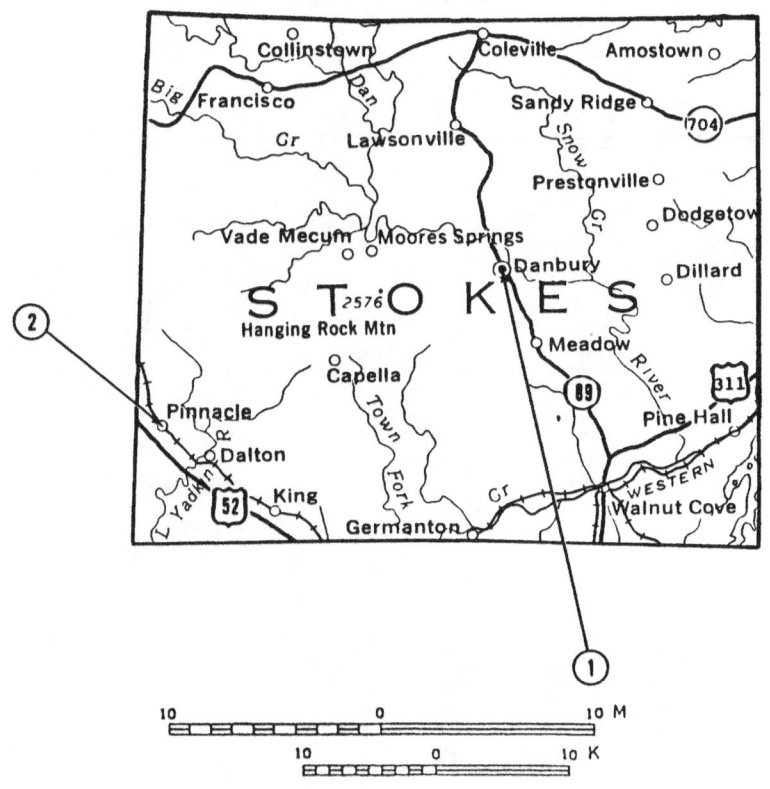

2. Pinnacle area: in the Lauratown Mountains; at Coffee Gap.
    Lazulite: massive. (5-quartz -16)

SURRY COUNTY

1. Burch area: 1.5 miles east of SR-268; on the C. Greenwood
   property.
    Jasper: blue. (1)

2. Burch area: along the Yadkin River.
    Jasper: blue. (7-18-20)

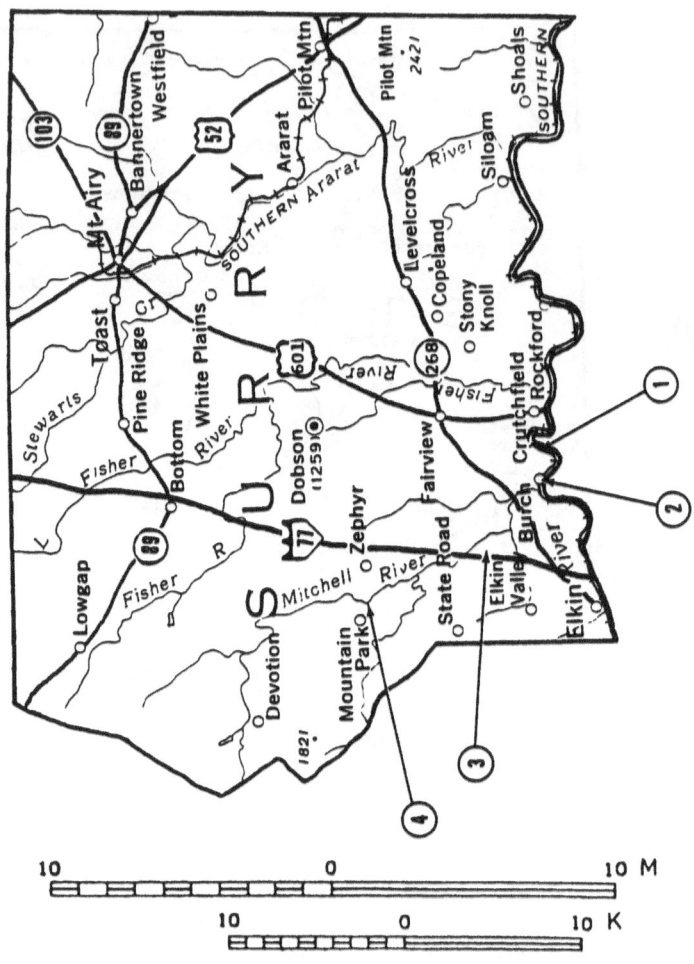

3. State Road area: 3 miles southeast; along the Mitchell River.

> Steatite (Soapstone): attractive cutting grade; boulders. (1-7-13-20)

4. Zephyr area: 1.7 miles west; off Mountain Park Road; near the Cross Road Church; in fields.

> Pyrite. (1-13)

SWAIN COUNTY

1. Bryson City area: 1.5 miles north of Deep Creek Campground.

   Kyanite. Staurolite. (27)

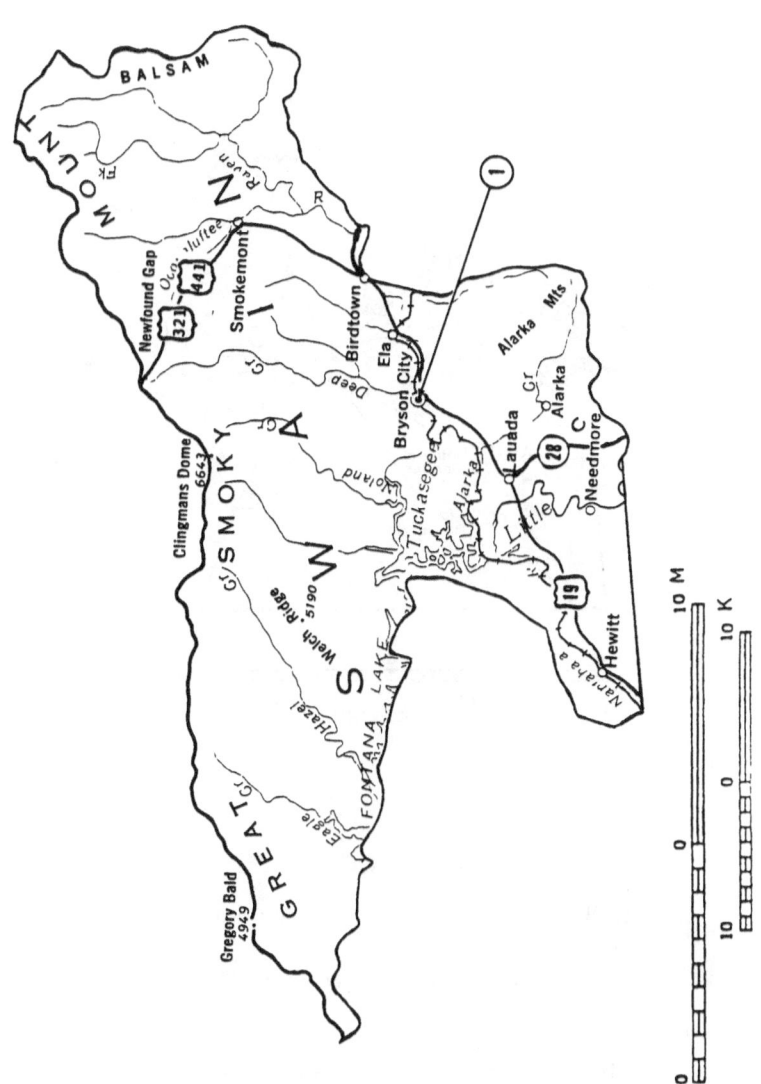

UNION COUNTY

1. Monroe area: to Potters Station; then 1.5 miles north; at the Bonnie Belle Mine (aka Washington Mine).

    Chalcopyrite. Gold. Pyrite. (2)

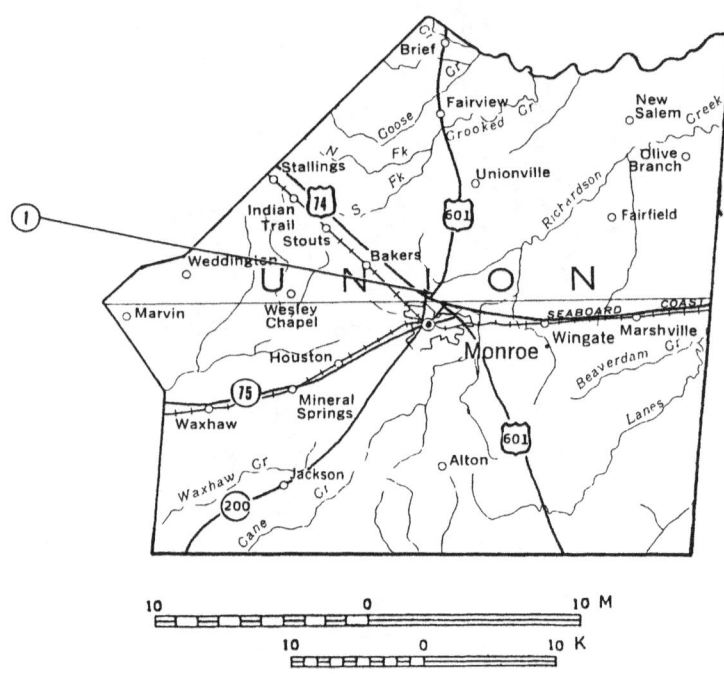

VANCE COUNTY

1. Kittrell area: north; in road cuts and granite outcroppings along US-1 Byp.

    Opal, hyalite: very fluorescent. (4-29-granite)

2. Townsville area.

    Huebernite: red-brown crystals. (19-quartz)

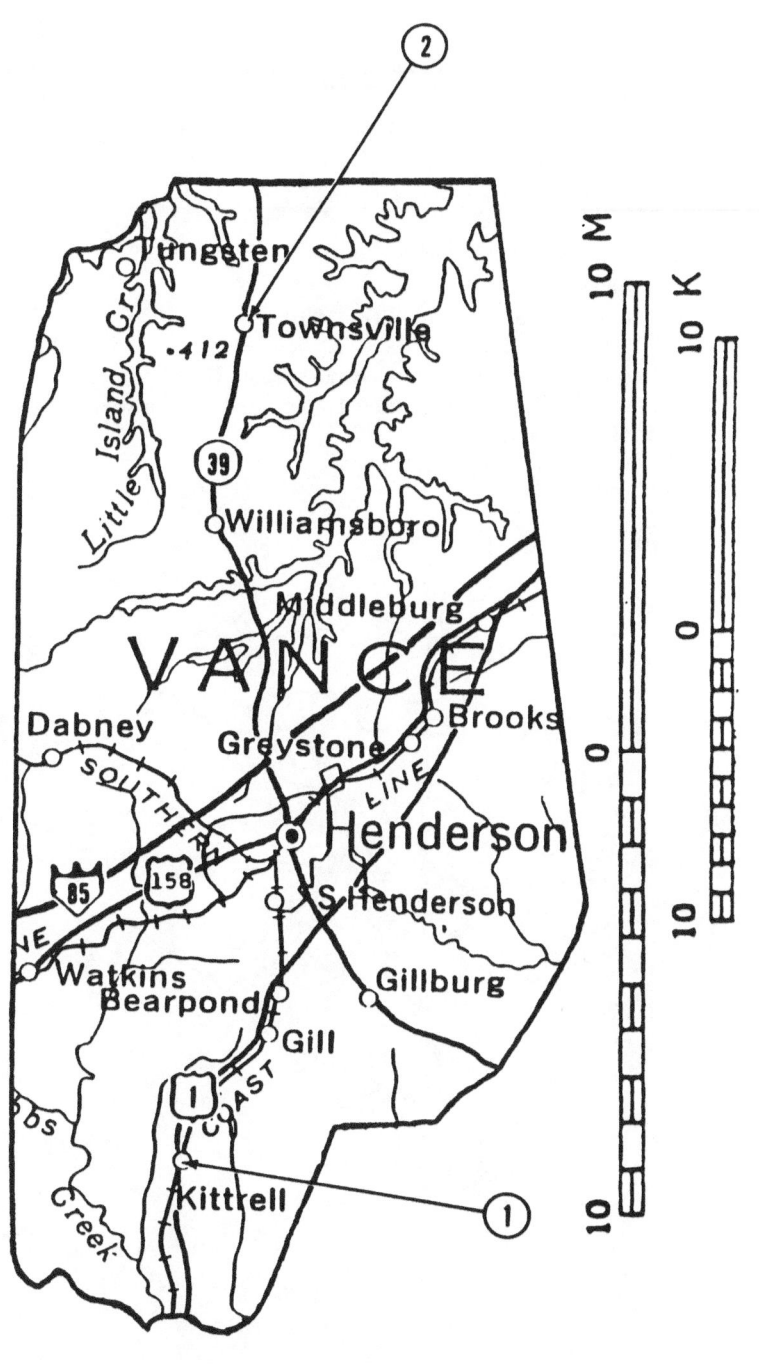

WAKE COUNTY

1. Bayleaf area: west to SR-50; then 2 miles to the Barton Creek bridge.

   Steatite (Soapstone). (1-13-20-29)

2. Millbrook area: 2.4 miles southeast; on the G. W. Partin property.

   Amethyst: gem quality. (19-quartz)

3. Raleigh area: 5.6 miles east-northeast; US-64 to 0.5 mile east of Wilders Grove; then 1 mile north to the west side of the Neuse River.

   Amethyst. (1-3-7-13-20-29)

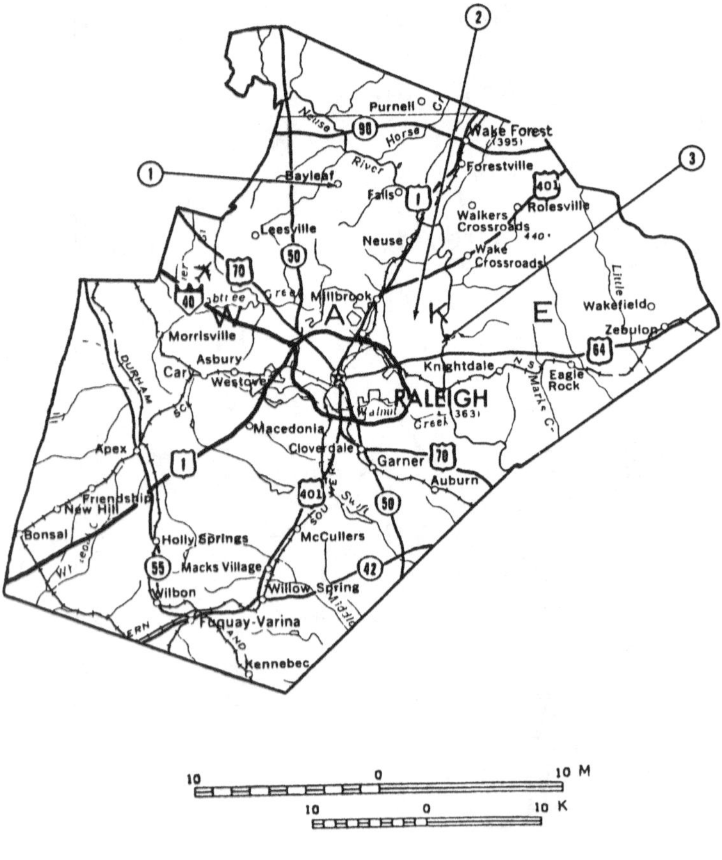

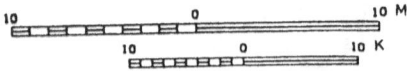

WARREN COUNTY

1. Inez area: 3 miles southwest; near Shocco Creek and Isinglass Creek; at the Alston Mine.

   Beryl. Muscovite: green. Perthite. Quartz, rock. (2-16)

2. Inez area: 2 miles south; on the J. B. Williams property; on a small hill 0.25 mile northwest of house.

   Amethyst: gem quality. (1)

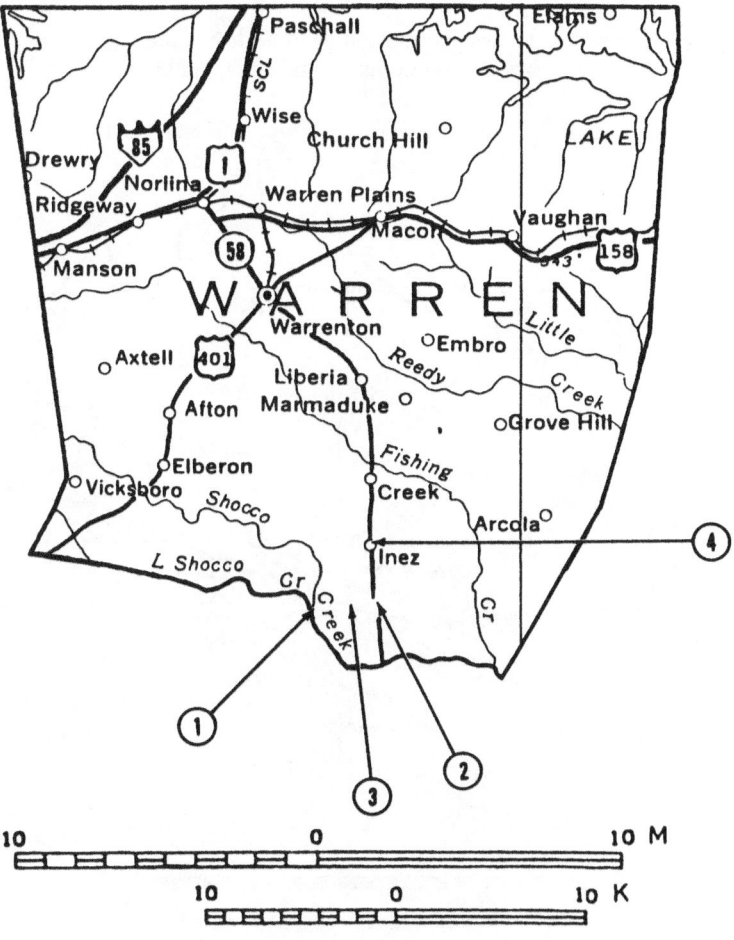

3. Inez area: 2 miles southwest; on the Connell property.
   Amethyst: gem quality. (1)

4. Inez area: on the Lathrop property.
   Amethyst: gem quality; colorless to deep purple; crystals to 3" thick. (1)

WILKES COUNTY

1. Pleasant Hill area: 1.5 miles on Pleasant Hill Road to unmarked road on right just before Pleasant Hill School; follow this to first unmarked road on left; take that for 1 mile; cross bridge and go just past the David Darnell property.

   Agate. Calcite (Honey Onyx var.). Jasper. (1-13)

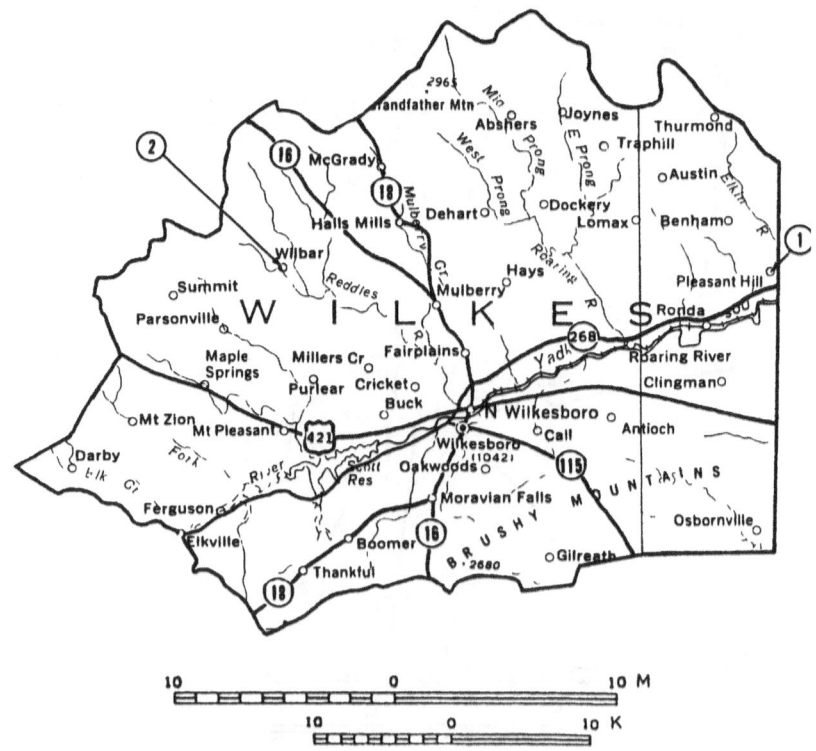

2. Wilbar area: Honey Creek headwaters.

   Amethyst: pale purple; faintly smoky. (1-16)

YANCEY COUNTY

1. Bowditch area: 1.9 miles east; on west bank of the South Toe River; at the Gibbs Mine.

   Oligoclase (Sunstone var.): transparent; colorless to pale blue-green. (2)

2. Burnsville area: 2 miles southeast; at the Mas Celo Mines.

   Albite. Apatite. Beryl. Biotite. Chalcocite. Chalcopyrite. Galena. Garnet. Kyanite: crystals to 4" long. Muscovite. Pyrite. Pyrrhotite. Sphalerite. (2)

3. Burnsville area: 1 mile southeast; at the Tantrough Mine.

   Allanite: small needle crystals; some bladed crystals to 6" long. (2)

4. Celo area: in the Toe River.

   Corundum. (7-18-20)

5. County-wide in streambeds.

   Beryl: fine; pale chartreuse; transparent. (7)

6. Micaville area: 2 miles east on US-19; cross the Toe River; turn south on Blue Rock Road and go 1 mile to the Fanny Gouge Mine.

   Aquamarine. Garnet. (2)

7. Micaville area: 2 miles east on US-19; cross the Toe River; turn south on Blue Rock Road and go 1 mile to the Spec Mine.

   Aquamarine. Garnet. (2)

8. Micaville area: 1 mile northeast; on ridge near the Googrock Mine.

   Asbestos, amphibole (aka Amphibolite). (1-13-29)

9. Micaville area: at the Hungerford Mine.

   Beryl. (2)

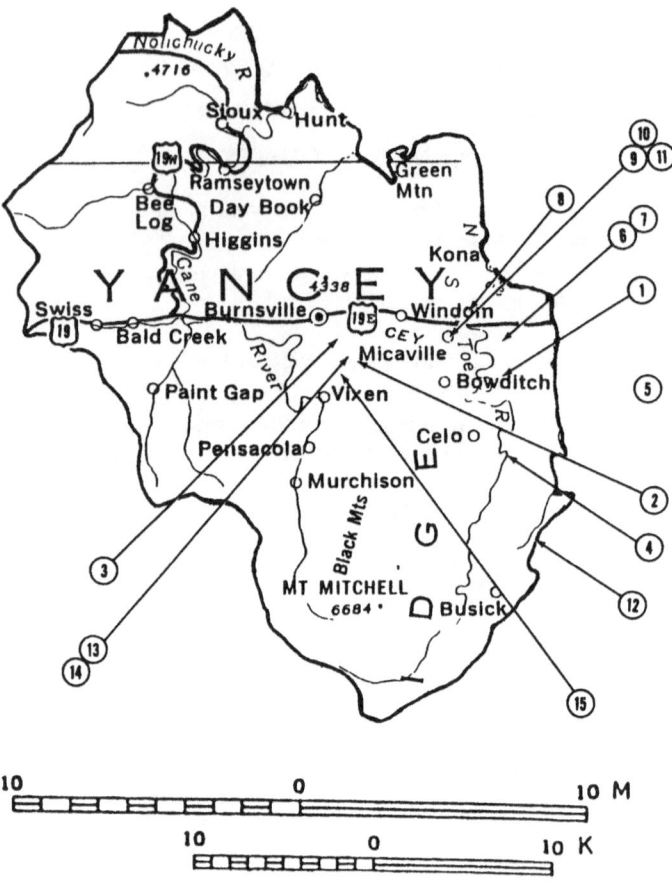

10. Micaville area: opposite the mouth of Blue Rock Branch on the northwest side of the Toe River.

   Amphibolite. (1-13-29)

11. Micaville area: take Blue Rock Road to the Spider Mine.

   Kyanite: gem quality; blue; bladed crystals. Thulite: small; fine-grained masses. (2)

12. Newdale area: 1.2 miles south; southeast of the bridge over the South Toe River; on the Thad Young property.

    Actinolite: needles. Epidote: granular; fine. Garnet: small crystals and massive. (16-milky quartz -21)

13. Vixen area: 1.7 miles northeast; opposite church is rough road; take it 1 mile to the Ray Mica Mine.

    Amazonite. Aquamarine. Beryl: golden, green; fine gem quality. Kyanite: gem quality; blue; bladed crystals. (2)

14. Vixen area: 1.7 miles northeast; in the Kyanite Mine.

    Garnet. Kyanite. (2)

15. Vixen area: 1 mile northeast; at the Yancey County Kyanite Mines.

    Albite. Beryl. Biotite. Bornite. Chalcocite. Chalcopyrite. Feldspar. Galena. Garnet. Kyanite: crystals to 4″ long. Muscovite: pale green; gem quality. Pyrite: fine brassy cubic crystals. Pyrrhotite. Sphalerite. (2)

SOUTH CAROLINA

Statewide total of 98 locations in the following counties:

Abbeville (4)	Florence (2)	McCormick (3)
Aiken (2)	Greenville (4)	Newberry (1)
Allendale (1)	Greenwood (11)	Oconee (2)
Anderson (13)	Horry (1)	Pickens (4)
Cherokee (7)	Kershaw (2)	Richland (1)
Chesterfield (2)	Lancaster (4)	Saluda (1)
Darlington (2)	Laurens (4)	Spartanburg (7)
Edgefield (2)	Lexington (1)	Union (2)
Fairfield (4)	Marlboro (1)	York (9)

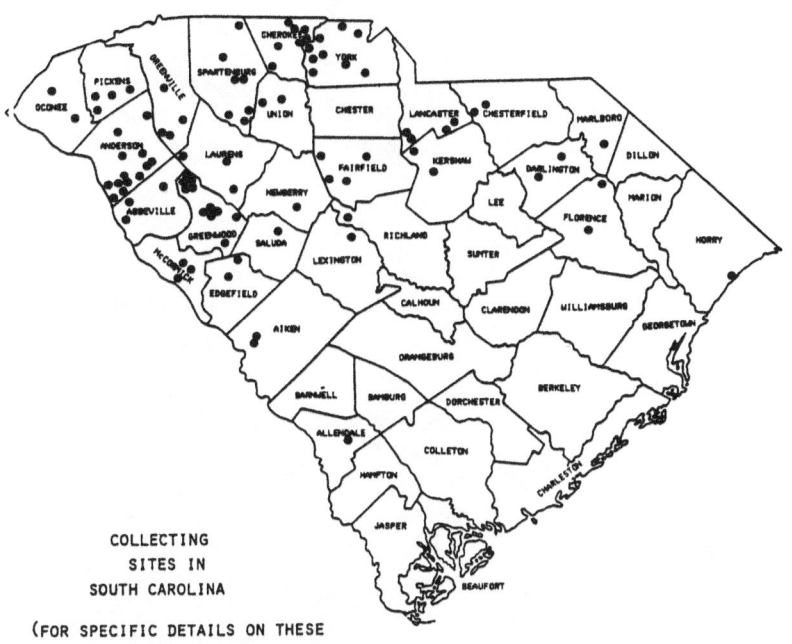

COLLECTING
SITES IN
SOUTH CAROLINA

(FOR SPECIFIC DETAILS ON THESE
LOCATIONS, SEE THE FOLLOWING
COUNTY MAPS AND RELATED TEXT)

ABBEVILLE COUNTY

1. Antreville area: 3 miles southwest on SR-284; then 1.75
   miles north on S-1-72, bearing right to house (at which pay
   CF); backtrack south to next road left and go east and
   north to mine.

   Amethyst. Quartz, smoky. (2)

2. Due West area: at the Ellis-Jones Amethyst Mine.

   Amethyst: superb gem quality; rich, velvety purple;
   color zoning is minimal; very large crystals and clusters (some clusters to 50 lb.); individual crystals short
   and to 3" thick; some crystals are enhydros. (2-25-
   yellow)

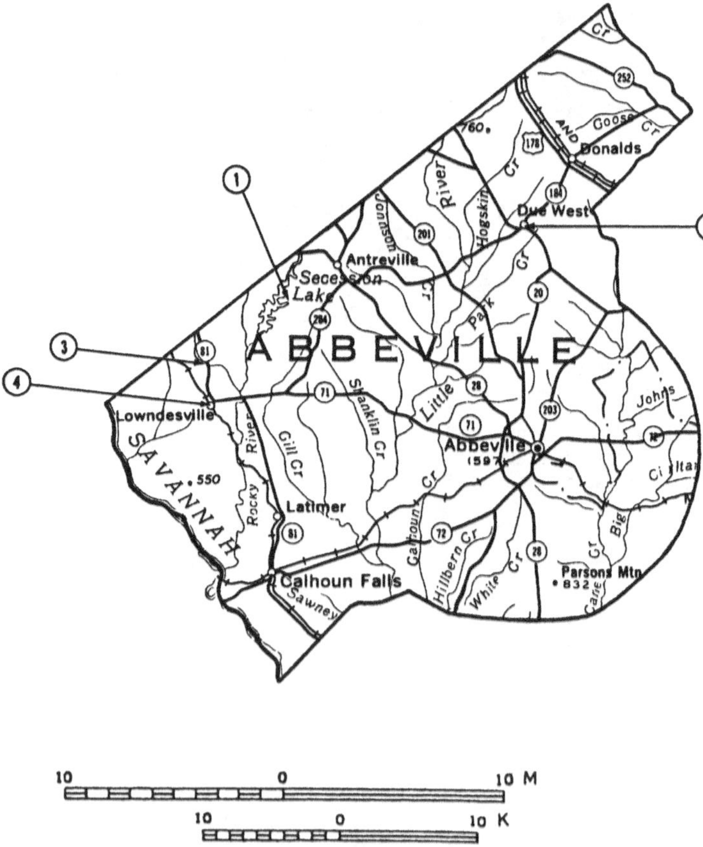

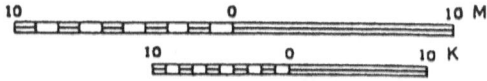

3. Lowndesville area: 1.8 miles north; at the Barnes property.
    Amethyst: superb gem quality. (1-13)

4. Lowndesville area: just east-southeast of town; at the McCalla property.
    Amethyst: gem quality. (1-13)

AIKEN COUNTY

1. Clearwater area: 1 mile from US-1 in Horse Creek Valley; at the Perry Gravel Company.
    Amethyst. Chalcedony. Ilmenite. Monzanite. Quartz, blue. Quartz, rock: some rutilated. Quartz, smoky. Rutile. Staurolite. Zircon. (10)

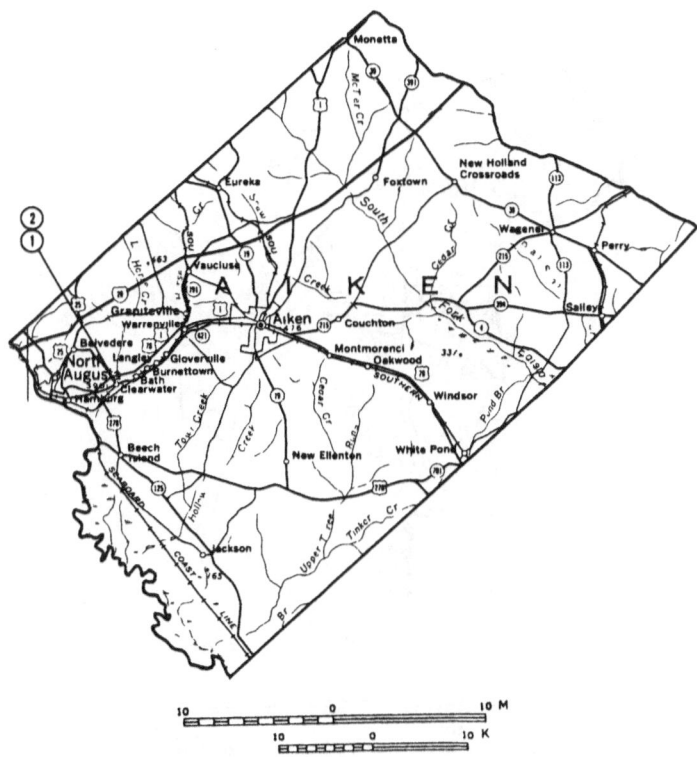

2. Clearwater area gravel pits.

> Amethyst. Chalcedony. Ilmenite. Monzanite. Quartz, rock. Quartz, rutilated. Quartz, smoky. Rutile: good crystals. Zircon. (10)

ALLENDALE COUNTY

1. Allendale area: 6.3 miles southwest on US-301 to where SR-3 becomes CR-26; then 7 miles south to just north of St. Pauls Church on east side of road; turn right on side road and go 1.5 miles to Red Bluff Point on the Savannah River; walk 1 mile south along river edge to mouth of Kings Creek.

> Chalcedony: nodules; blue, pink, white. Chert. Fossils: coral replaced by chalcedony. (1-13-20)

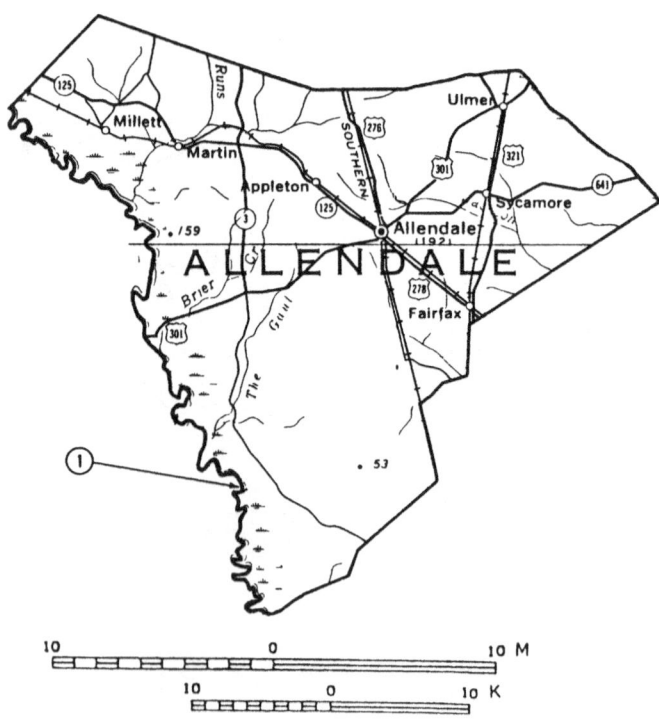

ANDERSON COUNTY

1. Anderson area: 2.3 miles southwest on SR-81 to right turn on SR-28; 9.6 miles south on SR-28, almost to border of Abbeville County; in road cuts through dike near Secession Lake.

    Amethyst. Garnet. (4-16)

2. Anderson area: 8.7 miles northeast on SR-81; then 2.1 miles northwest to near the Six and Twenty Creek Dam; at the Burgess Mica Mine.

    Feldspar: red. Garnet. Quartz, rock: asterism. (2)

3. Anderson area: 3 miles east; on the McConnell property; at the McConnell Mica Prospect.

    Aquamarine: bright green. Beryl: emerald green; gem quality; transparent. Emerald. Garnet, almandine: dark red; gem quality; large crystals. Limonite: pseudomorphs after pyrite. Pyrite. Quartz, rock. Quartz, smoky. Topaz. Tourmaline: black. (2-21)

4. Barnes Station area: at the J. B. Anderson property.

    Beryl: golden. (1-13-21)

5. Craytonville area: just northwest; at the Jackson property.

    Corundum: pale blue, pink; gem quality. (21)

6. Craytonville area: just northwest; at the Thomas property.

    Corundum: pale blue, pink; gem quality. Garnet: red; massive. Quartz, milky. Zircon: deep yellow. (21)

7. Iva area: 7.2 miles southwest; at the Sherard property.

    Amethyst. (1-13)

8. Iva area: 5.8 miles southwest on SR 184; along the northeast bank of the Savannah River; 0.25 mile from the mouth of Big Generostee Creek; at the Fretwell Prospect.

    Aquamarine. Beryl, golden. (21)

9. Iva area: 1.75 miles northeast; in Wilson Creek.

    Beryl. (1-7-13-20)

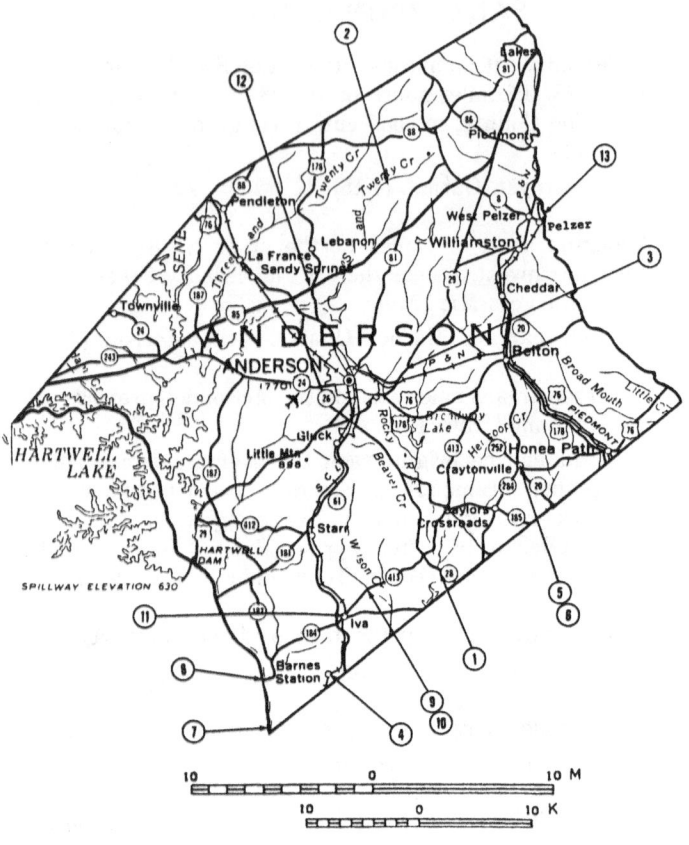

10. **Iva area:** 1.75 miles northeast on SR-413; at the Frank Pruitt property.

   Beryl, golden. (21)

11. **Iva area:** south near Moffettsville; on the W. T. Sherard property.

   Amethyst: superb crystals; single and in clusters; gem quality. (1-9-16-24-28-mica)

12. **Lebanon area:** 1.7 miles south on US-178; at the Martin-Blackwell-Ferguson Mine.

   Beryl: gem-quality; transparent. (2)

13. Pelzer area: from 1 mile south of town, north along border of Greenland County to 1.5 miles southwest of Piedmont.

Aquamarine: clear; blue; gem quality. Tourmaline (Indicolite var.): blue, green; gem quality. (1-13-16-24)

CHEROKEE COUNTY

1. Blacksburg area: 2.5 miles northwest on Buffalo Church Road; in pegmatites adjacent to the road; on the Andrew Moore property.

Emerald. Sapphire. (1-3-14-16-24)

2. Blacksburg area 2 miles north; near Earles Station.

Emerald. (6)

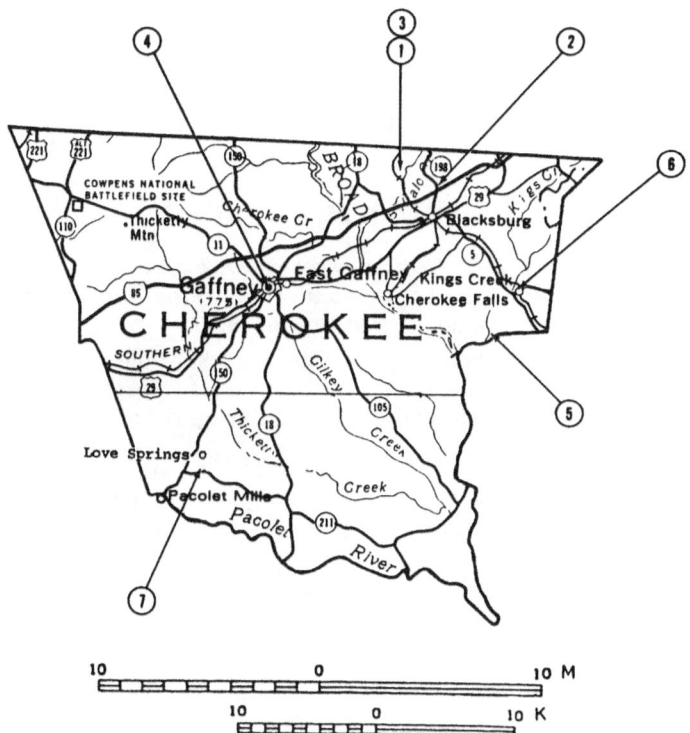

3. Blacksburg area: near the Buffalo Church on the W. T. Gibbons property and in adjacent streams.

> Amethyst: rich purple. Corundum. Rutile: very dark; lustrous. (1-7-13-20)

4. Gaffney area: north; in vicinity of Porters Hill; in the Broad River and its tributaries.

> Emerald; gem quality. Sapphire: grassy green; gem quality. Topaz: small crystals. Zircon: gem quality. (1-3-7-13-20)

5. Kings Creek area: 2 miles southwest; at the Barkat Mine.

> Amethyst. Azurite. Chalcopyrite. Copper, native. Galena. Gold. Hematite. Kyanite. Malachite. Pyrite. Quartz, rock. Steatite (Soapstone). (2)

6. Kings Creek area: mines southeast of town.

> Barite. Calcite. Galena. Gold. Pyrite. Sphalerite. (2)

7. Love Springs area: 0.5 mile south; at the abandoned Troy Blanton Mica Mine.

> Garnet. Tourmaline. (2)

CHESTERFIELD COUNTY

1. Jefferson area: 2.2 miles west on SR-265 to bridge; north along west side of Lynches River to Brewer Knob; at prospects and the abandoned Brewer Mine.

> Gold. Kyanite. Pyrite. Quartz, rock. Rutile. Staurolite. Tourmaline: massive, with areas of transparent gem-quality material in blue, champagne, and yellow. (1-2-7-13)

2. Jefferson area: just north of SR-265; at abandoned mines along the Lynches River.

> Topaz. (2)

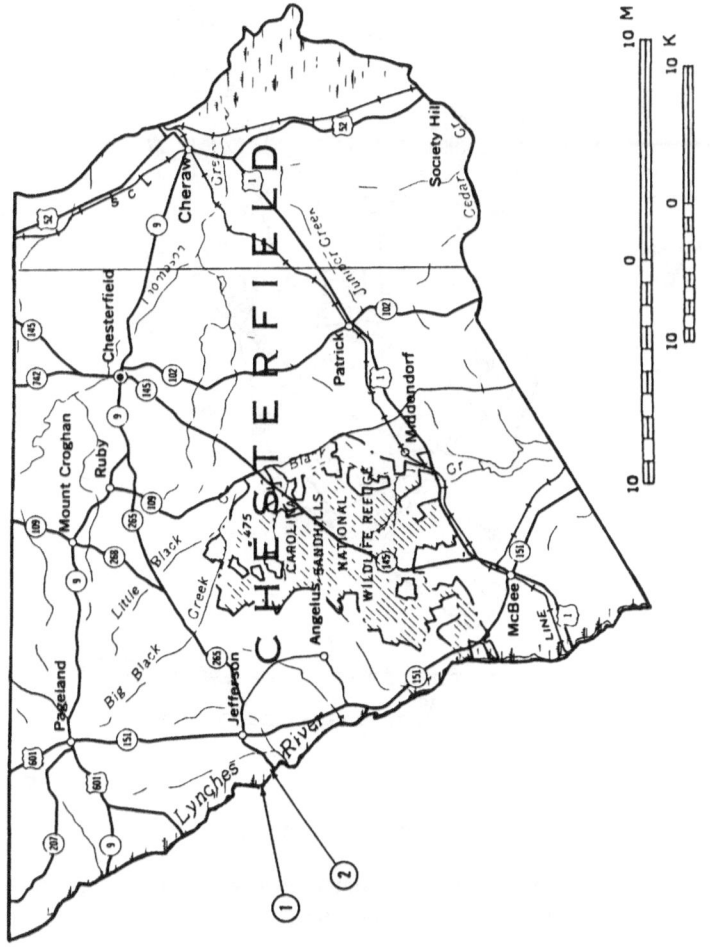

DARLINGTON COUNTY

1. County-wide in road, RR, and stream cuts.
   Petrified wood. (3-4-7-18-20)

2. Hartsville area: just south along Bellyache Creek.
   Petrified wood. (1-7-13-20)

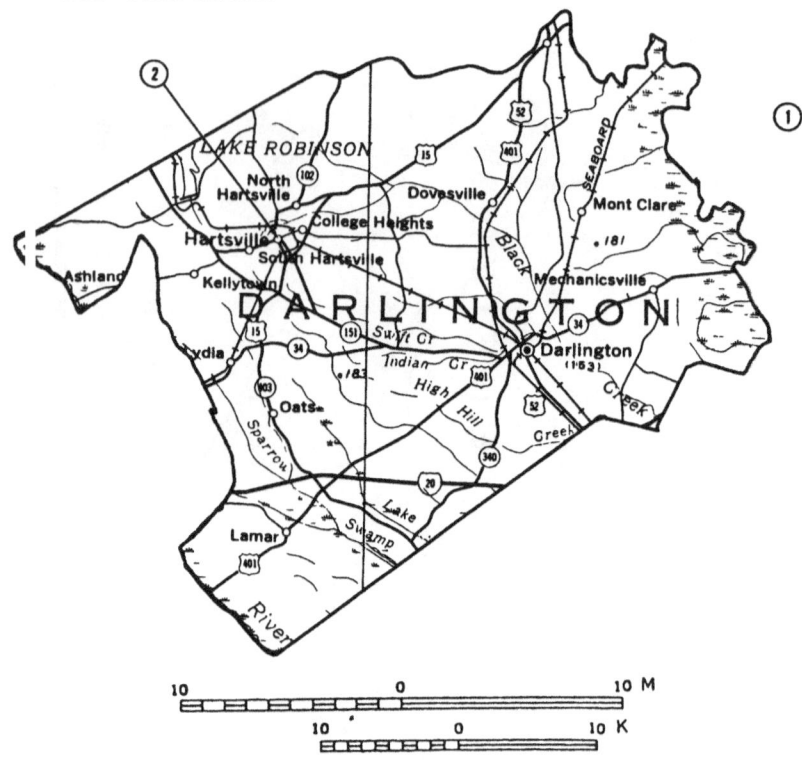

EDGEFIELD COUNTY

1. Edgefield area: 12 miles north-northwest; on the east side
   of Sleepy Creek; at numerous mines.
   > Gold. (2)

2. Edgefield area: 7 miles north on US-25; in Log Creek along
   road.
   > Serpentine. (7-20)

FAIRFIELD COUNTY

1. County-wide in gravel pits and sand pits.
   > Petrified wood. (10-11)

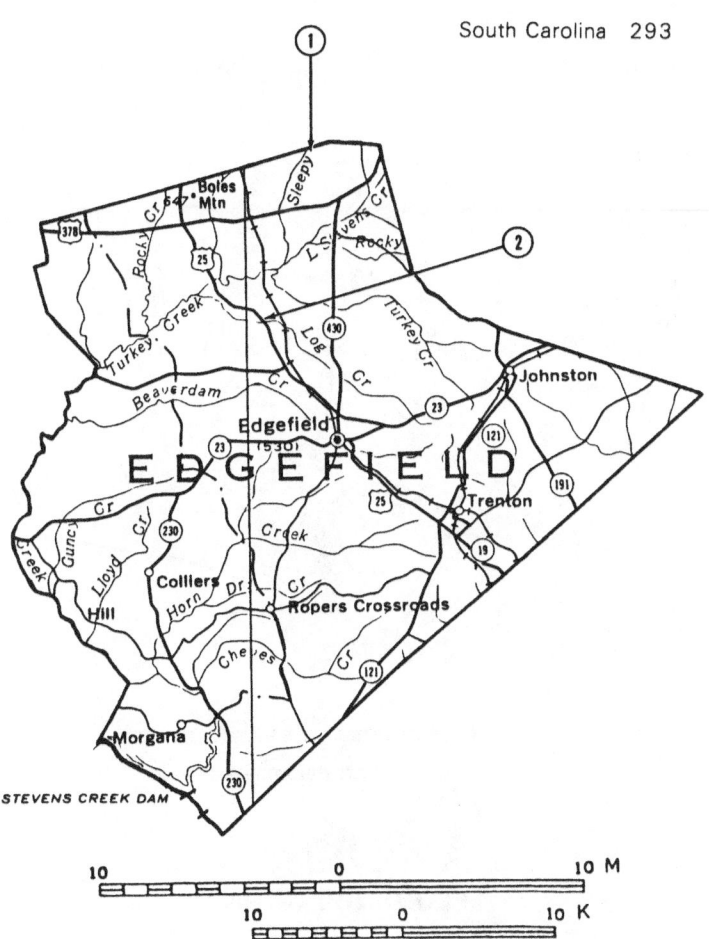

2. Jenkinsville area: south to the west side of the Broad
   River; at mine.
   Kyanite: gem quality. (2)

3. Monticello area: at the Anderson Quarry.
   Granite: high quality. Kyanite. Pyrite. (12)

4. Shelton area: along the Broad River; at abandoned kyanite
   mine.
   Kyanite. (2)

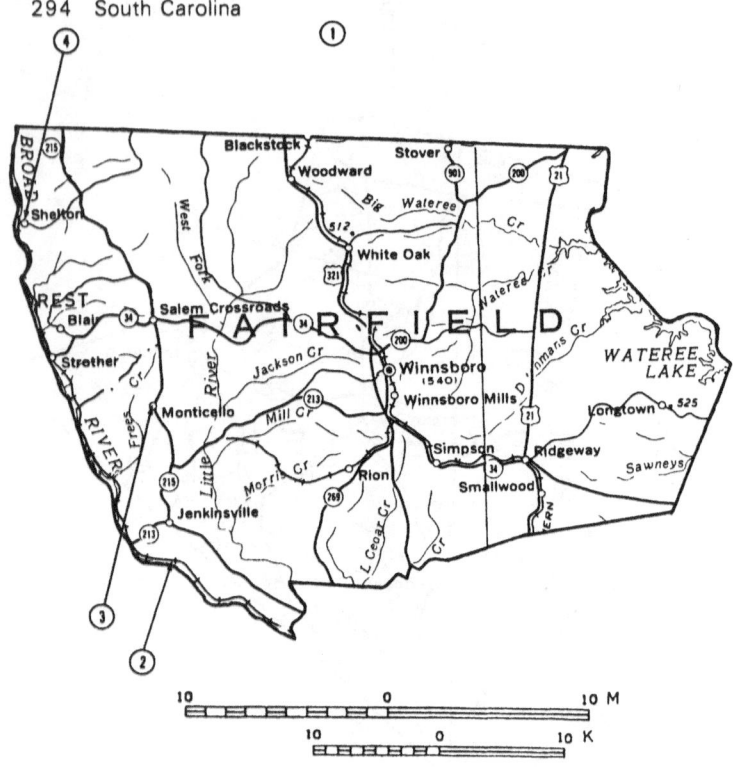

FLORENCE COUNTY

1. County-wide in road, RR, and stream cuts.
    Petrified wood. (3-4-7-18-20)

2. Quinby area: 1.5 miles north; along both shores (especially north shore) of the Pee Dee River.
    Petrified wood: some finely agatized. (1-13-20)

GREENVILLE COUNTY

1. Golden Grove area: 1.1 miles south; close to Piedmont (Anderson County); at the D. McNeely property.
    Beryl. (1-13)

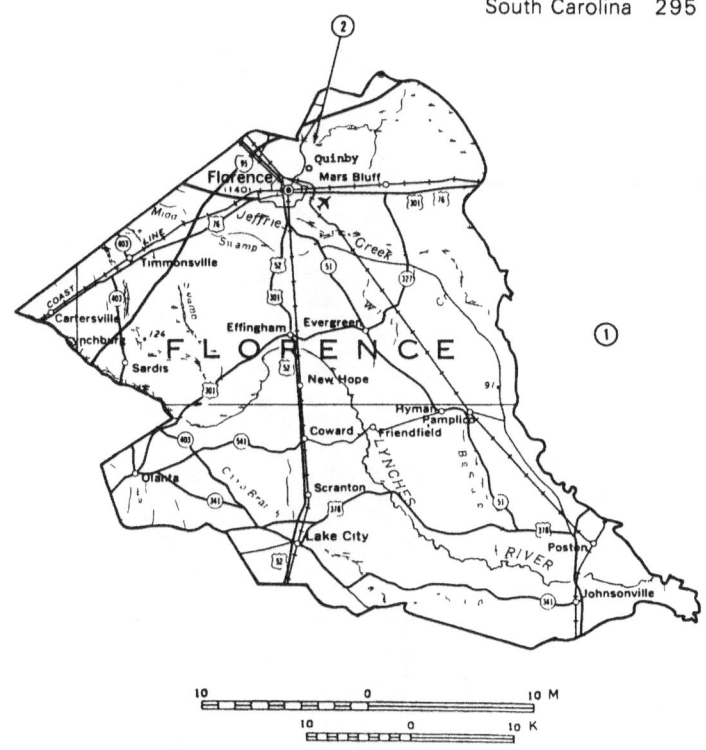

10   0   10 M

10   0   10 K

2. Golden Grove area: just to the west of SR-20; across the
   RR tracks; close to the Saluda River; at the Cleveland Mica
   Prospect.

   Beryl: clear. (21)

3. Greenville area: 5.9 miles north; close to the east border of
   Paris Mountain State Park; 350′ east of road to confluence
   of two brooks; at the Boling Prospect.

   Garnet: bright red. Quartz, rock: very clear. Quartz,
   smoky: gem quality. Sillimanite: brown. Tourmaline:
   black. (21-24)

4. Simpsonville area: at the Willimon Mine.

   Amazonite: pale blue-green. Feldspar: peach-colored;
   chatoyant. Kyanite: fine crystals. Rutile. Xenotime:
   bright yellow crystals. (2)

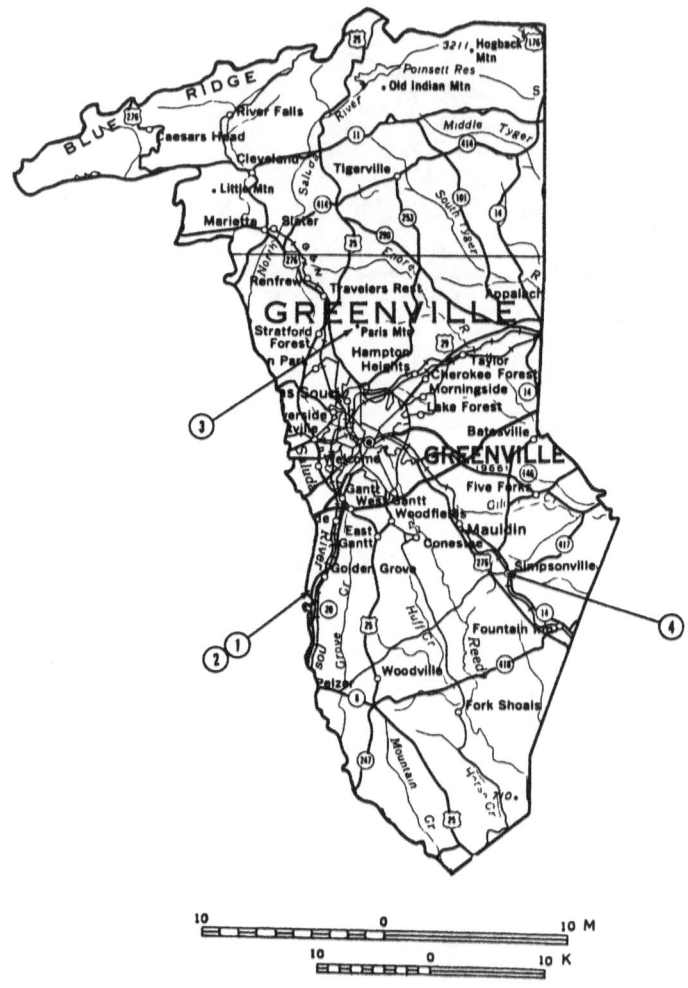

GREENWOOD COUNTY

1. Greenwood area: at the Harper property powerhouse.
   Chalcedony. (1-13)

2. Greenwood area: at the Milford property.
    Quartz, rock. Quartz, smoky. (1-13)

3. Greenwood area: at the Stockman Quarry.
    Garnet. (12)

4. Greenwood area: at the Wrenn property.
    Amethyst. (1-13)

5. Greenwood Shores area: along the west shore of Lake
   Greenwood.
    Unakite. (20)

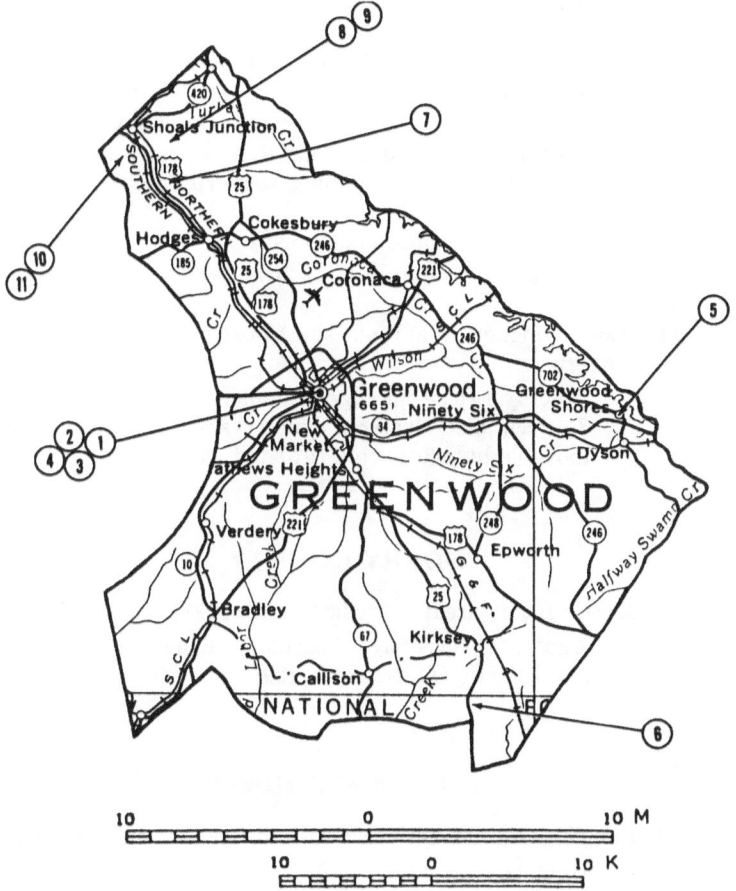

6. Kirksey area: south on US-25; in road cut about 0.5 mile north of border of Edgefield County.

> Limonite: pseudomorphs after pyrite; cubes to 2″ thick; dark; striated; some magnetic. (4)

7. Shoals Junction area: 2.5 miles southeast; at the J. T. Algary property.

> Amethyst: pale lilac; single crystals and clusters; individual crystals to 3″ thick; some enhydros with movable bubbles; some rutilated. (1-13-25)

8. Shoals Junction area: 1.5 miles southeast.

> Amethyst: gem quality. (1-3)

9. Shoals Junction area: 1.5 miles southeast; on the R. M. Haddon property.

> Amethyst: pale lilac; gem quality; single crystals and clusters; individual crystals to 3″ thick; some enhydros with movable bubbles; some rutilated. (1-13-25)

10. Shoals Junction: 1 mile southwest.

> Amethyst: gem quality. (1-3)

11. Shoals Junction area: 1 mile southwest; on the R. W. Dunn property.

> Amethyst: pale lilac; gem quality; single crystals to 3″ thick; also clusters; some crystals enhydros with movable bubbles; some rutilated. (1-13-25)

HORRY COUNTY

1. Myrtle Beach area: in seaside beach gravels.

> Agate. Chalcedony. Fossils, marine vertebrates: shark teeth. Quartz, rock. (20)

KERSHAW COUNTY

1. Camden area: 9 miles northwest; at the Lamar Gold Mine.

> Gold. (2)

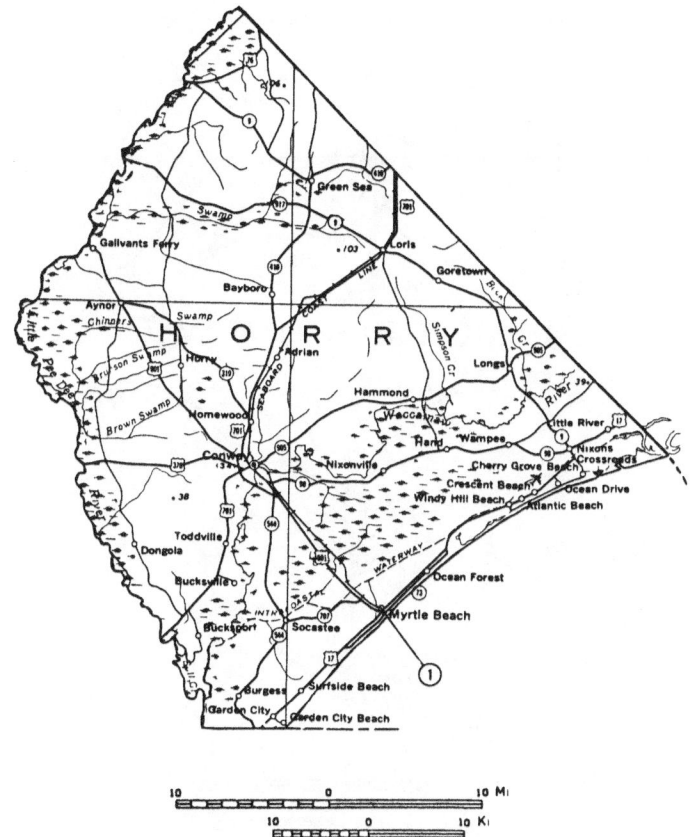

2. Liberty Hill area: tributary of the Catawba River; near the Wateree Reservoir.

> Quartz, smoky: gem quality; crystals to 2.5″ thick and 6″ long; well terminated. Zircon. (1-4-13-14-29)

LANCASTER COUNTY

1. Dearborn Dam area: just above the mouth of Cedar Creek.

> Quartz, smoky: fine gem quality; pale smoke to nearly black; sharply terminated crystals to 2″ thick and 6″ long; striated; color-zoned. Zircon. (1-7-13-16-20-24)

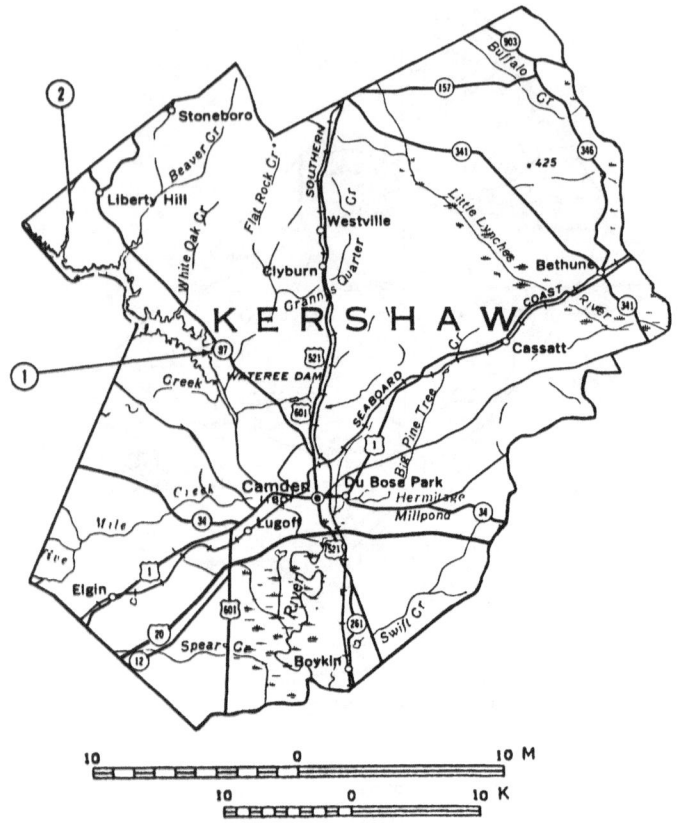

2. Dearborn Dam area; south; near Kershaw County border; at abandoned mine.

      Quartz, smoky. (2)

3. Kershaw area: 3 miles north; at the Haile Mine.

      Gold. Kyanite. Quartz, rock. Rutile. Staurolite. (2)

4. Kershaw area mines.

      Gold. Pyrite. Quartz, rock: drusy. (2-38)

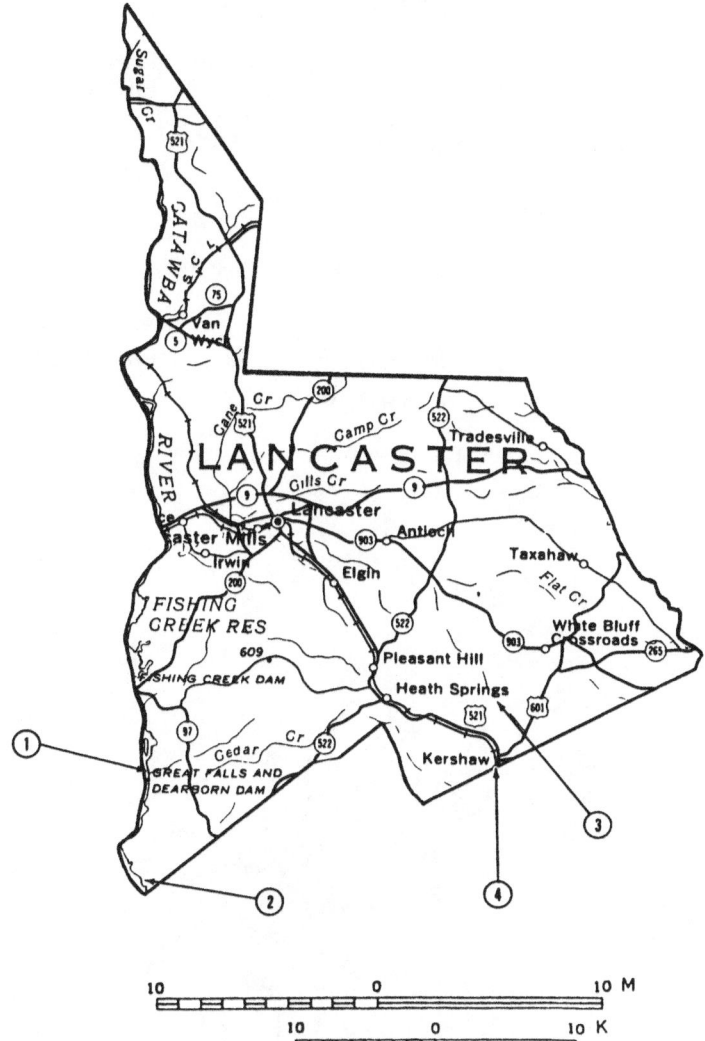

10   0   10 M

10   0   10 K

LAURENS COUNTY

1. Cross Hill area: south on SR-39 to unmarked road where old shack and large white house are on right; turn right on dirt road; follow to next right-angle bend. (Get permission from house on right.)

   Amethyst: gem quality; purple. Quartz, rock: fine; clear. (25)

2. Laurens area: on grounds of New Cemetery.
   Corundum. (1)

3. Laurens area: in city limits; at Dead Mans Cut on the RR.
   Corundum. Garnet, pyrope. (4)

4. Princeton area: 1.5 miles southwest; at spring.
   Amethyst. (1-13)

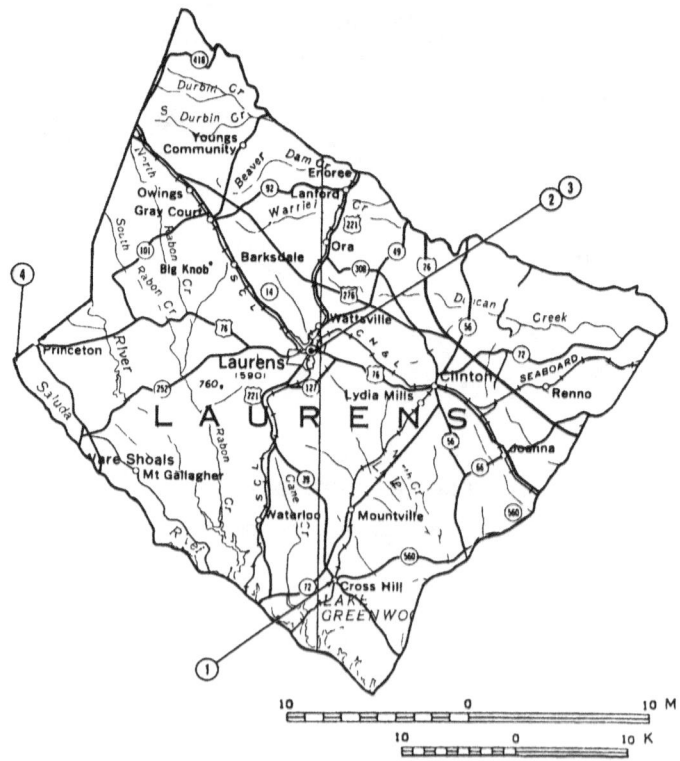

LEXINGTON COUNTY

1. Lexington area: 4.6 miles north; at Lake Murray Dam
   area; below dam on both sides of Congaree River.

   Garnet: dark red to brown; often fractured. Kyanite:
   sky blue to deep blue. (3-5-large rocks -20)

MARLBORO COUNTY

1. Blenheim area: 4 miles south.

   Fossils: cycads; silicified. (11)

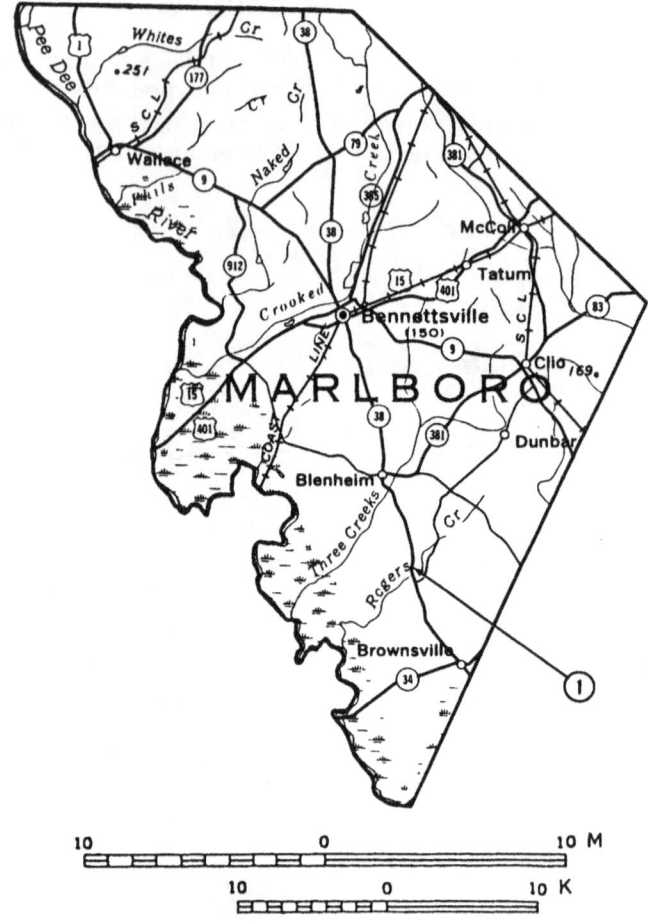

MCCORMICK COUNTY

1. McCormick area: 2.5 miles south; at abandoned gold mine.
   Gold. (2)

2. McCormick area: 2 miles southwest on US-378 to right
   turn onto Plum Branch Road; 2 miles to bridge over creek;
   in creekbed on right and in adjacent areas.

   Gold. Limonite: pseudomorphs after pyrite; cubes to
   2″ thick. (1-3-7-13-20)

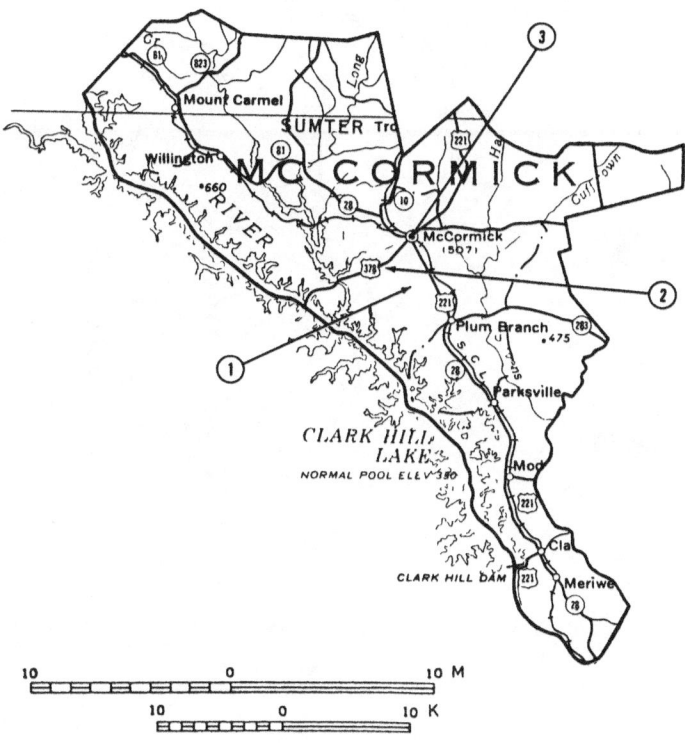

3. McCormick area streambeds.
   Gold. (17-21)

NEWBERRY COUNTY

1. Prosperity area.
   Rutile: fine single crystals and crystal clusters. (1-13-14-24)

OCONEE COUNTY

1. County-wide.
   Sillimanite: dark brown crystals to 2″ long. (1-2-13)

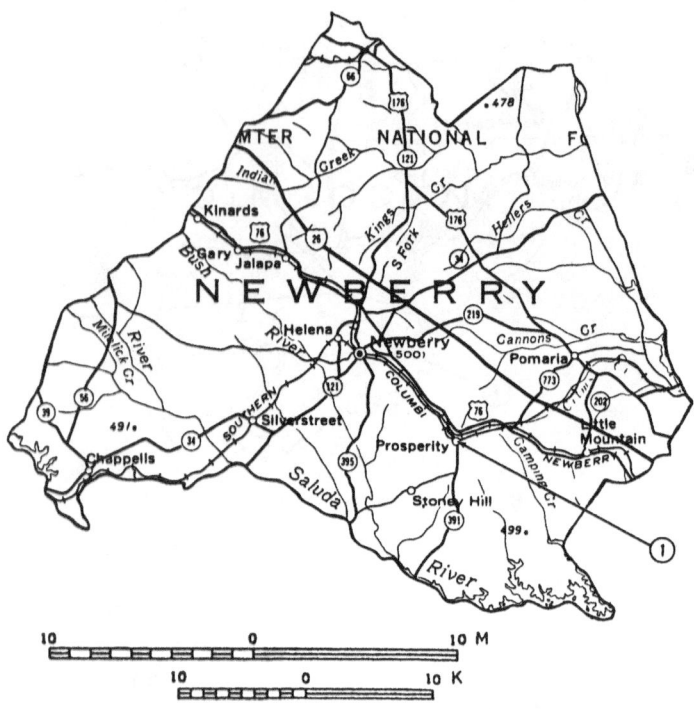

10   0   10 M

10   0   10 K

2. Seneca area: 2 miles north; on the Leroy property.

    Sillimanite: fine gem quality; cat's-eye; large crystals. (1-13-14-24)

PICKENS COUNTY

1. Clemson area: 4.5 miles north; 0.7 mile northwest of Twelve Mile Creek; at the Head Mica Prospect.

    Garnet. Quartz, rock. (21)

2. Easley area.

    Talc (Soapstone). (27)

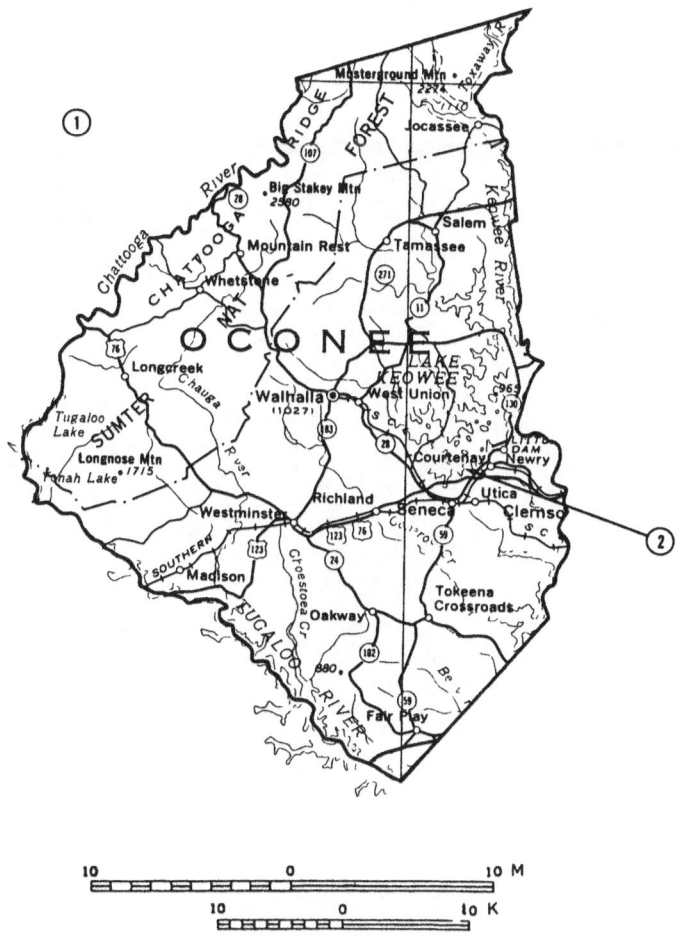

10   0   10 M

10   0   to K

3. Norris area: near Meece Mill; at the Meece Mine.
   Feldspar: pink; chatoyant. (2)

4. Six Mile area: 1.8 miles southwest; 1 mile east of the
   Seneca River; on Sixmile Creek; at the Davis Prospect.
   Feldspar: chatoyant. Garnet. Quartz, rock. Tourma-
   line. (21)

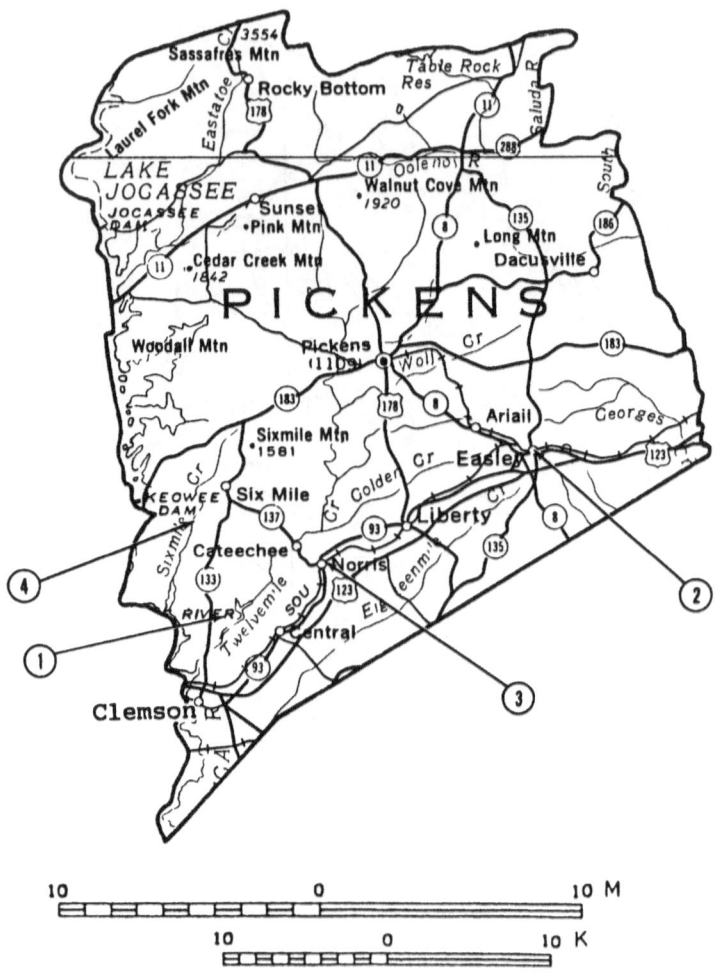

RICHLAND COUNTY

1. Ballentine area: along the east shore of Lake Murray.
   Amethyst. (1-13-20)

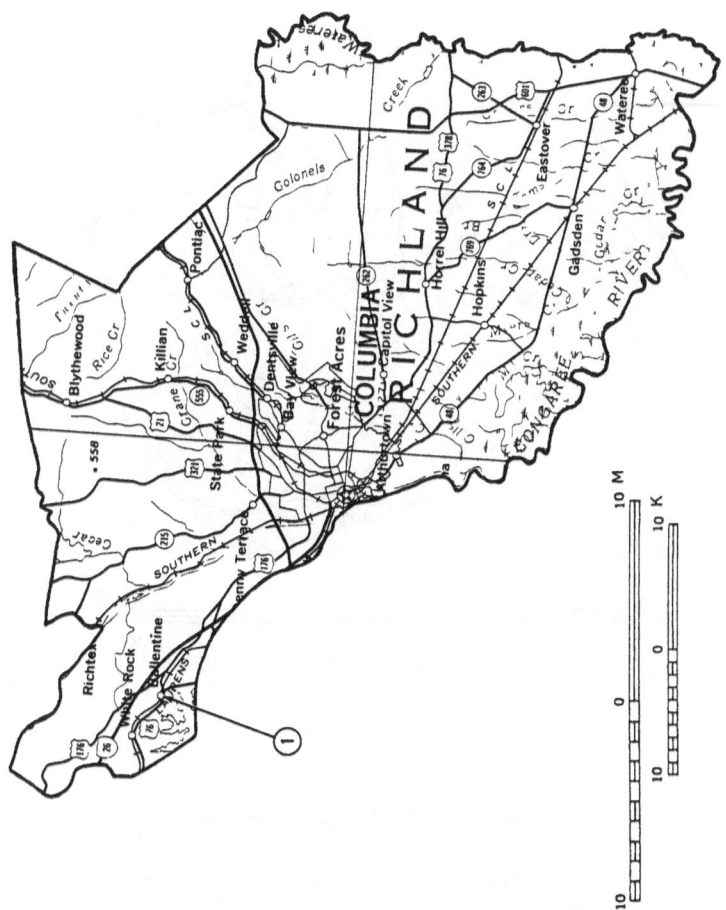

SALUDA COUNTY

1. Saluda area: 6 miles northeast; near the mouth of Big
   Creek; between that creek and the Little Saluda River; at
   the Culbreath Mine.

   Amphibolite. Chalcopyrite. Chlorite. Gold. Hornblende: crystals; green. Magnetite. Niccolite. Pyrite.
   (2)

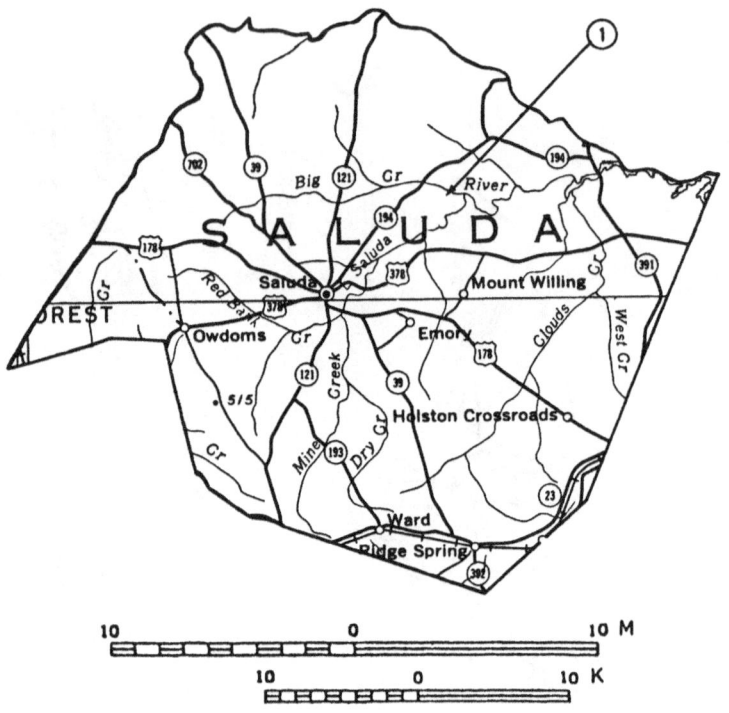

SPARTANBURG COUNTY

1. Chesnee area: at the abandoned Cowpens Mica Mine.
   Garnet. Quartz, milky: massive. Tourmaline. (2)

2. County-wide in streams.
   Diamond: rare; small. Gold. (3-7-17-18)

3. Cross Anchor area: at abandoned mica mine.
   Tourmaline: black; sharply terminated pencil crystals.
   (2)

4. Enoree area: where Enoree Road crosses Twomile Creek.
   Quartzite, red: gem quality. (1-4-13-14-18-20)

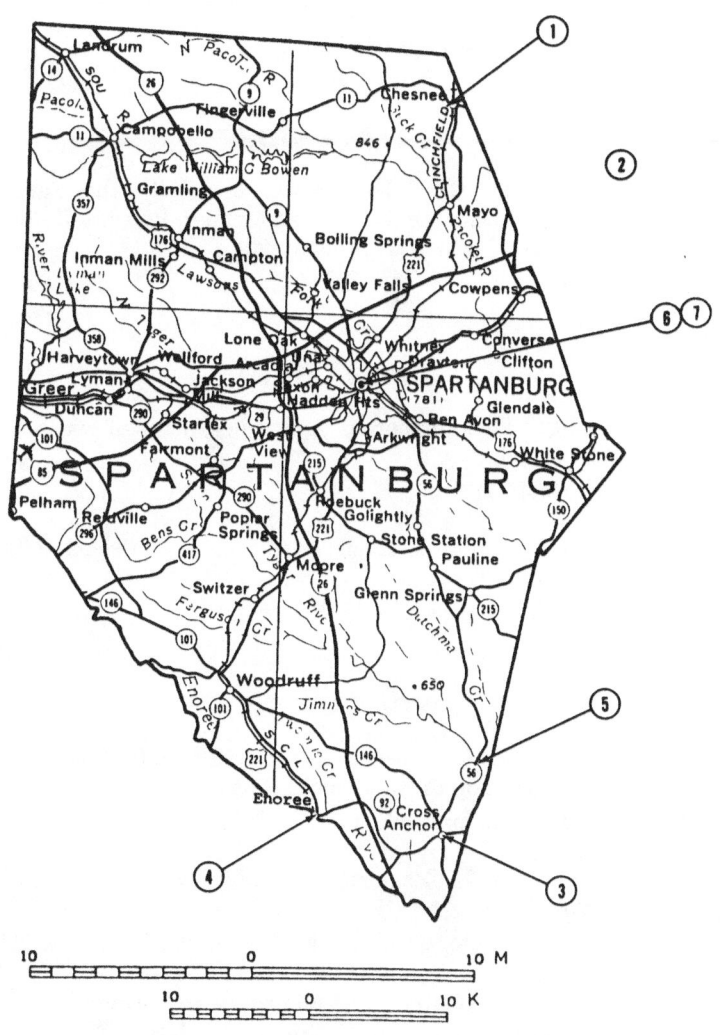

5. Glenn Springs area: 8 miles south; at mines.
     Psilomelane. (2)

6. Spartanburg area mines.
     Asbestos: crystals in massive. (2)

7. Spartanburg area streambeds.
   Gold. Zircon. (3-7-17-18-21)

STATEWIDE

1. In streams throughout state, especially the Santee flowage, Congaree River, and Cooper River and their tributaries, as well as in many lakes.
   Pearl. (7)

UNION COUNTY

1. County-wide in numerous gold mines, many of which are abandoned.
   Gold. (2)

2. West Springs area: adjacent to Fairforest Creek; at the Nott Mine.
   Copper, native. Gold. Pyrite. (2)

YORK COUNTY

1. Clover area: 6.3 miles southeast on SR-55; 0.25 mile west of the junction of SR-49 and SR-55.
   Corundum: black crystals. (1-13)

2. Clover area: 3 miles northwest on SR-508; on the slopes of Henry Knob.
   Andalusite. Kyanite. Lazulite. Staurolite. (1-4-13-29)

3. Hickory Grove area: west and southwest; at gold mines.
   Calcite. Galena. Gold. Pyrite. Sphalerite. (2)

4. Hickory Grove area: 2.2 miles west on Smiths Ford Road; at the Magnolia Mine.
   Gold. Malachite. (2)

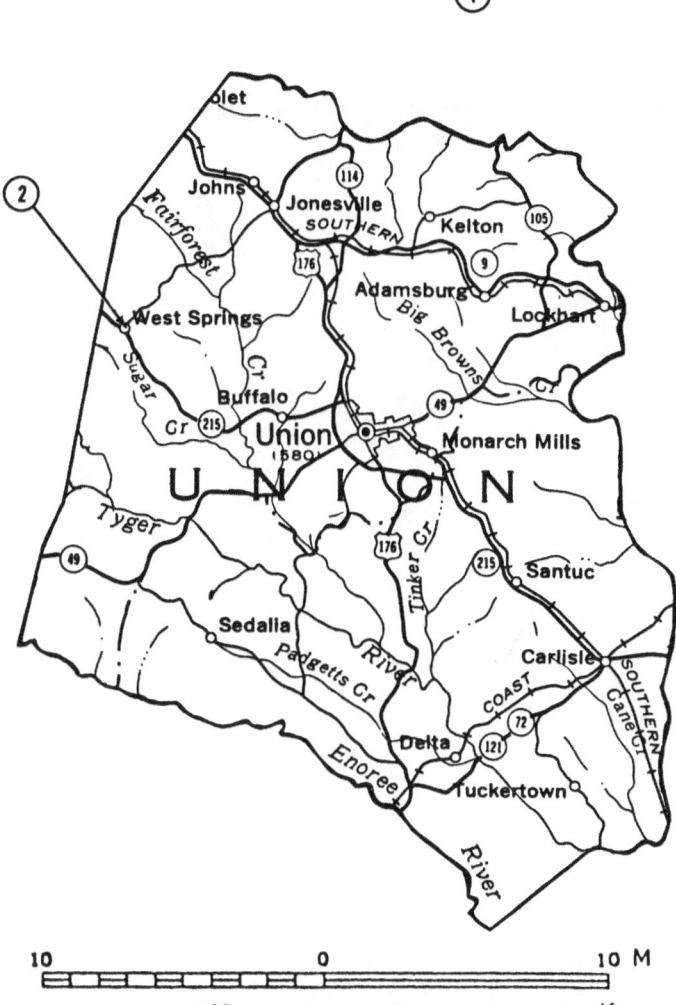

5. Hickory Grove area: 4 miles southwest on SR-211; 0.5 mile
   northwest of road; on slopes of Worth Mountain.
       Andalusite. Kyanite. Lazulite. Staurolite. (1-4-13)

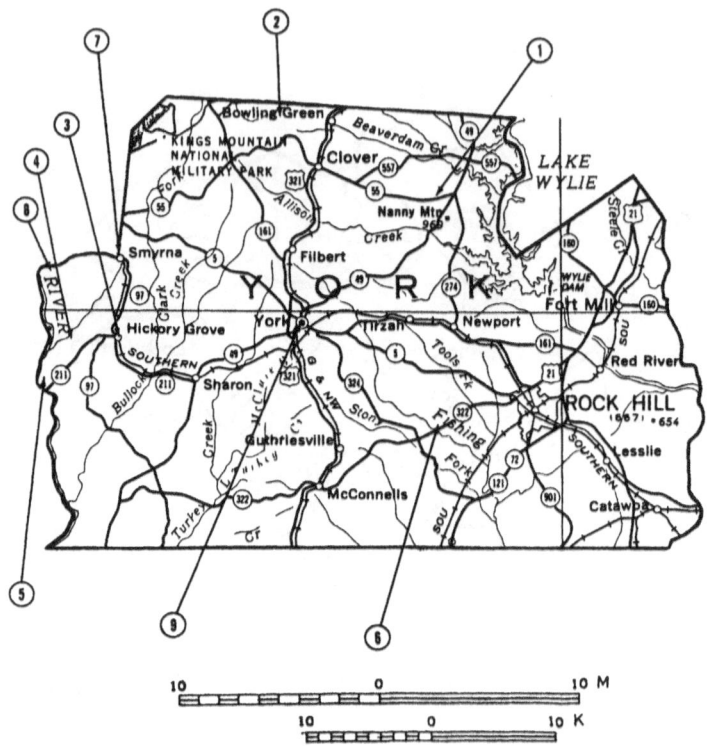

10   0   10 M

10   0   10 K

6. Rock Hill area: 5 miles west on SR-322; along the road in
   the area of Fishing Creek.
       Sillimanite. (1-13)

7. Smyrna area: north and west; at gold mines.
       Calcite. Galena. Gold. Pyrite. Sphalerite. (2)

8. Smyrna area: 3.5 miles west; at the Hull Mine.
       Tourmaline: black; flowerlike stellate crystals clusters
   on snow-white quartz; superb cabinet specimens. (2)

9. York area: just northeast; at abandoned copper mine.
       Azurite. Copper, native. Cuprite. Malachite. (2)

TENNESSEE

Statewide total of 96 locations in the following counties:

Bedford (5)	Hamblen (1)	Roane (1)
Blount (3)	Hamilton (1)	Robertson (1)
Bradley (1)	Hardin (1)	Rutherford (1)
Campbell (1)	Hawkins (2)	Sevier (1)
Cannon (1)	Jefferson (2)	Shelby (1)
Carter (4)	Knox (1)	Smith (3)
Claiborne (1)	Lawrence (1)	Sullivan (6)
Clay (1)	Marion (1)	Unicoi (5)
Cocke (8)	McMinn (1)	Union (4)
Cumberland (1)	Meigs (1)	Warren (2)
Davidson (1)	Monroe (6)	Washington (3)
Fentress (4)	Overton (2)	Wayne (1)
Greene (5)	Polk (5)	White (1)
Grundy (1)	Putnam (2)	Williamson (1)

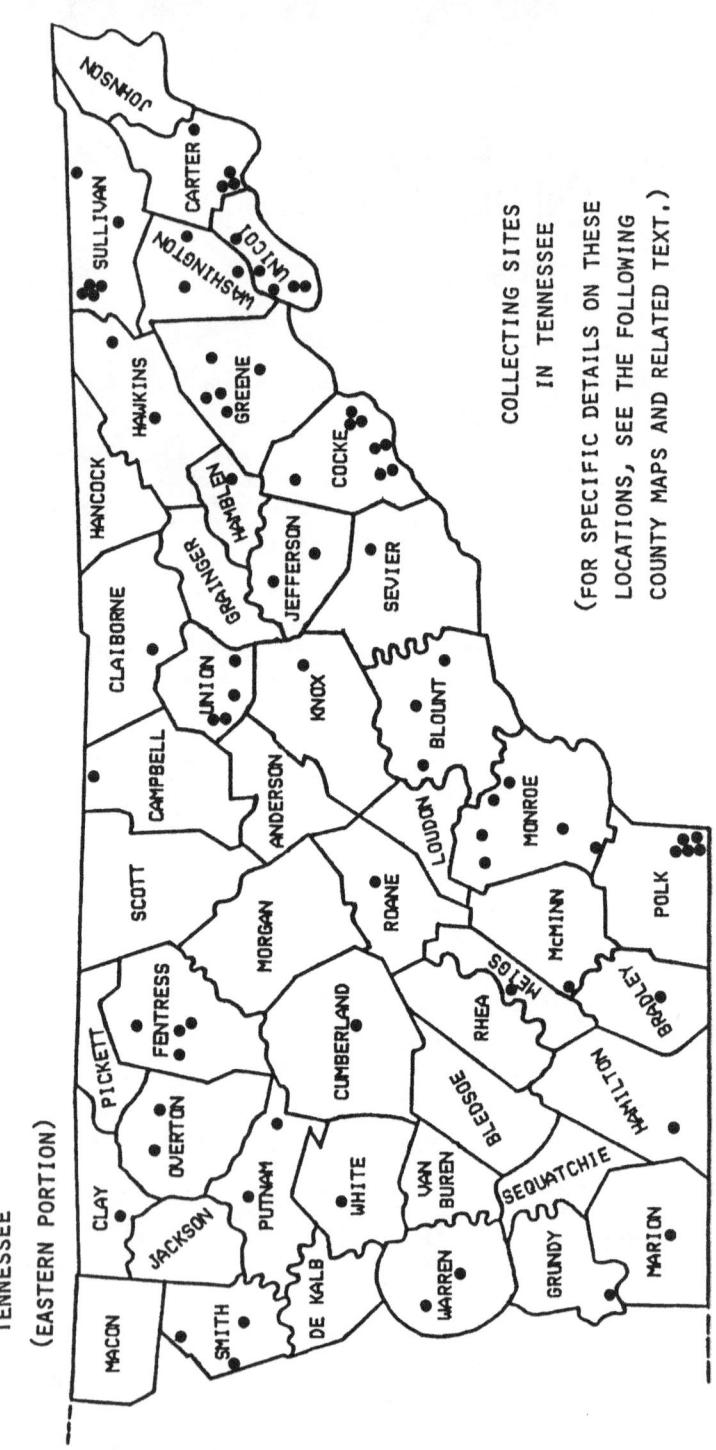

TENNESSEE
(EASTERN PORTION)

COLLECTING SITES
IN TENNESSEE

(FOR SPECIFIC DETAILS ON THESE
LOCATIONS, SEE THE FOLLOWING
COUNTY MAPS AND RELATED TEXT.)

TENNESSEE
(WESTERN PORTION)

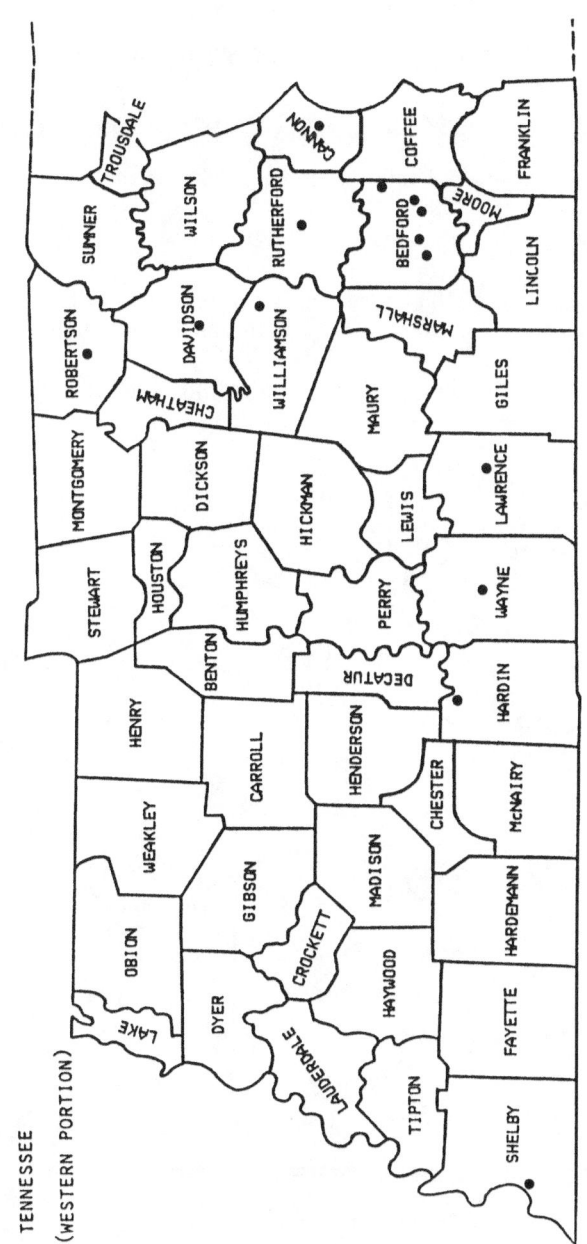

COLLECTING SITES IN TENNESSEE

(FOR SPECIFIC DETAILS ON·THESE LOCATIONS, SEE THE FOLLOWING COUNTY MAPS AND RELATED TEXT.)

BEDFORD COUNTY

1. Bell Buckle area: 6 miles east of SR-82; in area south and west of US-41.

    Agate: nodules to 4" diameter. (1-13)

2. Shelbyville area: 6 miles east on SR-64; at the Velmer Curvow farm (CF).

    Agate: iris. Carnelian. (1-13-39)

3. Shelbyville area: 3.3 miles southwest on SR-64 to Sugar Creek; upstream and down on both banks for 5 miles in each direction.

    Fossils: coral; agatized; fine; to 20 lb. (1-3-7-13-20)

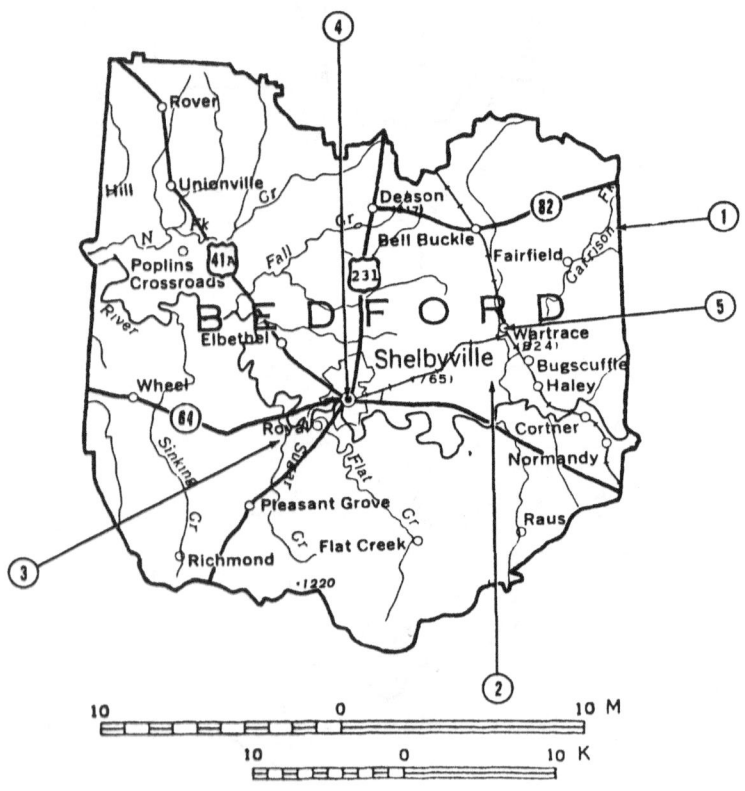

4. Shelbyville area: center of a virtually county-wide bed of agates; the best and most abundant come from a roughly 12-mile-diameter circular area whose center is on Horse Mountain Road halfway between Shelbyville and Wartrace.

> Agate (aka "Horse Mountain Agate" locally): superb varieties; banded to clear; transparent to translucent to opaque; water-clear through pale orange to bright salmon, blue, black, smoky, amber, gray; patternless through honeycomb, turtleback, dendritic; ranging in size up to 12" diameter and 16 lb. (1-3-7-13-20)

5. Wartrace area: at the W. E. Dye property (CF).

> Agate (aka "Horse Mountain Agate" locally). Fossils, marine: corals; agatized. (1-4-13)

BLOUNT COUNTY

1. Friendsville area.

> Marble. (12-30)

2. Maryville area: around Mountvale Springs.

> Gold. (3-7-17-18)

3. Townsend area: in road cuts and streams.

> Epidote. (4-7-20)

BRADLEY COUNTY

1. Cleveland area mines.

> Sphalerite. (2)

CAMPBELL COUNTY

1. Jellico area: southeast for several miles along US-25W; especially in road cuts.

> Agate: good quality. (1-4-7-13-20)

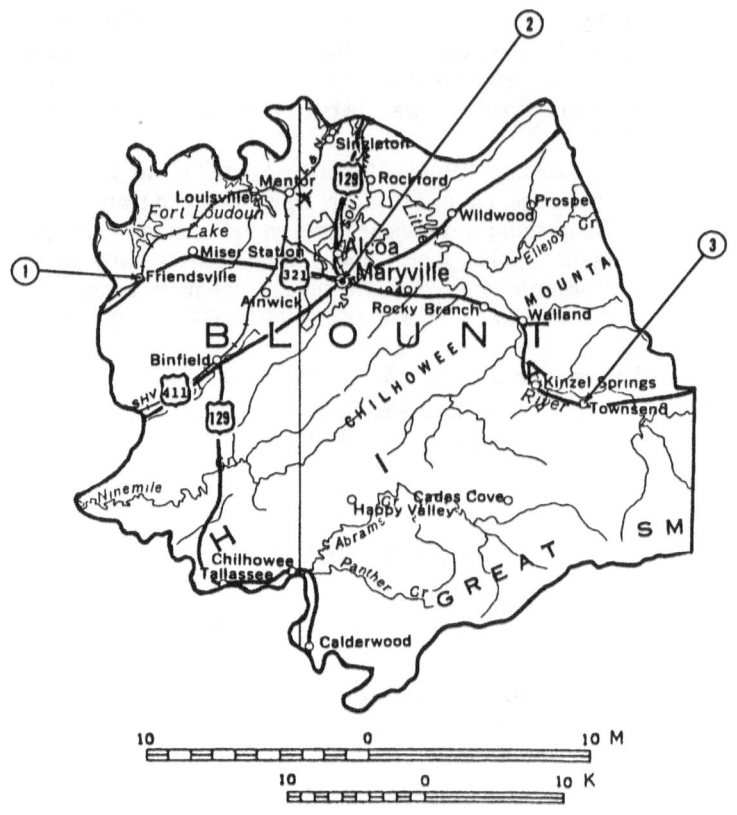

CANNON COUNTY

1. Woodbury area: 5 miles east on US-70S; in the valleys of the Stones River and its tributaries, even those hardly more than erosion ravines.

   Quartz: geodes to 14″ diameter; thin-shelled; warty outer surface; locally called "Cannonballs"; with interior inclusions of white calcite, botryoidal chalcedony, pink dolomite, fluorite in various colors, goethite, limonite, botryoidal linings of opal, pyrite, and rock quartz in glittering drusy linings and sharply terminated crystals; blue celestite in some. (1-4-7-10-12-13-14-20-37)

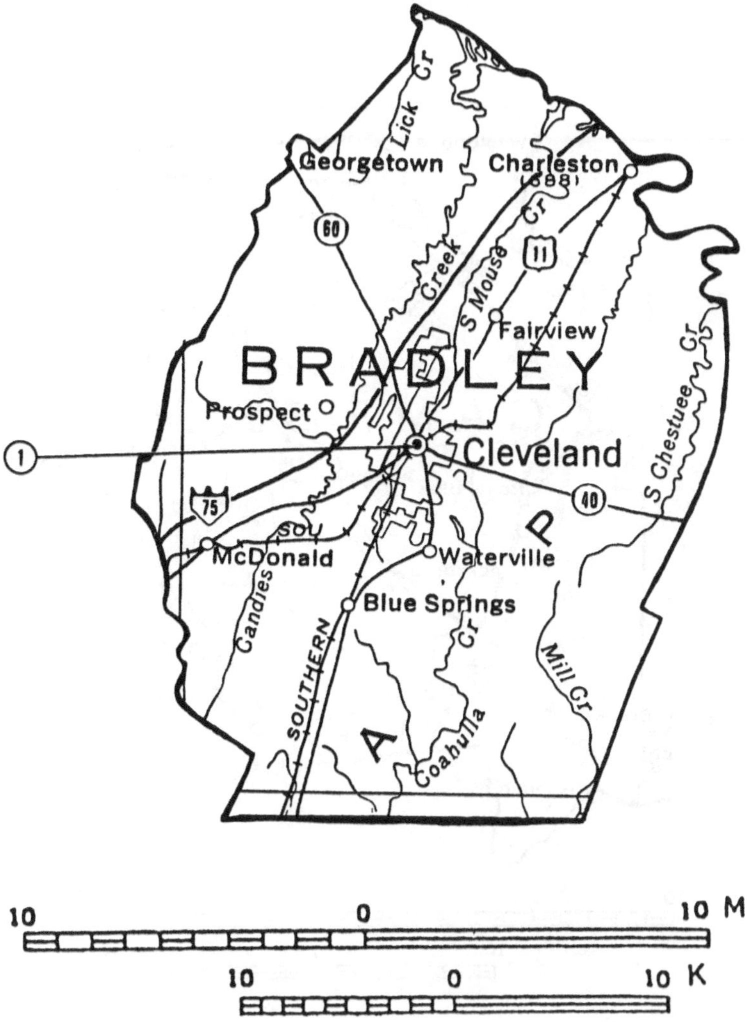

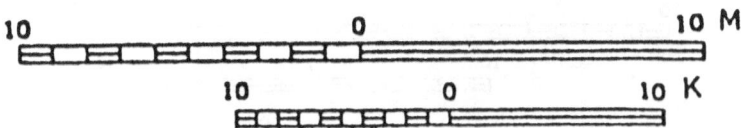

CARTER COUNTY

1. Elk Mills area: from US-321 bridge, go 1 mile south to road cut near top of hill.

    Dolomite: top quality. (4-37)

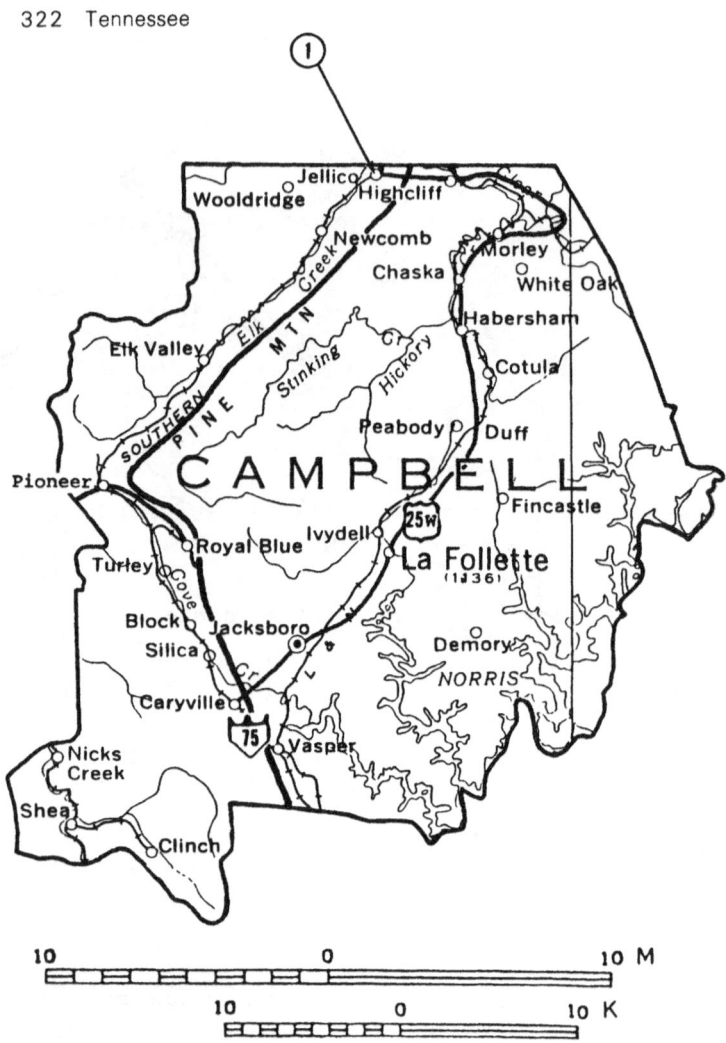

2. Roan High Knob area: at the Roan Mountain Flower Garden.

> Feldspar: green. Moonstone. Thulite. Unakite. (1)

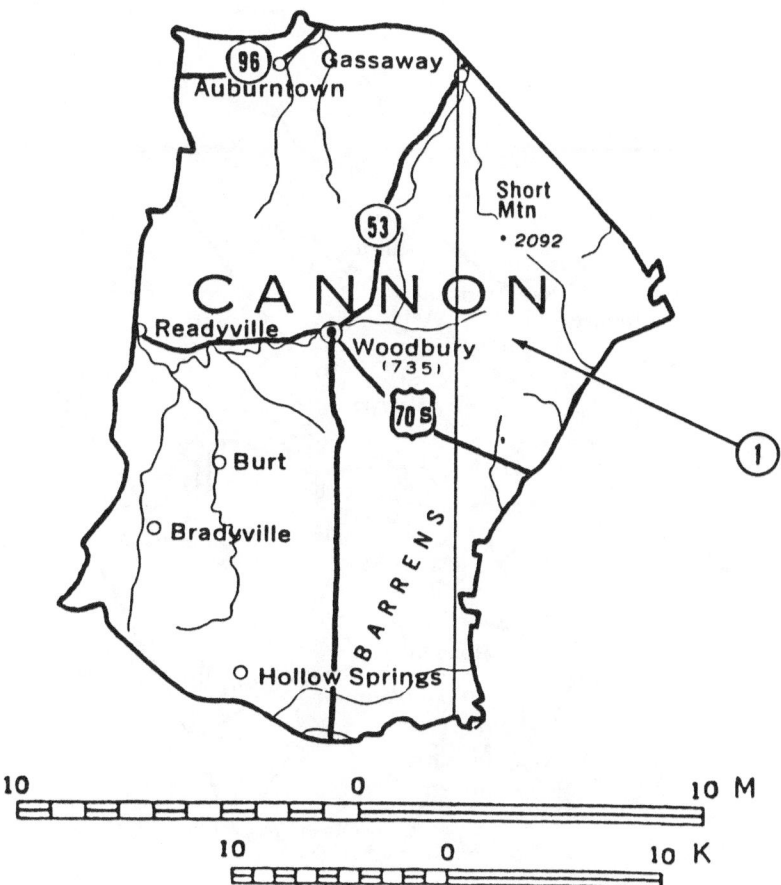

3. Roan High Knob area: on all slopes of Roan Mountain; along US-19 from Blevins to Elk Park.

   Unakite. (1-13-29)

4. Roan High Knob area: south on SR-143 to the next-to-last roadside table before the North Carolina border; on the Tennessee side of Roan Mountain; on both sides of Rock Creek, upstream and down.

   Unakite. (1-3-7-13-20)

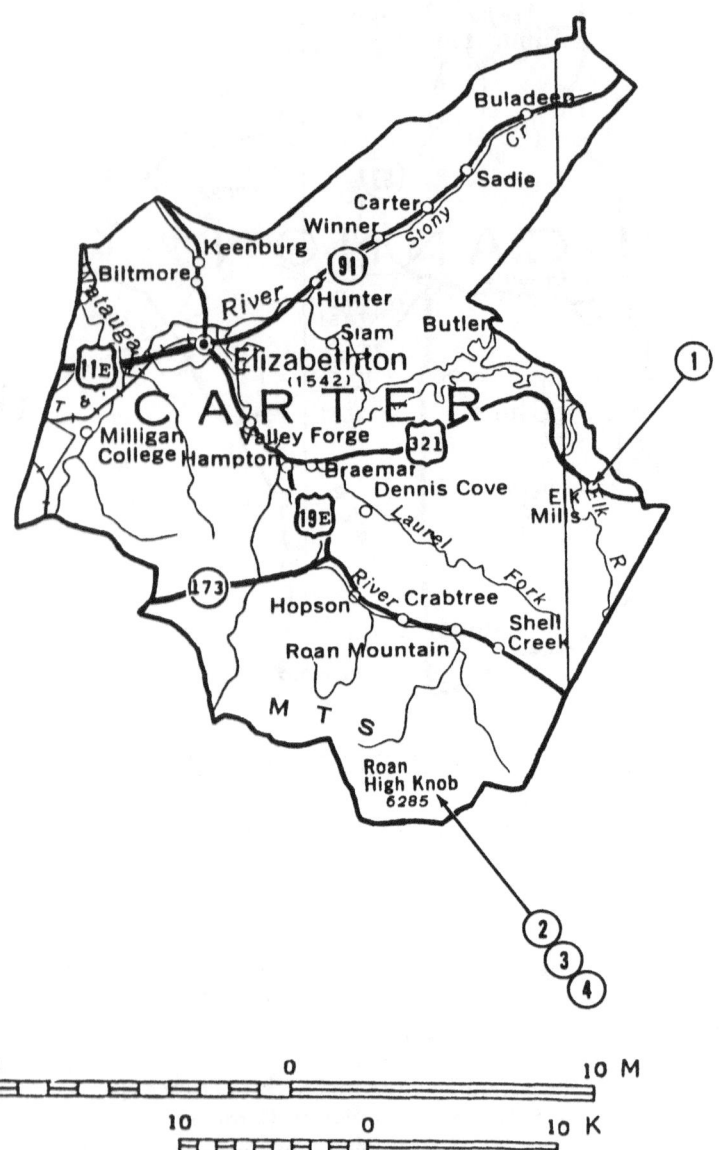

CLAIBORNE COUNTY

1. New Tazewell area: 5 miles southwest to Straight Creek; in area mines.

   Galena. Sphalerite. (2)

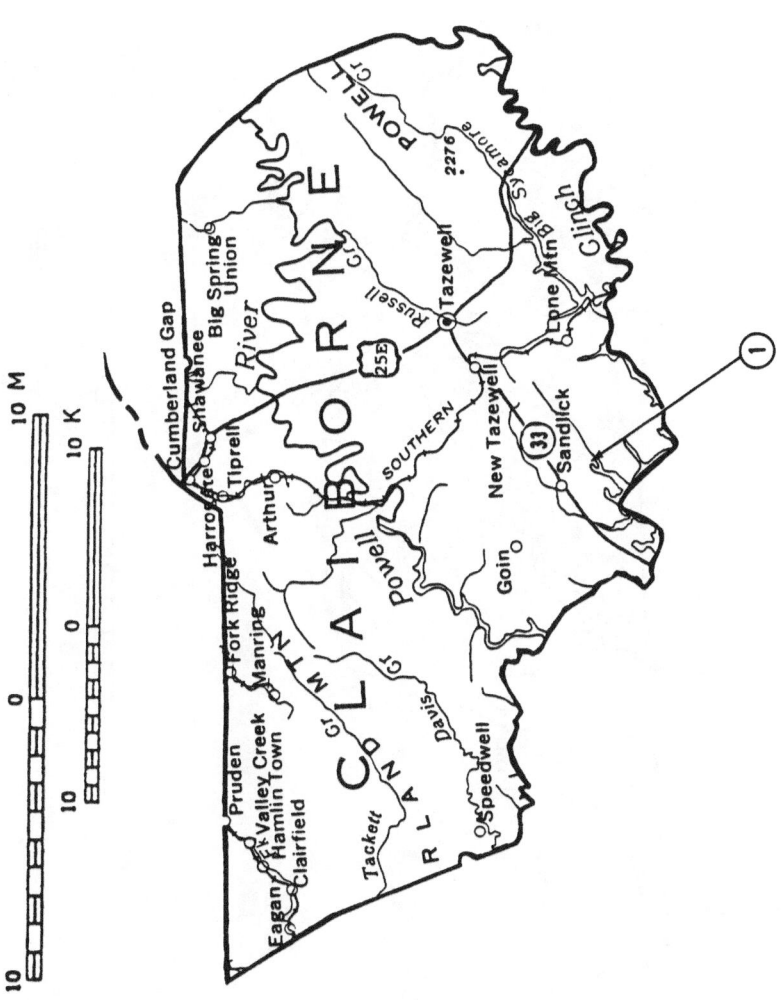

CLAY COUNTY

1. Dale Hollow Lake area: 15.3 miles northeast of Butlers Landing; most easily reached from Oakley (Overton County); from there, go 13 miles north to the south-central shore of Dale Hollow Lake.

   Quartz: geodes to 11″ diameter; thin-shelled; warty outer surface; locally called "Cannonballs"; with inte-

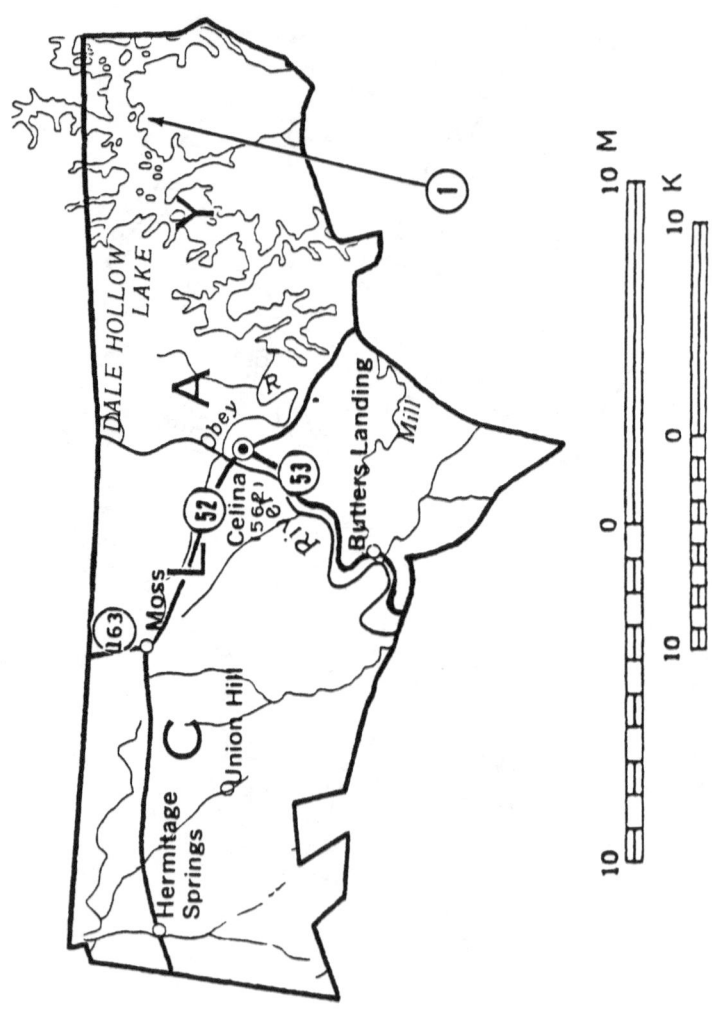

rior inclusions of white calcite, botryoidal chalcedony, pink dolomite, fluorite in various colors, goethite, limonite, botryoidal linings of opal, pyrite, and rock quartz in glittering drusy linings and sharply terminated crystals; blue celestite in some. (14-20)

COCKE COUNTY

1. Bybee area.
    Unakite: pebbles, boulders. (1-7-13-20-29-granite)

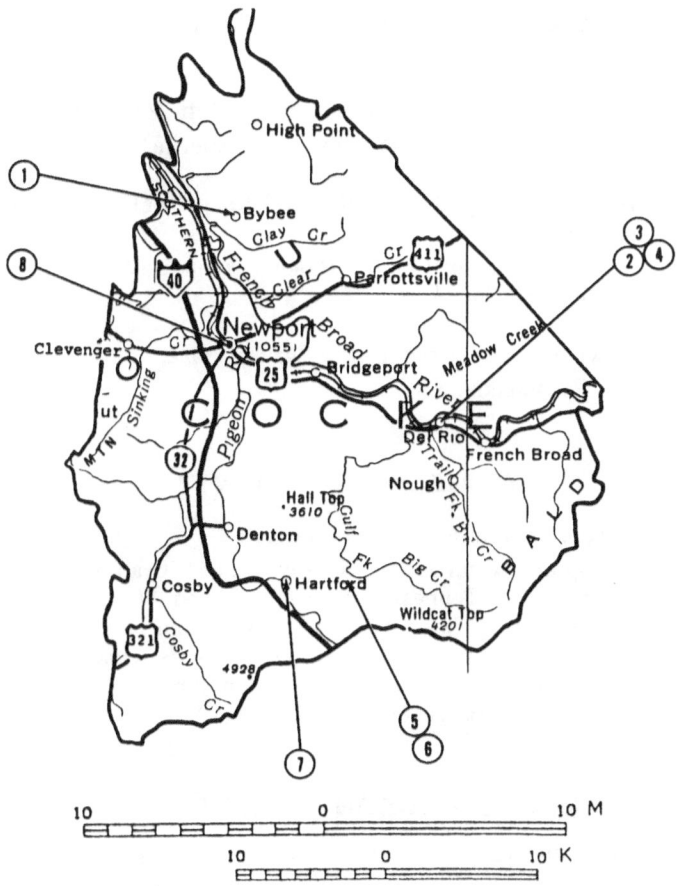

2. Del Rio area mines.
   Barite. Calcite. Celestite. (2)

3. Del Rio area stream valleys.
   Unakite: gem-quality crystals.

4. Del Rio area: to the Mine Ridge prospects.
   Azurite. Chalcopyrite. Galena. Hematite. Malachite. Pyrite. Zincite. (21)

5. Hartford area: 2.5 miles east; along Gulf Fork of Big Creek.
   Chalcopyrite. Galena. Pyrite. Sphalerite. (1-7-13-20-29-quartz)

6. Hartford area: 2.5 miles east; at the Gulf Fork mines.
   Chalcopyrite. Galena. Pyrite. Sphalerite. (2)

7. Hartford area: just south of Cogdill Chapel; at the Coggins property.
   Chalcopyrite. Copper. Galena. Hematite. Malachite. Pyrite. Zincite. (21)

8. Newport area: at the Yellow Springs Mine.
   Psilomelane. (2)

CUMBERLAND COUNTY

1. Dorton area: in a series of quarries.
   Quartzite-sandstone: locally called "Crab Orchard Stone," "Tennessee Quartzite," and "Cumberland Quartzite"; wide range of colors, from pale tints to vivid hues; patterned in swirls, bands, concentrics, dendritics, geometric shapes; in large sheets and blocks that are easily separated. (12)

DAVIDSON COUNTY

1. Nashville area: to the Haysborough lead mine.
   Barite. Galena. (2)

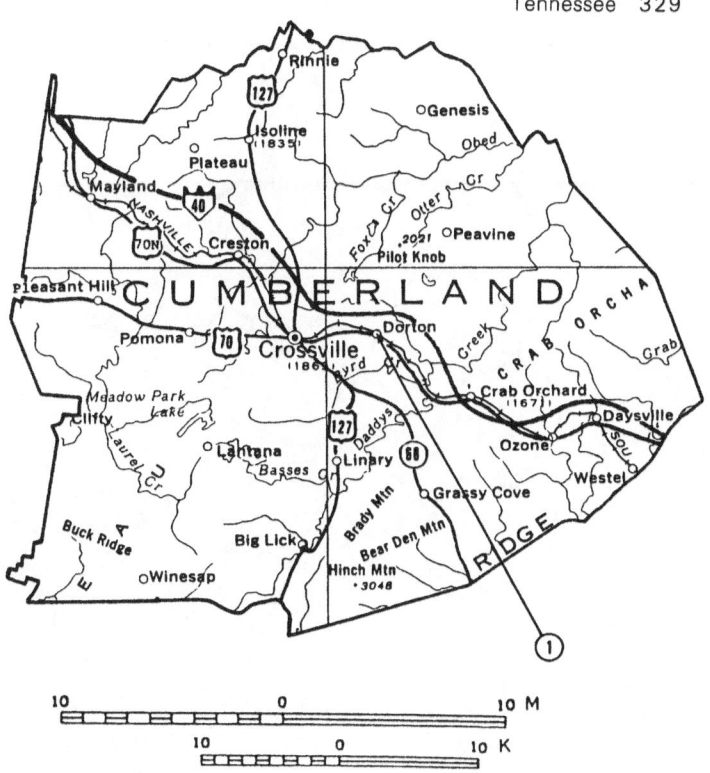

FENTRESS COUNTY

1. Boatland area: 1 mile from the church; on the north side of Boles Creek.

> Quartz: geodes to 11″ diameter; thin-shelled; warty outer surface; locally called "Cannonballs"; with interior inclusions of white calcite, botryoidal chalcedony, pink dolomite, fluorite in various colors, goethite, limonite, botryoidal linings of opal, pyrite, and rock quartz in glittering drusy linings and sharply terminated crystals; blue celestite in some. (1-7-13-20)

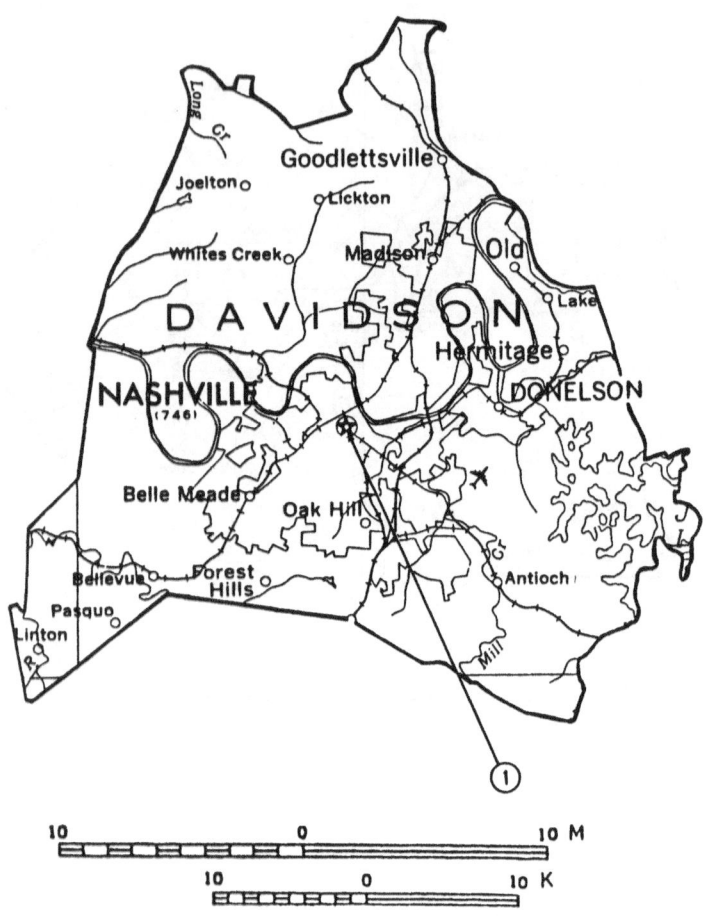

2. Jamestown area: 3.5 miles southwest; at Buffalo Cove.

Quartz: geodes to 11″ diameter; thin-shelled; warty outer surface; locally called "Cannonballs"; with interior inclusions of white calcite, botryoidal chalcedony, pink dolomite, fluorite in various colors, goethite, limonite, botryoidal linings of opal, pyrite, and rock quartz in glittering drusy linings and sharply terminated crystals; blue celestite in some. (1-7-13-20)

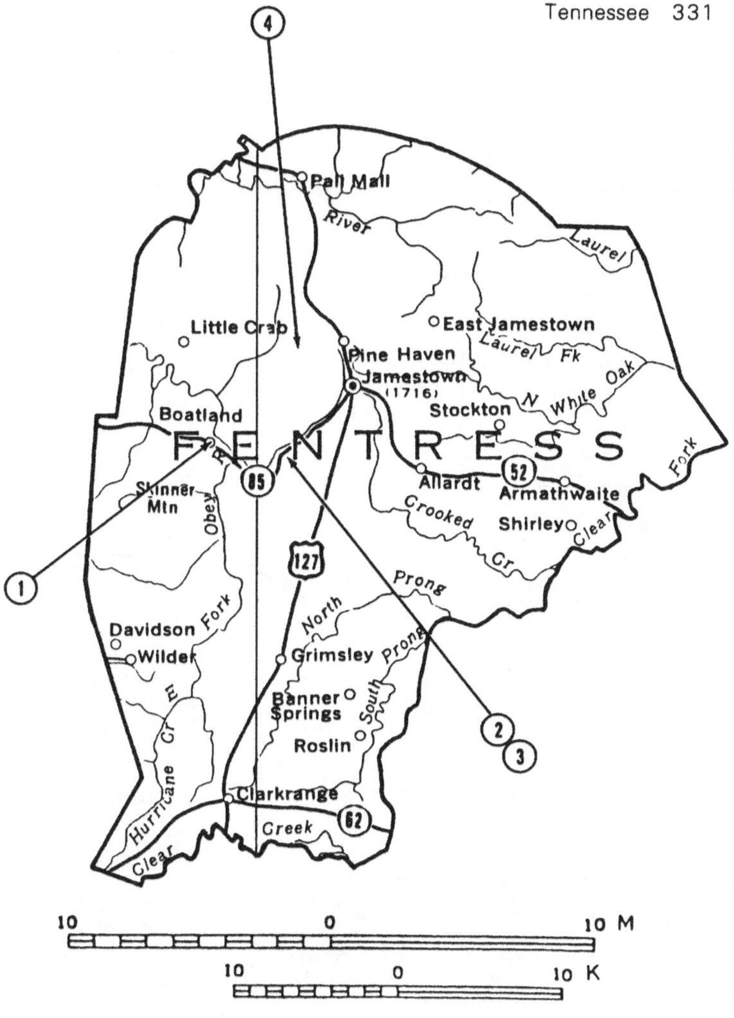

10   0   10 M

10   0   10 K

3. Jamestown area: 3.5 miles southwest; in Buffalo Cove area; on the Herbert Tipton farm (CF).

> Calcite (Onyx). (1-13)

4. Jamestown area: 2.5 miles northwest; north of SR-52; at Carpenter Hollow.

> Quartz: geodes to 11″ diameter; thin-shelled; warty outer surface; locally called "Cannonballs"; with inte-

rior inclusions of white calcite, botryoidal chalcedony, pink dolomite, fluorite in various colors, goethite, limonite, botryoidal linings of opal, pyrite, and rock quartz in glittering drusy linings and sharply terminated crystals; blue celestite in some. (1-7-13-20)

GREENE COUNTY

1. County-wide in road, RR, and stream cuts; also in quarries. Calcite. Celestite. Dolomite: crystals. Fluorite. Jasper: golden. Pyrite. Quartz, rock. (4-12)

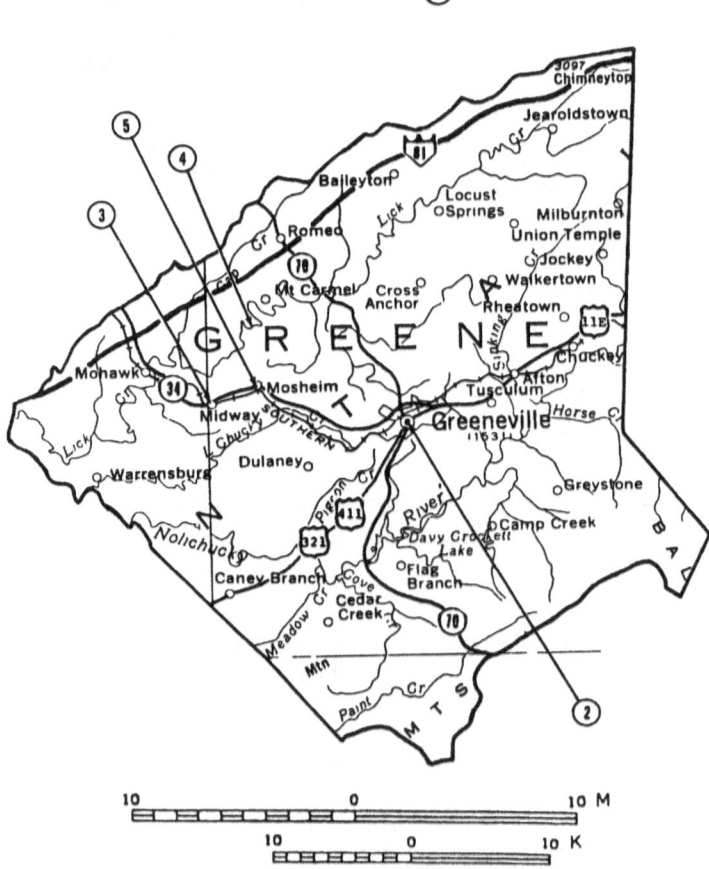

2. Greeneville area lead mines.

Sphalerite. (2)

3. Midway area: to Chestnut Ridge.

Quartz: geodes to 14″ diameter; thin-shelled; warty outer surface; locally called "Cannonballs"; with interior inclusions of white calcite, botryoidal chalcedony, pink dolomite, fluorite in various colors, goethite, limonite, botryoidal linings of opal, pyrite, and rock quartz in glittering drusy linings and sharply terminated crystals; blue celestite in some. (1-7-13-20)

4. Mosheim area: 3 miles north to gravel road on left; follow this west to Gethsemane Church area, where dirt road goes right; go 0.5 mile north on the dirt road to the Brown-Tipton Mine.

Calcite. Cerussite. Chert: high quality. Dolomite: crystals. Galena. Pyrite. Smithsonite. Sphalerite. (2)

5. Mosheim area.

Quartz: geodes to 9″ diameter; thin-shelled; warty outer surface; locally called "Cannonballs"; with interior inclusions of white calcite, botryoidal chalcedony, pink dolomite, fluorite in various colors, goethite, limonite, opal, pyrite, and rock quartz in glittering drusy linings and sharply terminated crystals; blue celestite in some. (1-7-13-20)

GRUNDY COUNTY

1. Monteagle area: just south of city in road cuts.

Calcite (Onyx): transparent to peach and pale pink; some banded. Hematite. Jasper. (4)

HAMBLEN COUNTY

1. Russellville area: in all road, RR, and stream cuts; also in quarries.

Quartz geodes to 12″ diameter. (4-12)

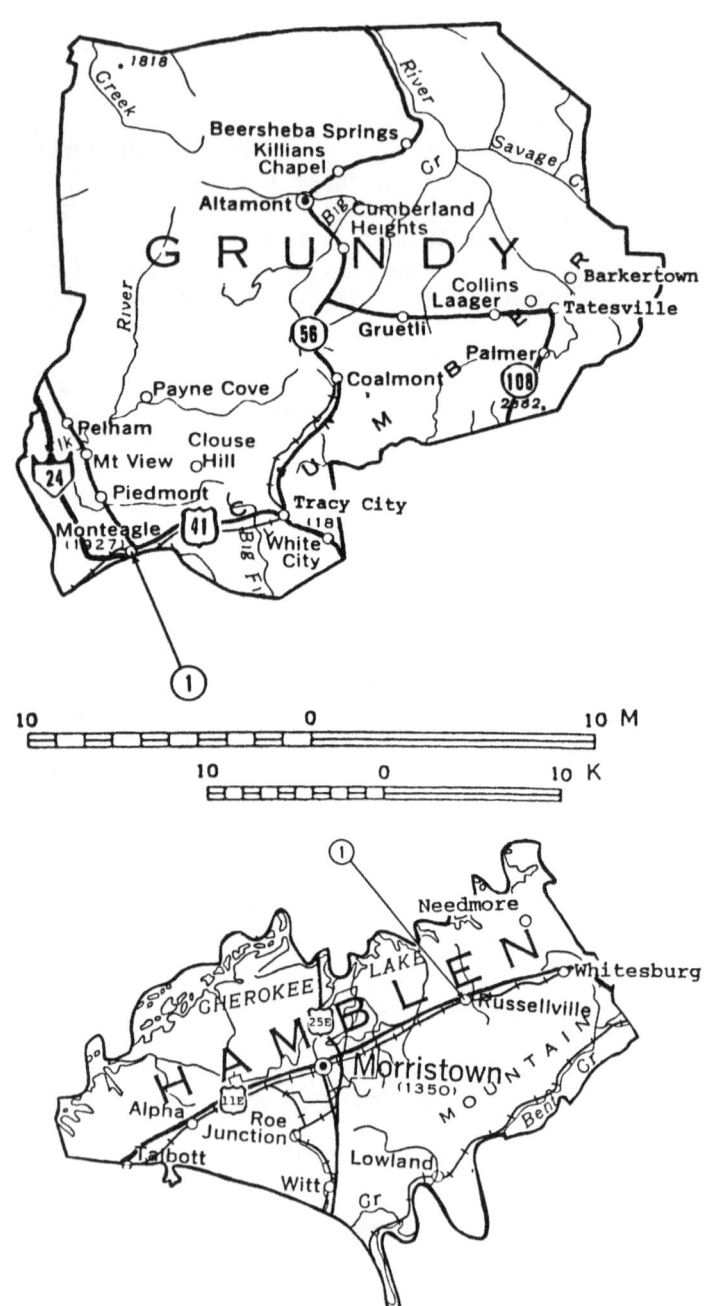

HAMILTON COUNTY

1. Chattanooga area: at the Missionary Ridge mines.
   Bauxite: crystals. (2)

HARDIN COUNTY

1. Saltillo area: near Milledgeville; on the Tennessee River banks at Coffee Bluff; most abundant in sands of bluffs, from which the material weathers.

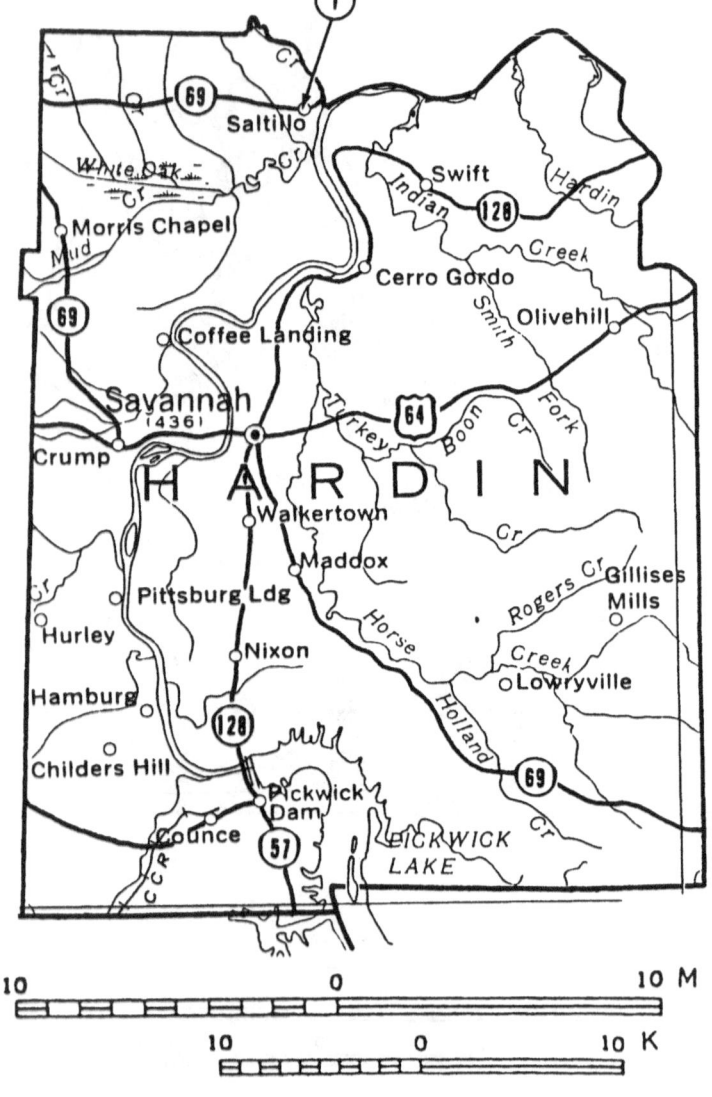

Amber (Succinite): reddish brown and faintly cloudy; nodules to 2" thick. (1-8-13-14-20-26)

HAWKINS COUNTY

1. County-wide in all road, RR, and stream cuts; also in quarries.

   Calcite. Celestite. Dolomite: crystals. Fluorite. Jasper: golden. Pyrite. Quartz, rock. (4-12)

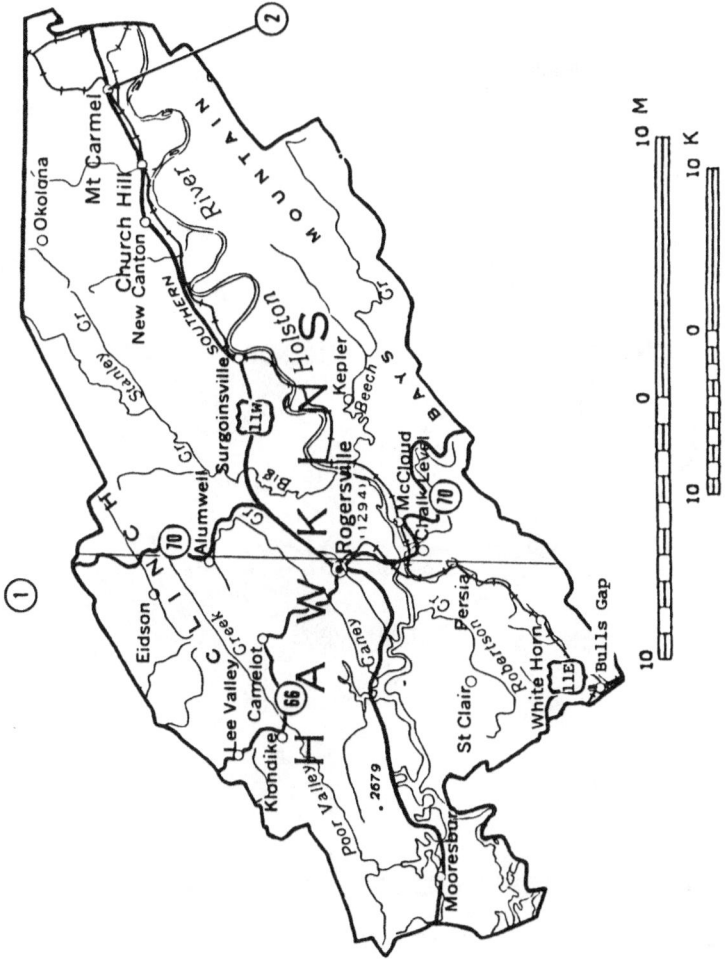

2. Mount Carmel area: 5 miles upstream from where US-11W crosses North Fork; along Senebaugh Branch.

> Barite: honey-colored; platy crystals to 3″ × 2″ × 0.5″; facet quality. (1-7-13-20)

JEFFERSON COUNTY

1. Dandridge area: at the Mossy Creek area mines.

> Calamine. Smithsonite (aka "Dry Bone" locally). Sphalerite. (2)

2. New Market area mines.

> Smithsonite (aka "Dry Bone" locally). Sphalerite. (2)

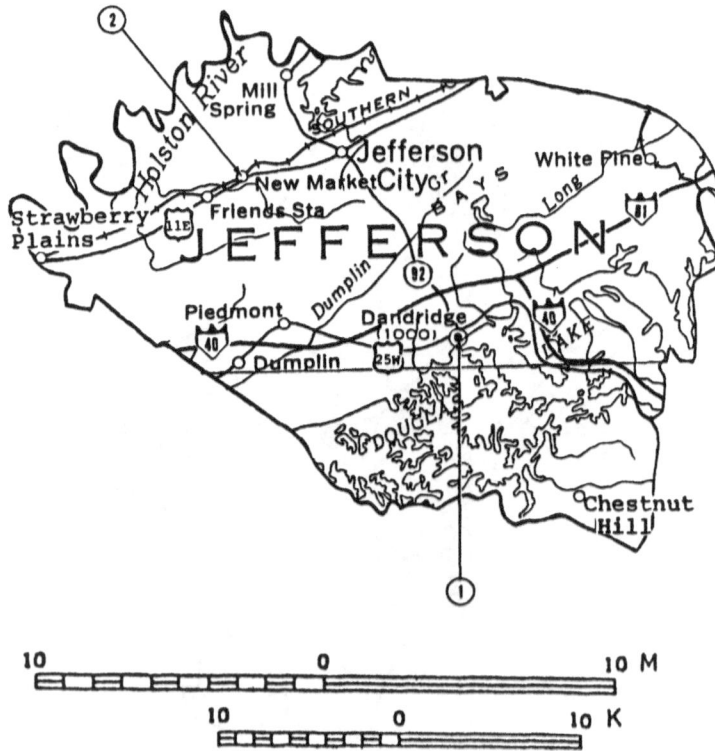

KNOX COUNTY

1. Mascot area mines.
    Sphalerite. (2)

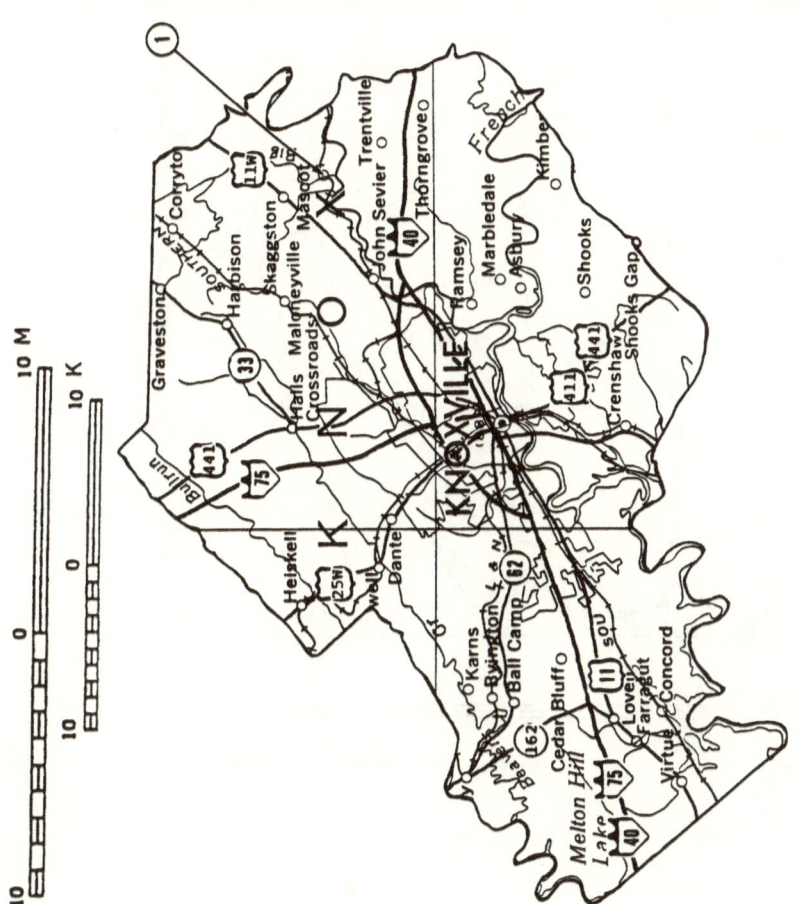

LAWRENCE COUNTY

1. Lawrenceburg area: northwest along US-64 to about 2
    miles beyond David Crockett State Park.
    Chalcedony. Chert. Quartz: geodes to 6″ diameter;

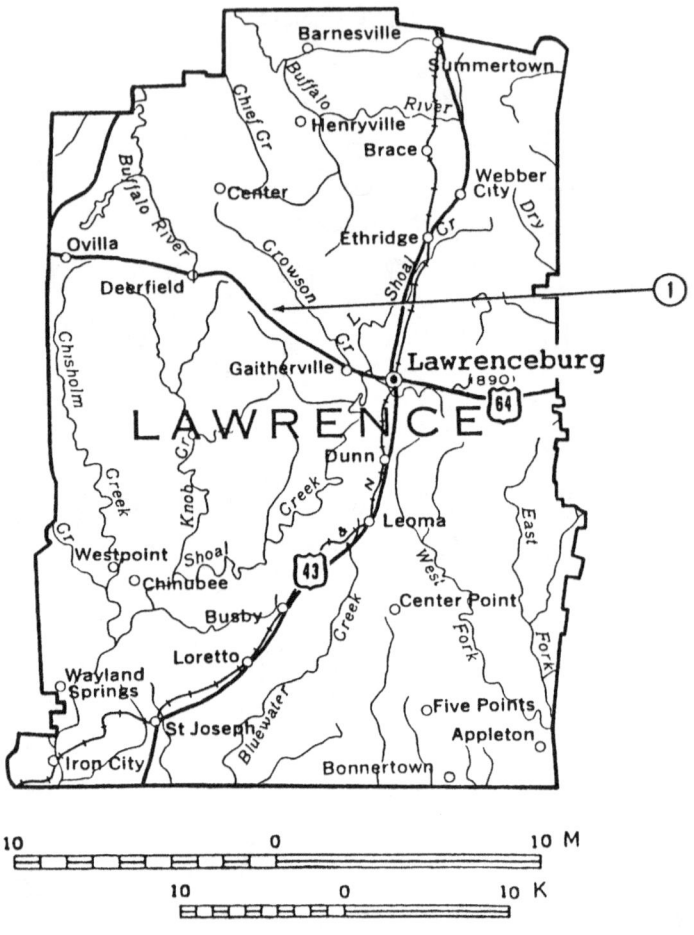

thin-shelled; warty outer surface; locally called "Cannonballs"; with interior inclusions of white calcite, botryoidal chalcedony, pink dolomite, fluorite in various colors, goethite, limonite, botryoidal linings of opal, pyrite, and rock quartz in glittering drusy linings and sharply terminated crystals; also, in rare instances, crystal inclusions of blue celestite and dolomite. (4)

MARION COUNTY

1. Jasper area: at the Chattanooga Shale Quarry.
   Barite. Calcite. Fluorite. Galena. Pyrite. Selenite. (12)

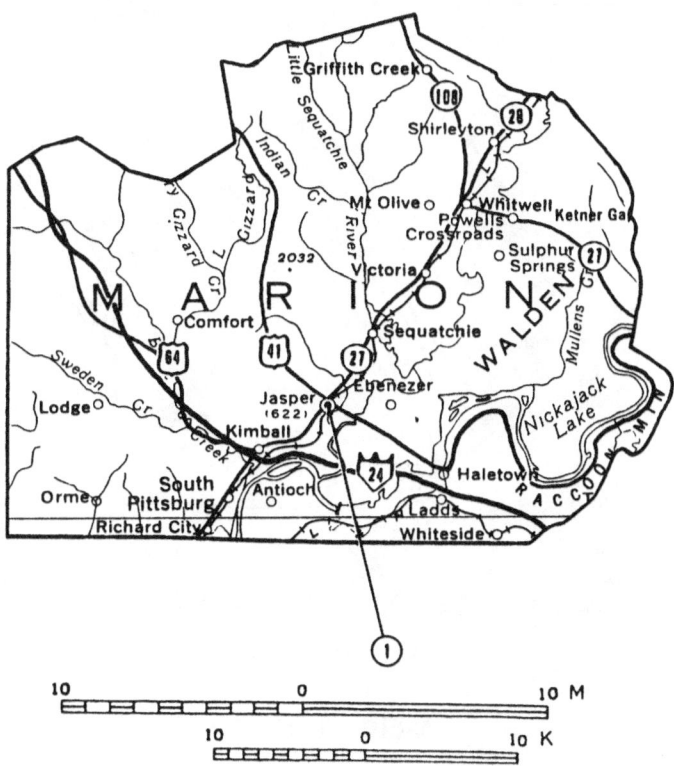

MCMINN COUNTY

1. Calhoun area: 6.9 miles northwest to the McMinn Ridge
   area; on southeast slope; at a small, unnamed mine.
   Barite: quality crystals; faceting grade. (2)

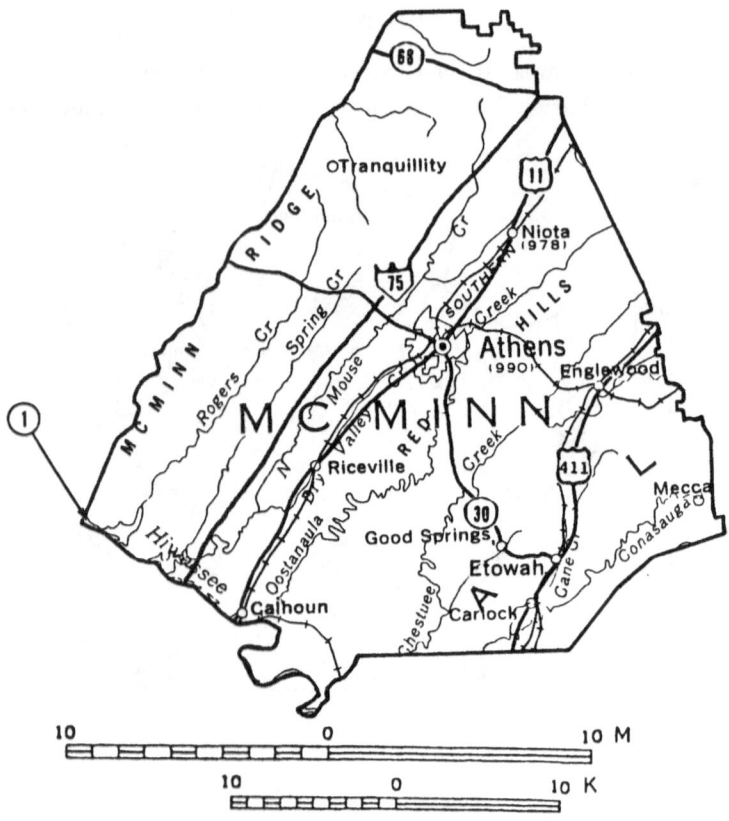

MEIGS COUNTY

1. Decatur area: from junction of US-58 and SR-30, 2 blocks
   north on US-58; turn left and leave town limits; collect on
   both sides of road for the next several miles.

    Chert: banded; gray. (1-13-37-39)

MONROE COUNTY

1. Citico Beach area: 1.25 miles east-northeast; on shores and
   in gravels of the Little Tennessee River near the mouth of
   Citico Creek.

    Quartz, rock: small, bright crystals. (1-7-13-20)

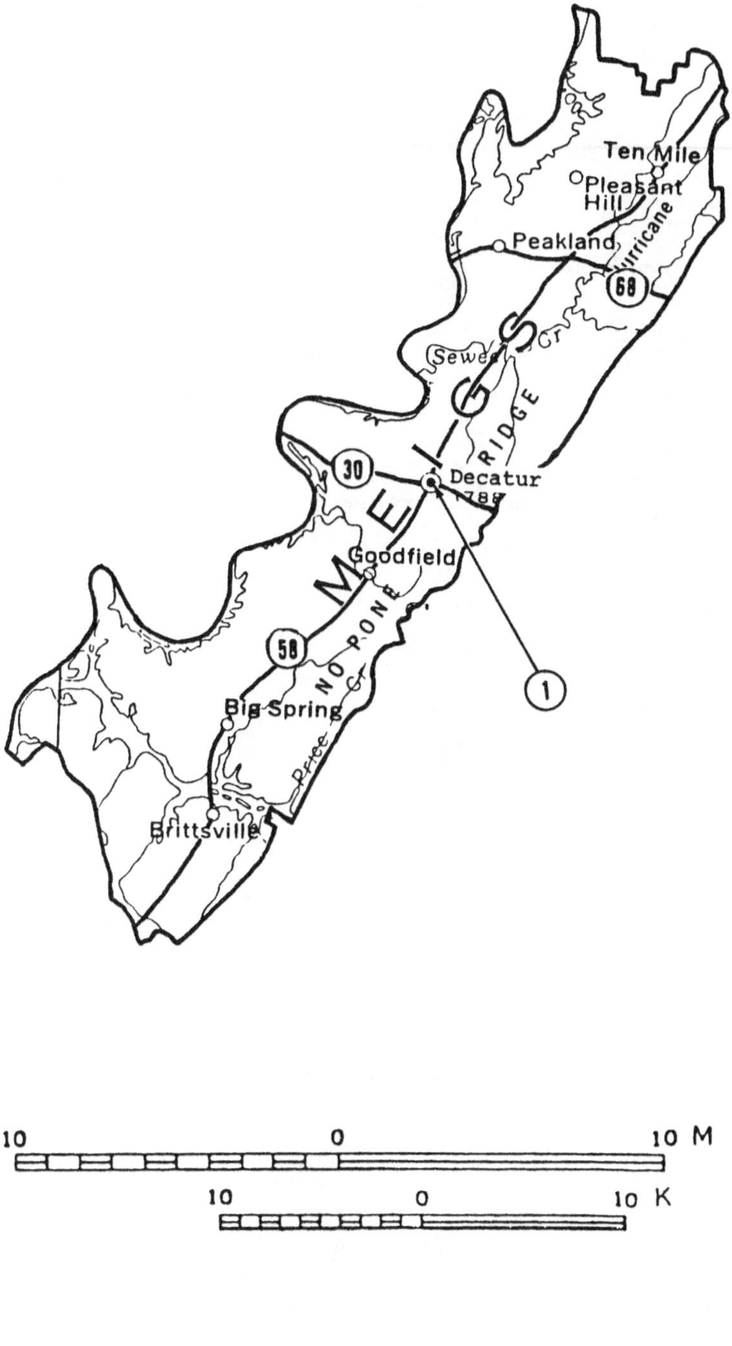

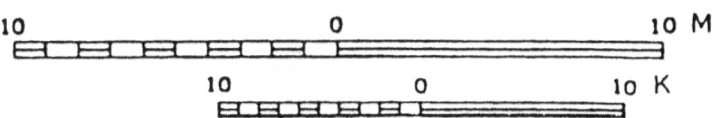

Ten Mile
Pleasant
Hill
Peakland
Sewee Cr
RIDGE
MEIGS
30
Decatur
788
Goodfield
58
NO PONE
Price Cr
Big Spring
Brittsville

10   0   10 M

10   0   10 K

2. Coker Creek area: 4.2 miles southwest; in Coker Creek.

Diamond: 3 stones; details not specified; found during panning. (17)

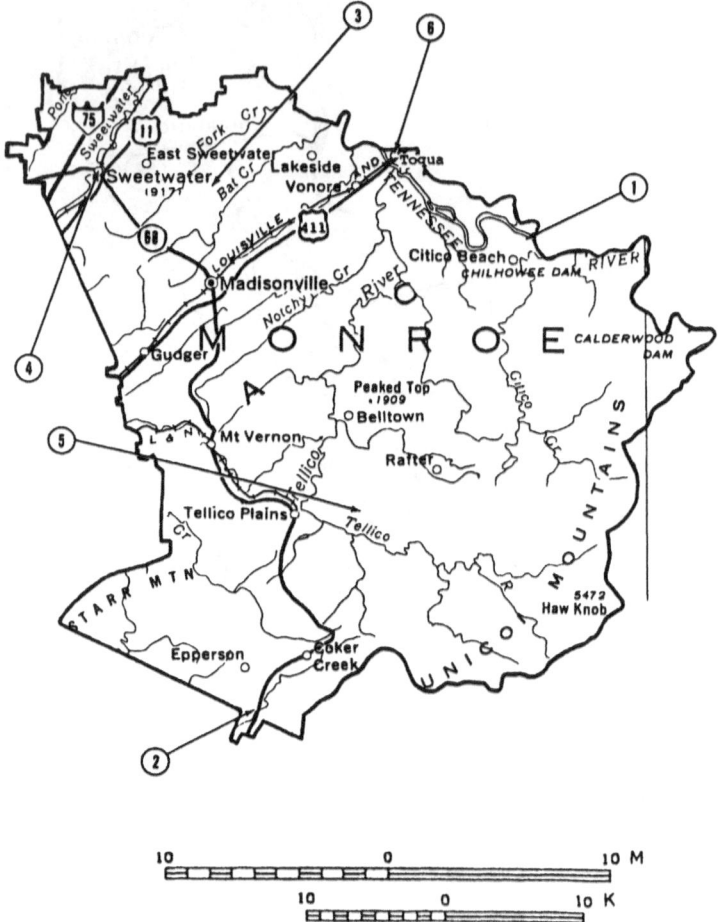

3. Sweetwater area: 5 miles east; near Rocky; at the Bullard Barite Mine.

Barite: glittering rosettes. Fluorite: green, purple, blue. Sphalerite: yellow. (2)

4. Sweetwater area mines and quarries.

> Barite. (2-12)

5. Tellico Plains area: from Murphy's Store on SR-68, go 3 miles east on gravel road to white brick store with red trim; turn left past store and continue left to crossroad and small graveyard; collect in graveyard and to right of it.

> Limonite: pseudomorphs after pyrite; cubes to 3". (1-13)

6. Toqua area: in the clay banks of the Little Tennessee River.

> Quartz, rock: fine crystals; well defined; some doubly terminated. (1-14-25-29)

OVERTON COUNTY

1. Allons area: west of SR-52.

> Quartz: geodes to 11" diameter; thin-shelled; warty outer surface; locally called "Cannonballs"; with interior inclusions of white calcite, botryoidal chalcedony, pink dolomite, fluorite in various colors, goethite, limonite, botryoidal linings of opal, pyrite, and rock quartz in glittering drusy linings and sharply terminated crystals; blue celestite in some. (1-7-13-20)

2. Monroe area: 0.75 mile southeast; on northeast slope of Pilot Knob.

> Quartz: geodes to 11" diameter; thin-shelled; warty outer surface; locally called "Cannonballs"; with interior inclusions of white calcite, botryoidal chalcedony, pink dolomite, fluorite in various colors, goethite, limonite, botryoidal linings of opal, pyrite, and rock quartz in glittering drusy linings and sharply terminated crystals; blue celestite in some. (1-7-13-20)

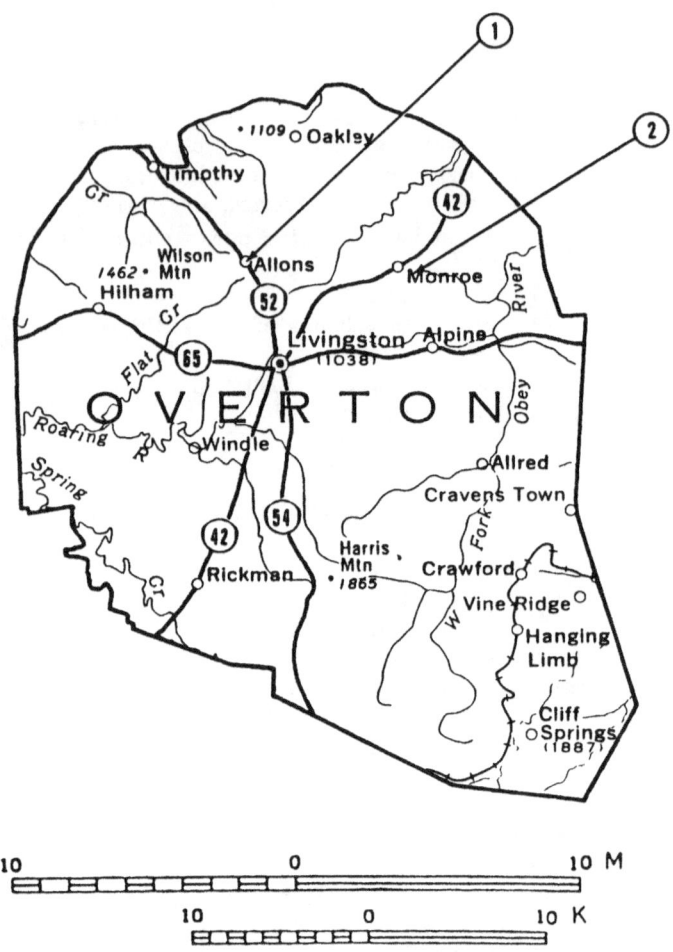

POLK COUNTY

1. Copperhill area: 1 mile north on SR-68; in eroded area along right side of road.

  Staurolite. (1-4-14-from bank -37)

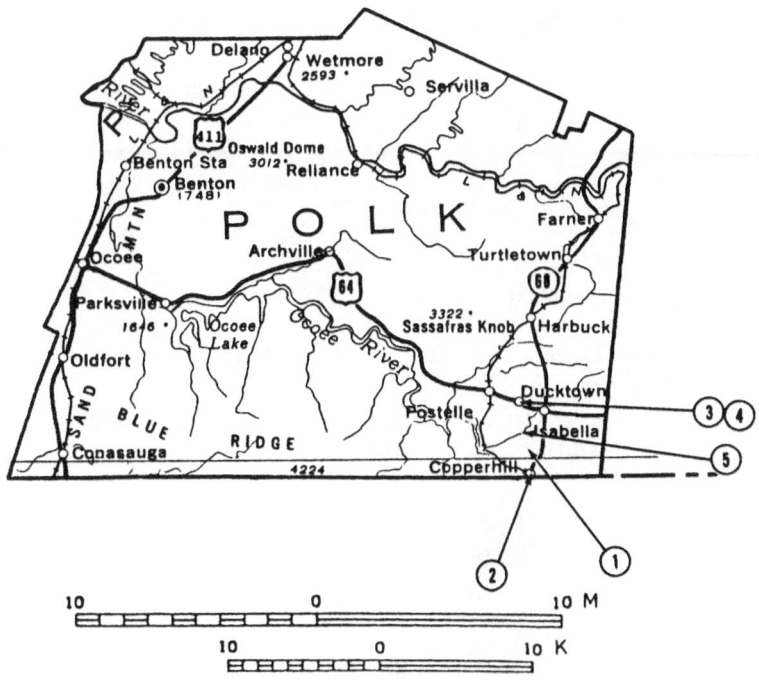

2. Copperhill area: north on SR-68 to power transformer on right; in parking lot behind the transformer and in adjacent areas.

> Garnet. Spodumene: large crystals. (1-13)

3. Ducktown area: in zinc mines.

> Melanterite (Copperas). Pyrrhotite: crystals. Zinc. Zoisite: brown crystals. (2)

4. Ducktown area: west on US-64 to Ocoee River; continue downriver to second red brick powerhouse; make U-turn here and return 1 mile to sharp curve left; park at end of guard post; in area toward ravine; also upstream from here along river, collecting for 1 mile.

> Azurite. Chalcanthite. Chalcopyrite. Copper. Garnet. Gold. Malachite. Pyrite. Pyrrhotite. (2)

5. Ducktown-Copperhill area: at five major copper mines of the district.

> Azurite. Chalcanthite. Chalcopyrite. Copper. Garnet. Gold. Malachite. Pyrite. Pyrrhotite. (2)

PUTNAM COUNTY

1. Cookeville area streams.
   > Agate. Jasper: nodules. (7-20)

2. Monterey area: 2 miles on US-70N; at abandoned quarry (now a dump).
   > Calcite. Celestite: blue crystals. Dolomite: pale pink crystals. Fluorite. Quartz, rock. (12)

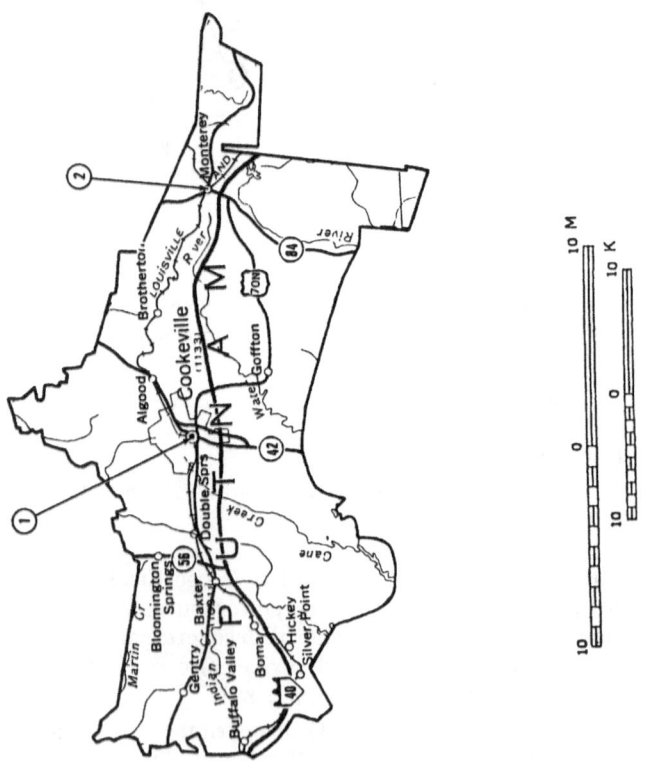

ROANE COUNTY

1. Kingston area: upstream on the Clinch River; near Indian
   Mound on the south bank.

   Diamond: 1 stone; 3 ct; cut to excellent gem of 1.25 ct;
   found in 1899. (3)

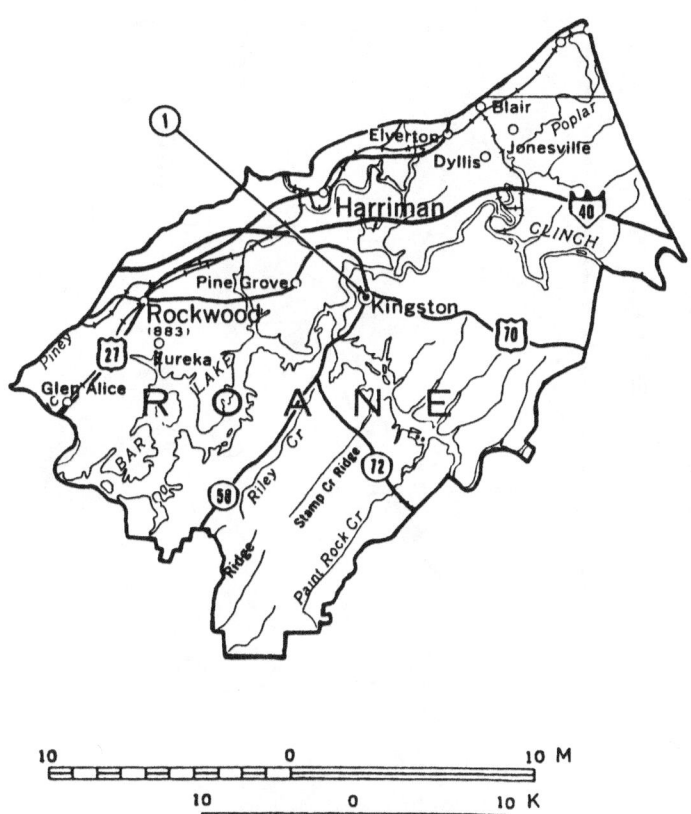

ROBERTSON COUNTY

1. Springfield area.
   Agate: good quality. (1-7-13-20)

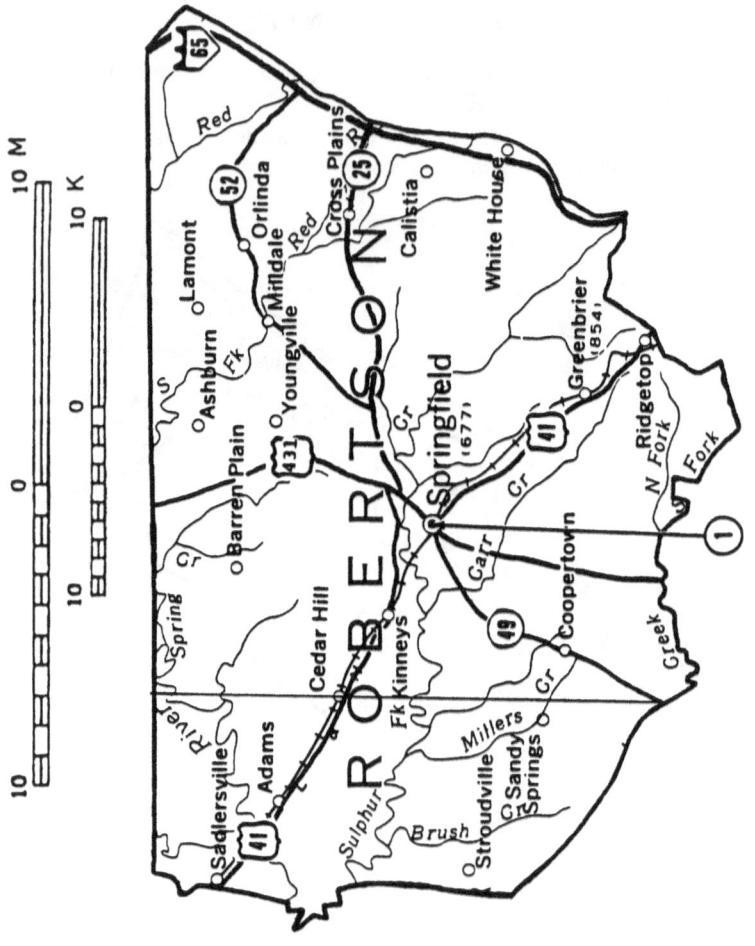

RUTHERFORD COUNTY

1. Murfreesboro area: road cuts and ditches.
Agate: nodules. Chert. (1-4-13)

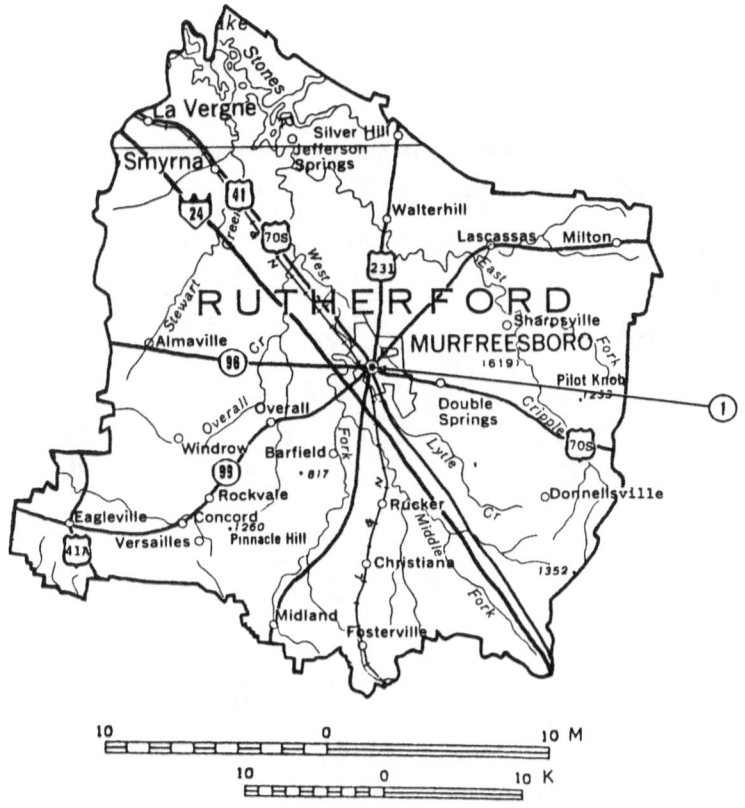

SEVIER COUNTY

1. Harrisburg area: 3.5 miles east; at the Nuns Cove mines.
    Barite. Calcite. Chert, blue. Sphalerite: crystals; faintly yellow. (2)

SHELBY COUNTY

1. Memphis area: vicinity of Richardsons Landing; at gravel dredging operations.
    Agate. Fossils: marine; including agatized corals and sponges. (1-3-10-18-20)

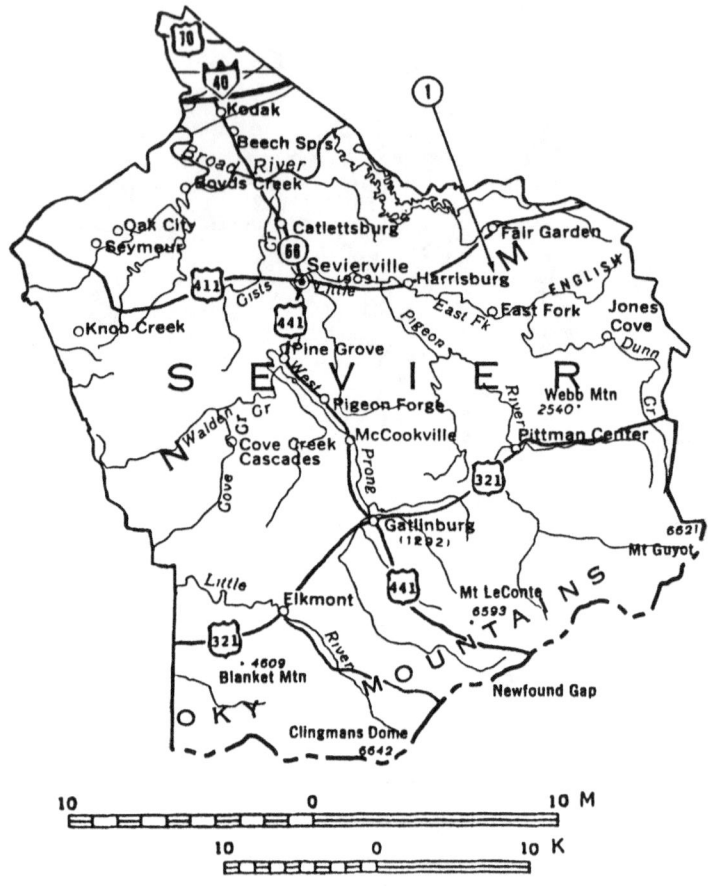

SMITH COUNTY

1. Dixon Springs area: at area mines; continuing northwest into Trousdale County.

    Calcite. Fluorite. Sphalerite. (2)

2. Elmwood area: at the Elmwood Mine.

    Calcite: superb brown-blue, doubly terminated crystals. (2)

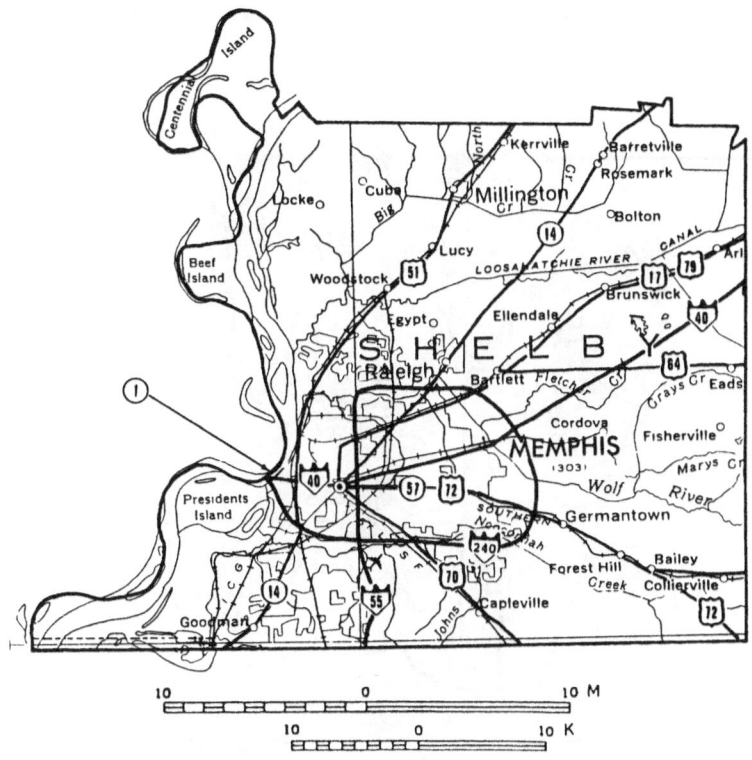

3. Rome area: 1.8 miles west-southwest on US-70N; at the
   Foley Mine.

   Fluorite. (2)

STATEWIDE

In such streams as the Elk, Clinch, Watauga, Duck,
Harpeth, Powell, Cumberland, Tennessee, Stone, Holston,
Calf Killer, Big Pigeon, Tellico, French Broad, Caney, and
Hiawassee and their tributaries, as well as in some lakes.

   Pearl. (7)

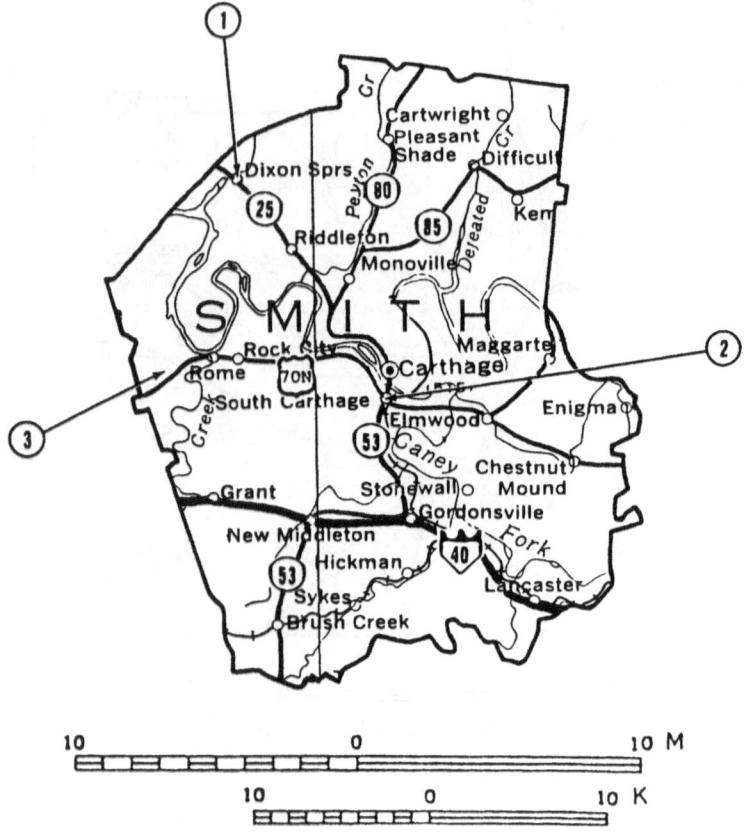

SULLIVAN COUNTY

1. Bristol area: in road cuts on US-421 Byp, just outside city limits.

    Calcite (Onyx): black with snow-white veinings. (4)

2. County-wide in all road, RR, and stream cuts; also in quarries.

    Calcite. Celestite. Dolomite: crystals. Fluorite. Jasper: golden. Pyrite. Quartz, rock. (4-12)

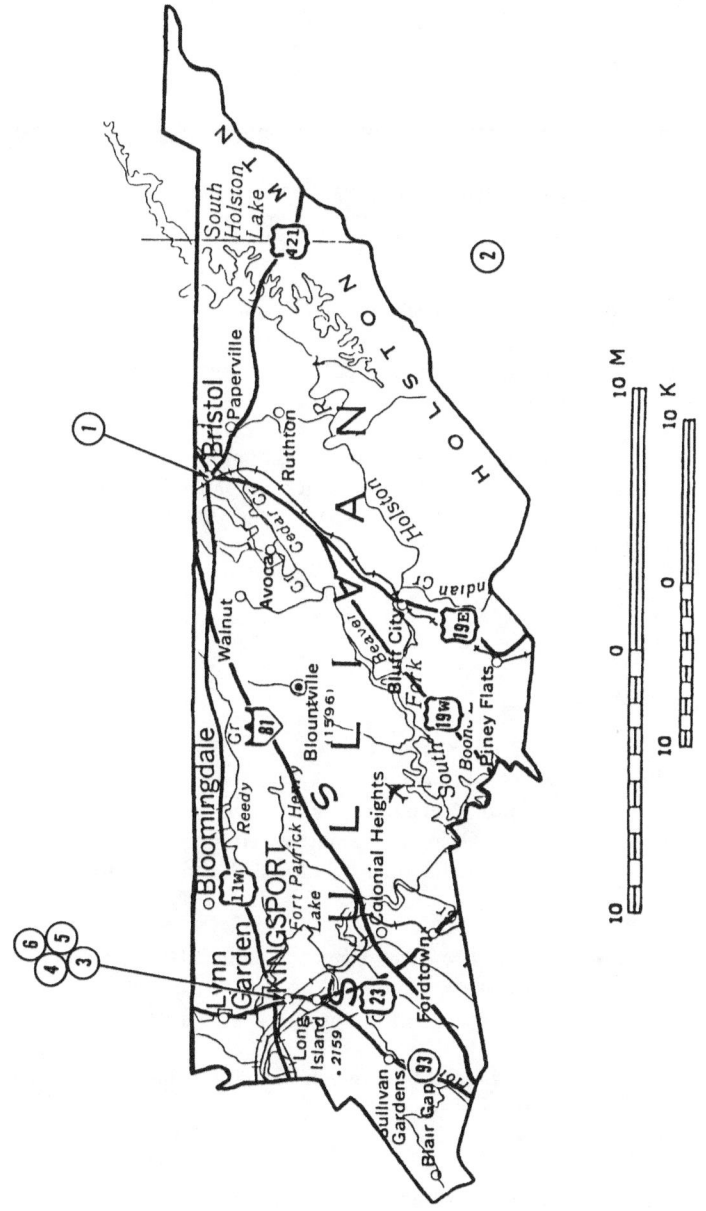

3. Kingsport area: at the Rockway Quarry.

> Calcite. Celestite. Dolomite: superb; pink. Fluorite: crystals; various colors. Pyrite. Quartz, rock. (12)

4. Kingsport area: at barite mines and iron mines, mainly abandoned.

> Barite: brown, green; fine crystals and massive. Quartz, rock; excellent; some crystals doubly terminated. Quartz, smoky. (2)

5. Kingsport area: at the Lambert Quarry.

> Calcite. Celestite. Dolomite: superb, pink. Fluorite: crystals of various colors. Jasper. Pyrite. Quartz, rock. (12)

6. Kingsport area: northeast on Bloomingdale Pike to Arcadia; watch for American Zinc Company sign on left and go 1 mile beyond to next sign on right; at the old Arcadia Mine.

> Sphalerite. (2)

UNICOI COUNTY

1. Bumpus Cove area mines.

> Anglesite. Cerrusite. Chalcedony: blue; to 25 lb. Chalcopyrite. Galena. Hematite. Hemimorphite. Jasper: tan; dark inclusions. Psilomelane. Smithsonite. Sphalerite. (2)

2. Erwin area: in the various Brooks Sand & Gravel Company pits and quarries.

> Chalcedony: good quality. Jasper. (10-12)

3. Rocky Fork area: 1.9 miles northwest; at the old Chandler Barite Mine dump.

> Barite. Epidote. Unakite. (2)

4. Rocky Fork area mines and quarries.

> Barite. Hematite. (2-12)

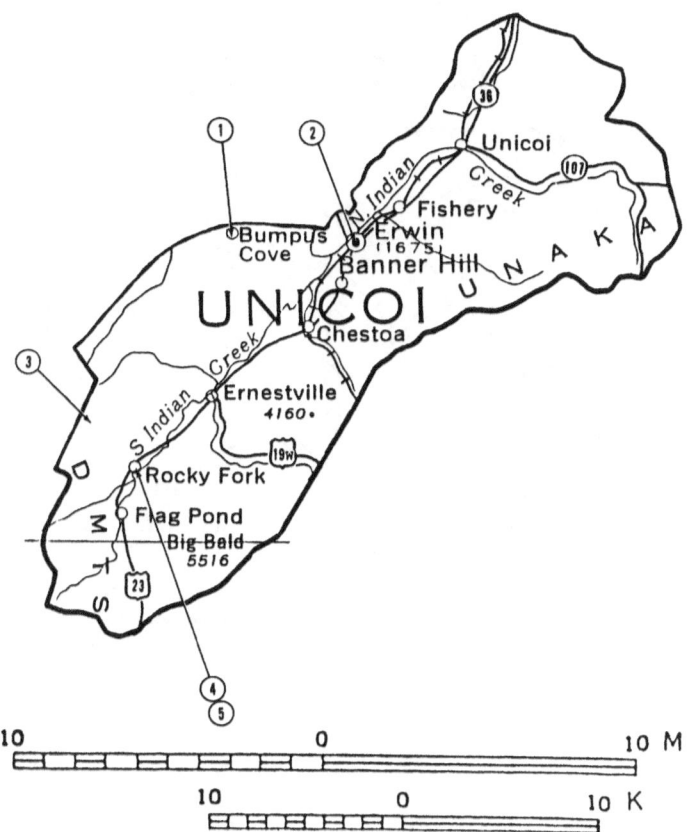

5. Rocky Fork area: in deep road cut on US-23; also upstream
   and down along creek to east.

   Unakite: fine material; green, coral, black. (1-4-13-20-
   30-37)

UNION COUNTY

1. Luttrell area: on west bank of Flat Creek.

   Diamond: 2 stones; one of 1.81 ct; colorless; flawless;
   found during panning; another of 1.69 ct found on
   bank in 1900. (20)

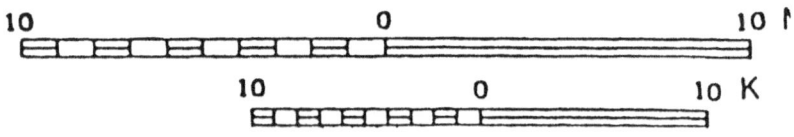

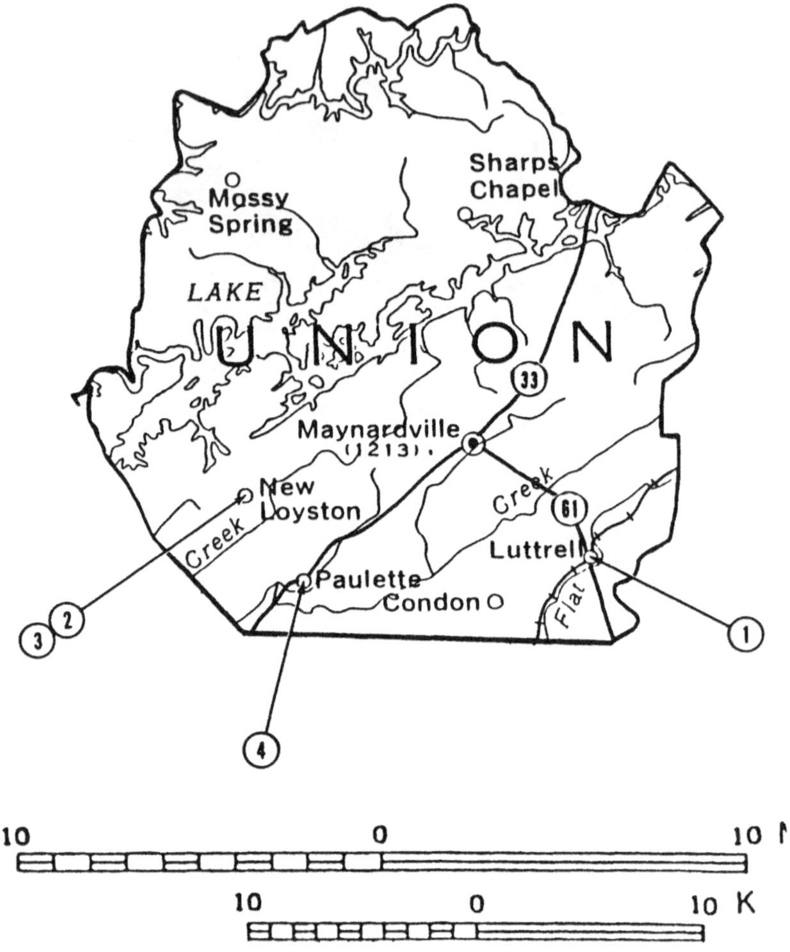

2. New Loyston area mines.
   Calamine. Smithsonite. Sphalerite. (2)

3. New Loyston area: at the Stiner Zinc Mine.
   Calamine. Smithsonite. Sphalerite. (2)

4. Paulette area mines.
   Calamine. Smithsonite. Sphalerite. (2)

WARREN COUNTY

1. Centertown area.

    Jasper: good quality; many colors. (1-13)

2. McMinnville area: south; between Bennett Hollow and McCorkle Hollow; on the west slope of Ben Lomond Mountain.

    Quartz: geodes to 11″ diameter; thin-shelled; warty outer surface; locally called "Cannonballs"; with interior inclusions of white calcite, botryoidal chalcedony, pink dolomite, fluorite in various colors, goethite, limonite, botryoidal linings of opal, pyrite, and rock quartz in glittering drusy linings and sharply terminated crystals; blue celestite in some. (1-7-13-20)

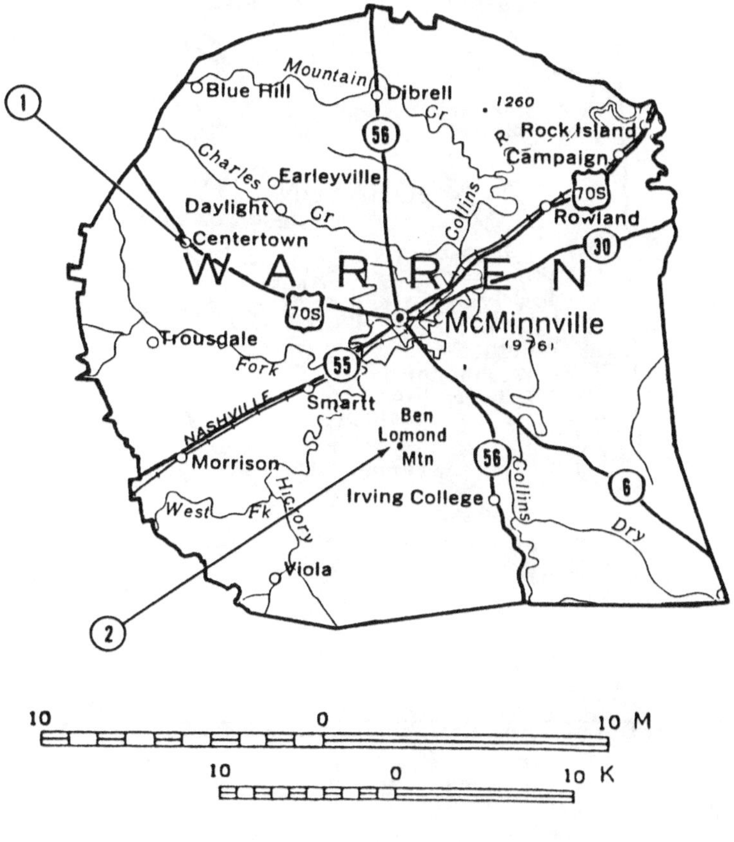

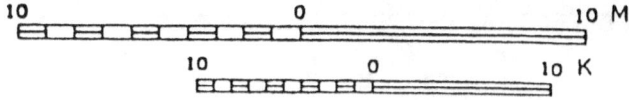

WASHINGTON COUNTY

1. County-wide in all road, RR, and stream cuts; also in quarries.

> Calcite. Celestite. Dolomite: crystals. Fluorite. Jasper: golden. Pyrite. Quartz, rock. (4-12)

2. Embreeville area: southwest; just north of Bumpus Cove (Unicoi County).

> Cerrusite. Hematite. Jasper. Manganite. Pyrolusite. Septaria (skeletal sphalerite). (1-2-13)

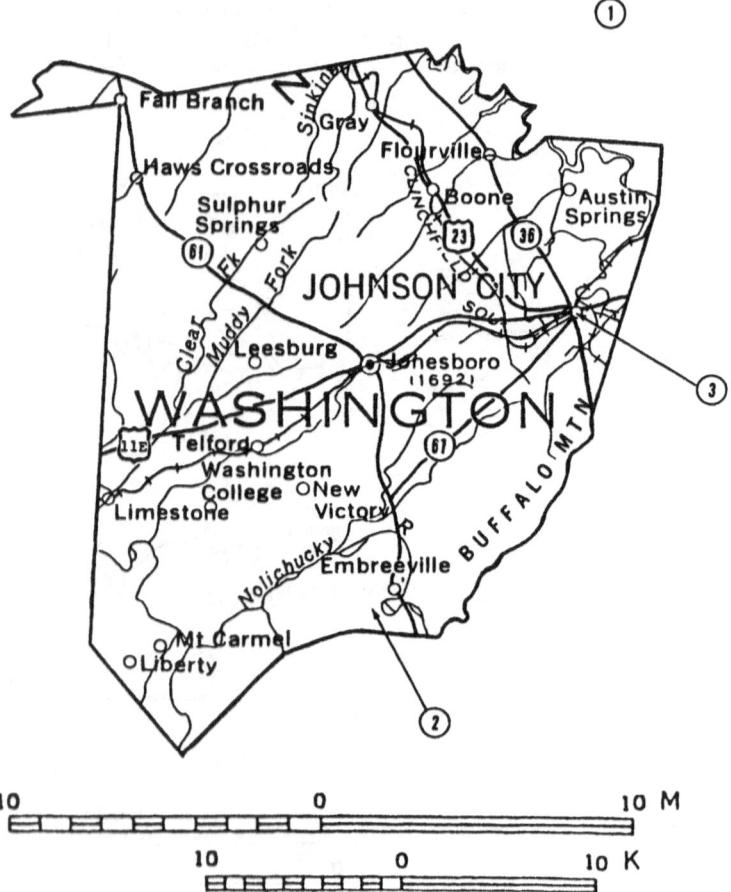

3. Johnson City area: at the Boone Tree Quarry.

> Calcite. Celestite. Dolomite: fine; pink. Fluorite: excellent crystals; various colors. Pyrite. Quartz, rock. (12)

WAYNE COUNTY

1. County-wide.

> Chert (Flint): fine grade; closely grained; often swirl-patterned. Chalcedony. Fossils: marine; many agatized. (1-7-13-20)

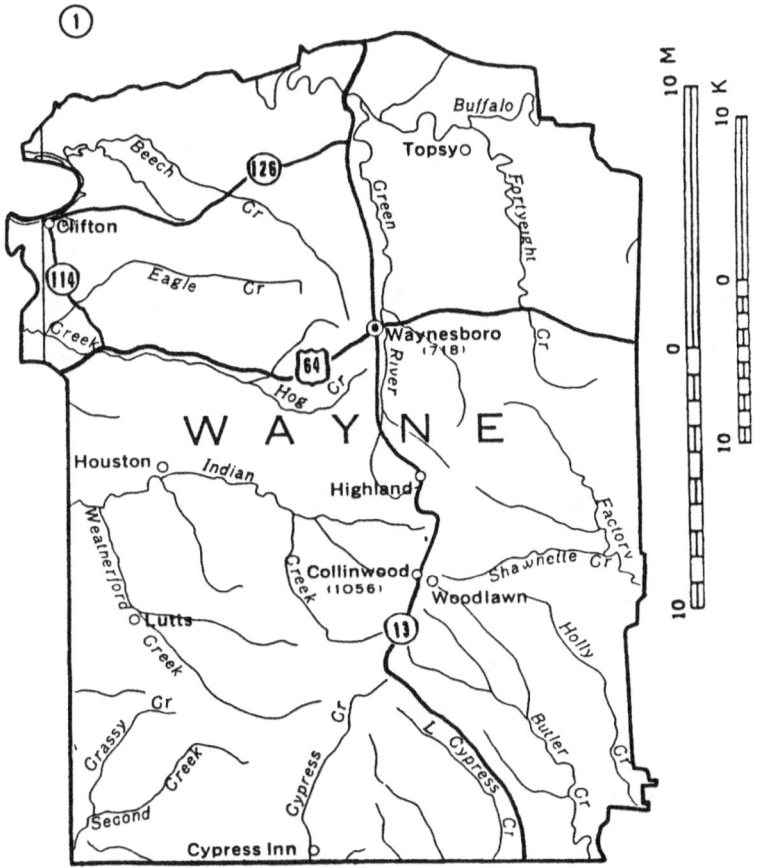

WHITE COUNTY

1. Sparta area: 1 mile southwest on US-70; at the White Company Limestone Quarry.

    Jasper. (12)

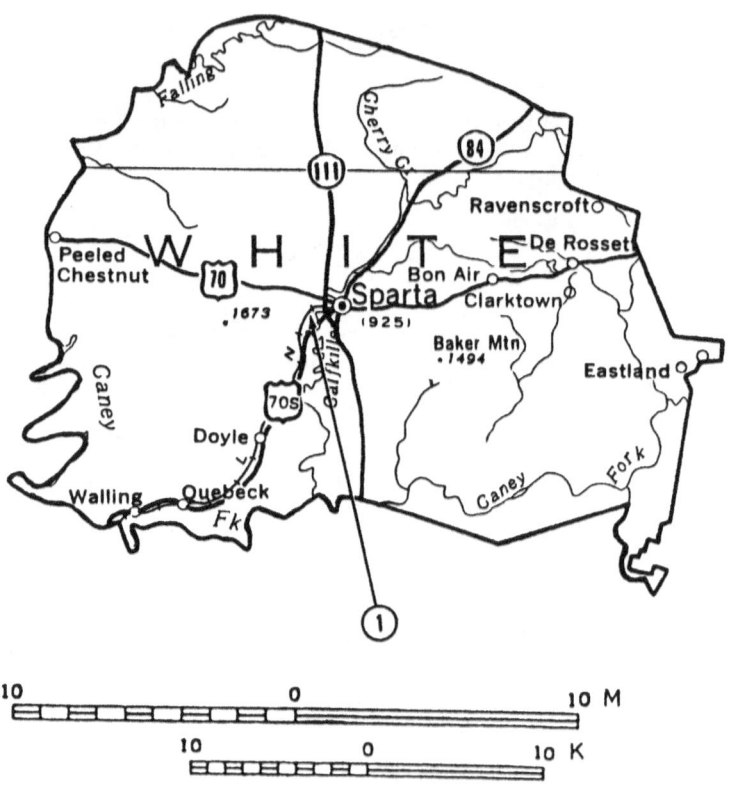

WILLIAMSON COUNTY

1. Nolensville area mines.

    Galena. (2)

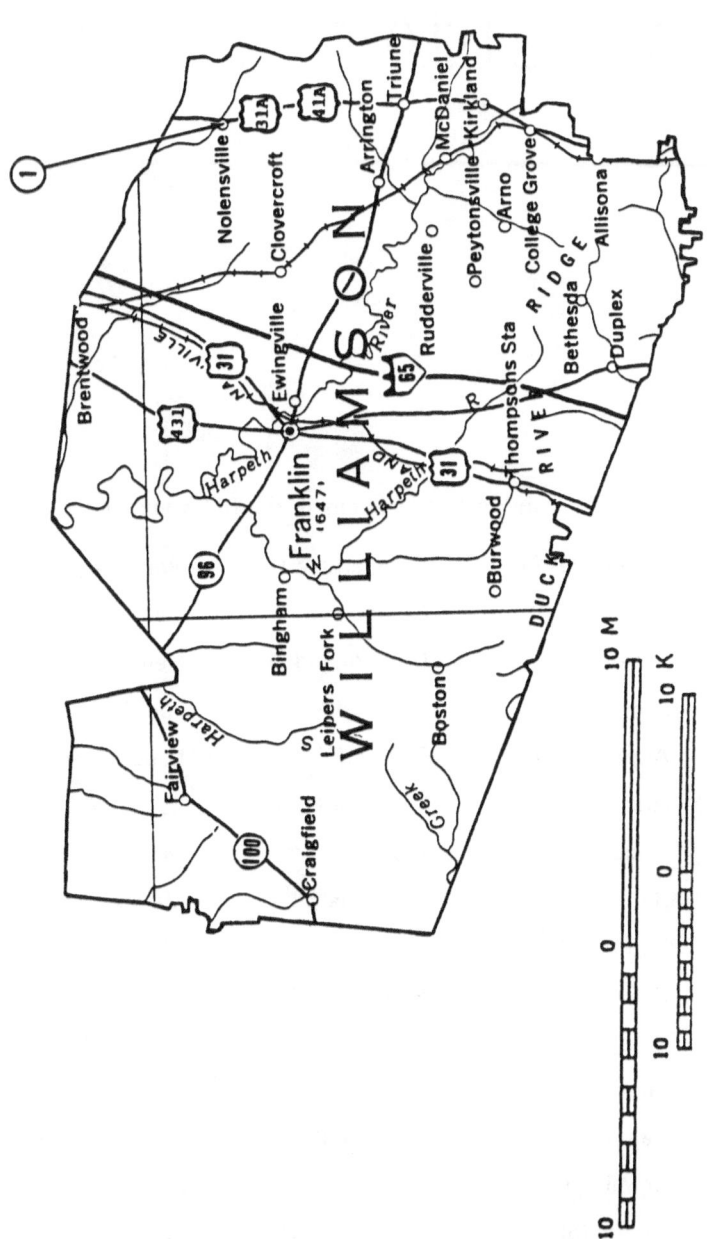

VIRGINIA

Statewide total of 244 locations in the following counties:

Albemarle (9)	Chesterfield (2)	Hanover (4)
Alleghany (1)	Culpeper (2)	Henrico (1)
Amelia (45)	Cumberland (1)	Henry (6)
Amherst (4)	Dinwiddie (1)	Lee (1)
Appomattox (2)	Fairfax (5)	Loudoun (8)
Arlington (1)	Fauquier (1)	Louisa (4)
Augusta (2)	Floyd (4)	Madison (2)
Bath (1)	Fluvanna (2)	Mecklenburg (1)
Bedford (8)	Franklin (6)	Montgomery (2)
Bland (2)	Frederick (1)	Nelson (10)
Botetourt (1)	Giles (1)	Orange (2)
Buckingham (1)	Goochland (1)	Page (7)
Campbell (10)	Grayson (9)	Patrick (5)
Caroline (1)	Greene (2)	Pittsylvania (7)
Carroll (6)	Greensville (1)	Powhatan (3)
Charlotte (8)	Halifax (1)	Prince Edward (4)

Prince George (1)

Prince William (1)

Pulaski (1)

Rockbridge (6)

Rockingham (3)

Russell (1)

Shenandoah (2)

Smyth (5)

Spotsylvania (3)

Stafford (2)

Tazewell (1)

Warren (4)

Washington (1)

Wise (1)

Wythe (2)

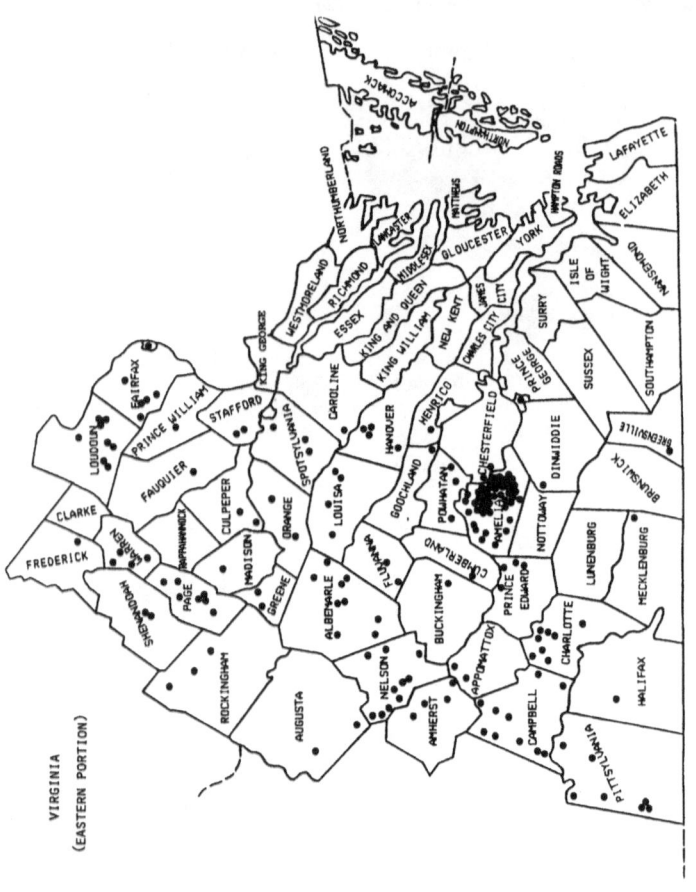

VIRGINIA
(EASTERN PORTION)

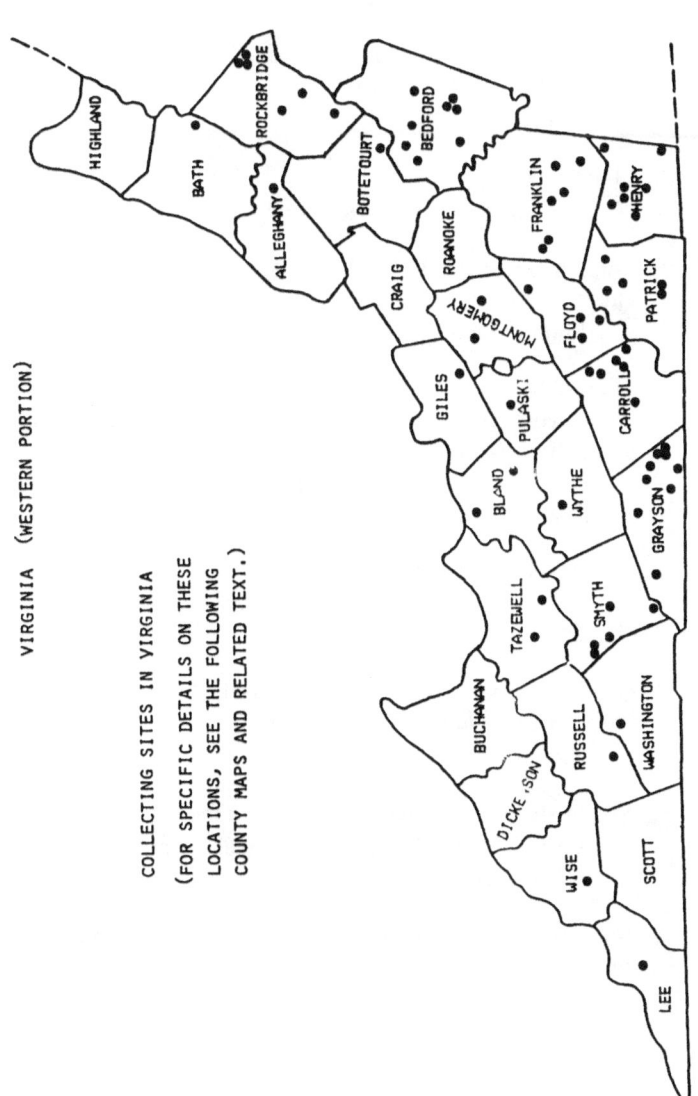

VIRGINIA   (WESTERN PORTION)

COLLECTING SITES IN VIRGINIA

(FOR SPECIFIC DETAILS ON THESE
LOCATIONS, SEE THE FOLLOWING
COUNTY MAPS AND RELATED TEXT.)

ALBEMARLE COUNTY

1. Charlottesville area: on Hydraulic Road; west of US-29; behind the 7-Eleven Store.

     Quartz, rock. (1)

2. Charlottesville area: road cut on I-64; just west of US-29.

     Quartz, rock. (4-5-greenstone)

3. Cobham area: 1 mile northeast; on southwest side of Mechum Creek.

     Limonite. (1-13-29)

4. Covesville area: 3.8 miles south; 1.9 miles northeast of Faber (Nelson County); between Shiloh Mountain and Appleberry Mountain.

     Calcite: fluorescent. Cerrusite: small needles. Galena. Zinc. (1-13-29)

5. Esmont area: at slate quarry.

     Limonite: pseudomorphs after pyrite; cubes to 3″. (12-29)

6. Shadwell area: 1 mile west on SR-250; 100 yards south of road on east slope of hill.

     Barite. (19-quartz)

7. Shadwell area: at the Albemarle Crushed Stone Quarry.

     Calcite. Dolomite. Fluorite: purple, white. Hematite. Quartz, rock. (5-rhyolite -12-16)

8. Stony Point area: along CR-600 for 0.1 mile beginning 0.3 mile from SR-20.

     Limonite. (1-13-29)

9. Yancey Mills area: just south; at Hillsboro Quarry.

     Quartz, blue: asterism. (12)

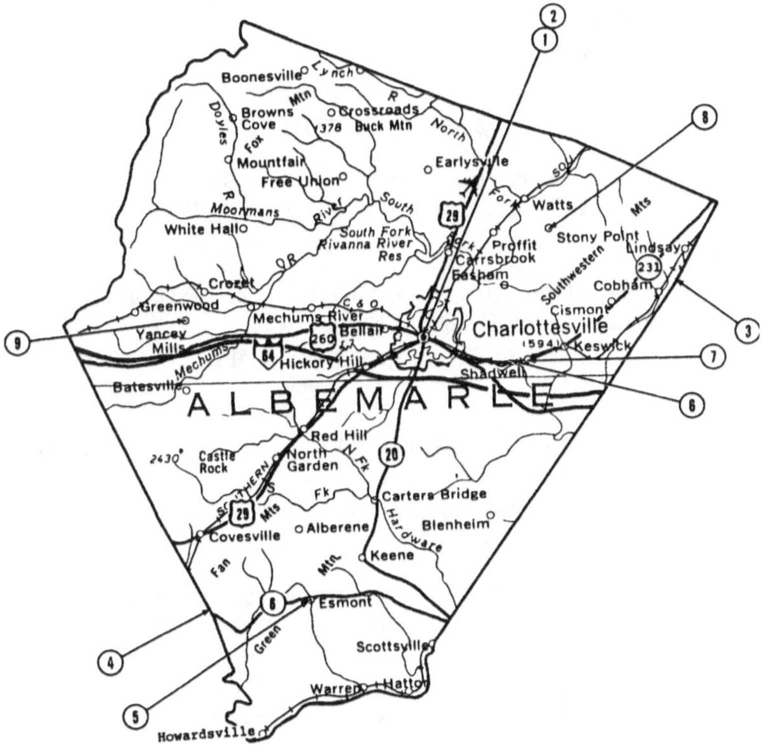

Boonesville

Browns
Cove

Crossroads
Buck Mtn
1378

Mountfair

Earlysville

Free Union

White Hall

South Fork
Rivanna River
Res

Watts

Stony Point

Proffit
Carrsbrook
Pantops

Cismont

Cobham

Crozet

Greenwood

Mechums River

Charlottesville

Keswick

Yancey
Mills

Bellair

Hickory Hill

Shadwell

Batesville

ALBEMARLE

Red Hill

Castle
Rock
2430°

North
Garden

Carters Bridge

Blenheim

Covesville

Alberene

Keene

Esmont

Scottsville

Warren Hatton

Howardsville

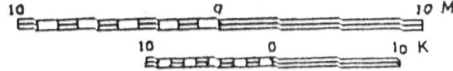

ALLEGHANY COUNTY

1. Clifton Forge area: in Island Ford Cave.
   Hydrozincite. (32)

AMELIA COUNTY

1. Amelia Court House area: 3.5 miles northeast; 0.3 mile
   west of CR-628; at the H. T. Flippen Mica-Beryl Prospect.
   Amethyst. Beryl. Mica. Moonstone. Tantalite. (21)

2. Amelia Court House area: 3.5 miles northeast on US-360;
   then south on CR-628 (CF).
   Amazonite: bright green. Beryl. Mica. Muscovite. (2-
   21)

3. Amelia Court House area: 3.4 miles east-northeast; at the
   Dobbin Prospect.
   Amazonite. Amethyst. Beryl. Mica. Muscovite. Tour-
   maline. (21)

4. Amelia Court House area: 3.4 miles east-northeast on US-
   360; then 0.75 mile southeast on CR-628; at the Morefield
   Mine.
   Amazonite. Beryl. Phenakite. Topaz: clear to blue;
   crystals and formless masses to 500 lb. (1-2)

5. Amelia Court House area: 3.35 miles northeast; at the
   Pinchbeck No. 1 Mine.
   Amethyst. Beryl. Cassiterite. Garnet. Mica. Perthite.
   (2)

6. Amelia Court House area: 3.2 miles southeast; at the Mott-
   ley Mine.
   Albite. Amazonite. Amethyst. Apatite. Beryl. Garnet.
   Mica. Moonstone. Muscovite. Topaz: blue; massive.
   Tourmaline. (2)

7. Amelia Court House area: 3.1 miles north-northeast; at the
   Maria Mine (aka the Old Pinchbeck Mine).
   Amethyst. Beryl: blue; gem quality. Cassiterite. Mica.
   Moonstone. Muscovite. (2)

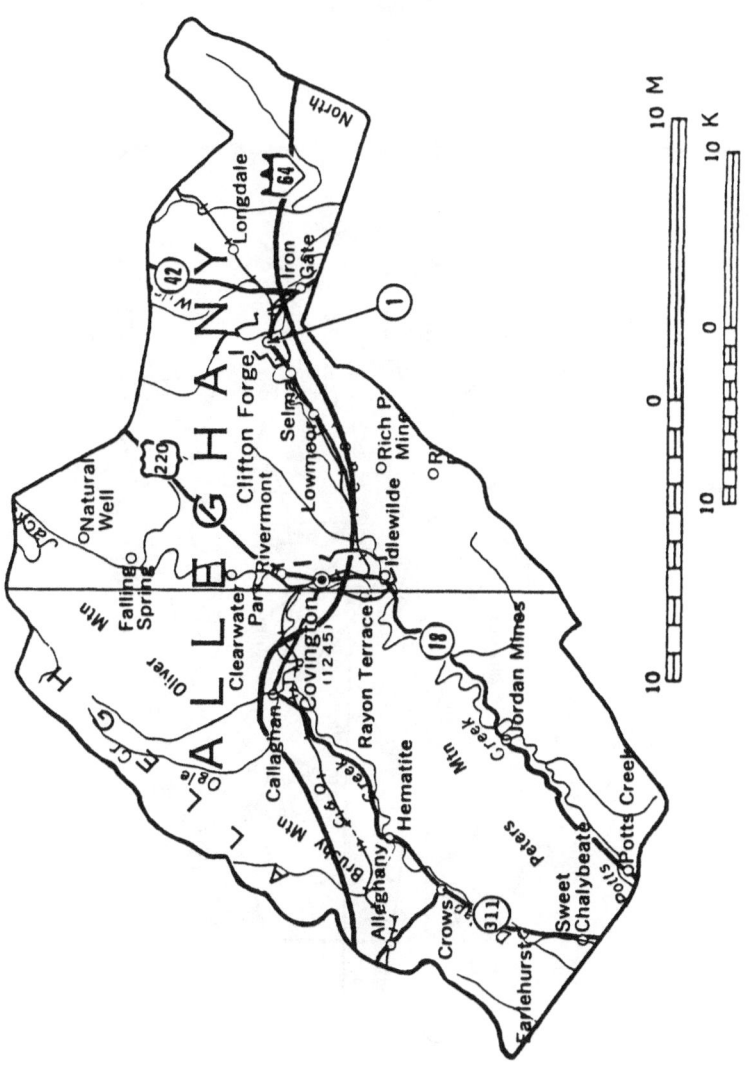

AMELIA
Amelia C.H.
(1366)
Scotts Fork
Mannboro
Ammon
Earls
Denaro
Jetersville
Deatonville
Rodophil
Painesville
Morven
Ladore
Truxillo
Winterham
Chula
Creek
Creek
West
Nibbs
SOUTHERN
Flat
Cr
Cr
10 0 10 M
10 0 10 K

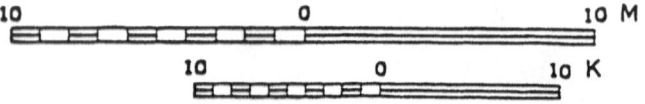

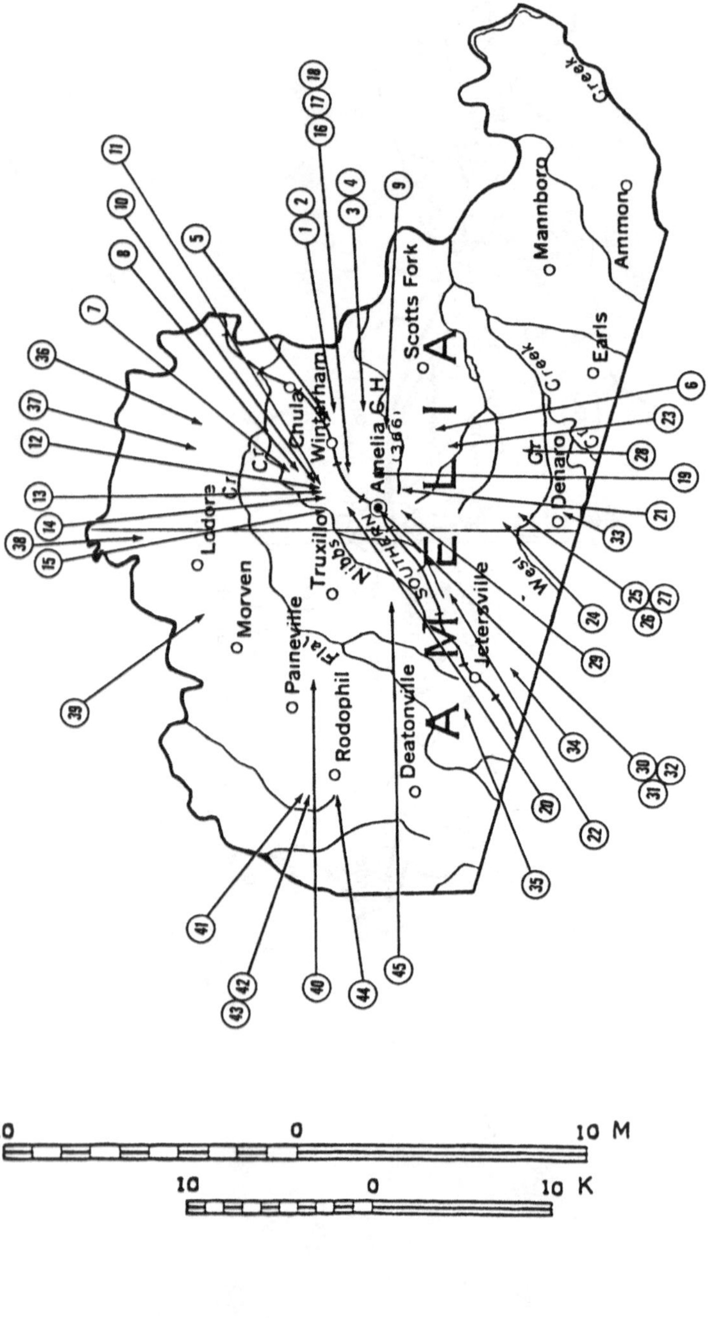

8. Amelia Court House area: 2.8 miles north-northeast; at the Nettie Taylor Mine.

> Amazonite. Beryl. Mica. Muscovite. Quartz, rock. Tourmaline: green. (2)

9. Amelia Court House area: 2.75 miles east; west side of CR-627; at the Vaughan Beryl Prospect.

> Beryl. (2)

10. Amelia Court House area: 2.3 miles north-northeast; at the International Mine.

> Amazonite. Amethyst. Beryl. Garnet. Mica. Quartz, rock. (2)

11. Amelia Court House area: 2.1 miles north-northeast; at the Patterson Mine.

> Amethyst. Apatite: gem quality. Beryl: some gem quality. Cassiterite. Mica. Muscovite. Perthite. Tantalite. (2)

12. Amelia Court House area: 2 miles north-northeast; at the Captain Tim Prospect.

> Amazonite. Amethyst. Apatite. Beryl. Cassiterite. Garnet. Mica. Moonstone. Muscovite. Perthite. Plagioclase. Quartz, rock. Tantalite. Tourmaline. (21)

13. Amelia Court House area: 1.8 miles north-northeast; at the O'Neil Prospect.

> Amazonite. Amethyst: gem quality. Apatite: gem quality. Beryl. Garnet. Topaz: blue; massive. (21)

14. Amelia Court House area: 1.75 miles north-northeast; at the Marshall Mine.

> Mica. Tourmaline. (2)

15. Amelia Court House area: 1.75 miles north on SR-49; east of road near Nibbs Creek; at the Trueheart Prospect.

> Aquamarine: tiny crystals. Moonstone. Tourmaline: black. (21)

16. Amelia Court House area: 1.55 miles northeast; at the Winston Mine.

> Beryl. Mica. Muscovite. (2)

17. Amelia Court House area: 1.5 miles northeast; at the Penn Prospect.

> Albite. Amazonite. Amethyst. Garnet. Mica. Muscovite. Tantalite. (21)

18. Amelia Court House area: 1.4 miles northeast; along the Richmond & Danville Line of the Southern RR; in remains of pit 15' deep, 35' diameter; at the Richeson Mica Mine.

> Amazonite. Amethyst. Beryl. Mica. (21)

19. Amelia Court House area: 1.2 miles east-northeast; at the Berry Mine.

> Beryl. Cassiterite. Garnet. Mica. Muscovite. Quartz, rock. Tourmaline. (2)

20. Amelia Court House area: 1 mile north; at the Rutherford Mines.

> Albite. Amazonite. Amethyst. Beryl. Cassiterite. Garnet. Mica. Moonstone. Muscovite. Perthite. Plagioclase. Quartz, rock. Tantalite. Tourmaline. (2)

21. Amelia Court House area: 1 mile southeast; at the Abner Pinchbeck Mine.

> Beryl: clear; gem quality. Mica. Muscovite. Perthite. Plagioclase. Quartz, rock. Tourmaline: parallel crystal aggregates. (2)

22. Amelia Court House area: 0.75 mile south on SR-38 to CR-614; right on CR-614; 2 miles south on CR-614 to CR-603; right on CR-603 and continue 2.5 miles to Golden Prospect on the north side of the road.

> Amazonite. Beryl. Mica. (21)

23. Amelia Court House area: 0.75 mile south on SR-38 to CR-614; right on CR-614; 2 miles south on CR-614 to CR-603; left on CR-603 and continue 2.8 miles to end of road; walk 0.3 mile southwest to the D. E. Kraft Prospect.

> Apatite. Beryl. Mica. Tourmaline. (21)

24. Amelia Court House area: 0.75 mile south on SR-38 to CR-614; right on CR-614; 3.7 miles south on CR-614;

walk 0.3 mile straight right off end of road to the Wyatt Vaughan Prospect.

Amazonite. Amethyst. Beryl. Garnet. Mica. Tourmaline. (21)

25. Amelia Court House area: 0.75 mile south on SR-38 to CR-614; right on CR-614; 4.7 miles south on CR-614 to CR-608 on left; directly opposite this junction on right of CR-614 is the Ponton Mine.

Amazonite. Beryl. Mica. Perthite. Tantalite. (2)

26. Amelia Court House area: 0.75 mile south on SR-38 to CR-614; right on CR-614; 4.7 miles south on CR-614 to CR-608; left on CR-608 for 0.5 mile east; on north side of road; at the Harrison Venable Prospect.

Amethyst. Beryl. Mica. Quartz, rock. (21)

27. Amelia Court House area: 0.75 mile south on SR-38 to CR-614; right on CR-614; 4.7 miles south on CR-614 to CR-608; turn left on CR-608 and go 0.5 mile; on south side of the road; at the Mays Mine.

Amazonite. Amethyst. Apatite. Beryl. Mica. Tourmaline. (2)

28. Amelia Court House area: 0.75 mile south on SR-38 to CR-614; right on CR-614; 4.7 miles south on CR-614 to CR-608; turn left and go 1.8 miles east on CR-608 to CR-653; turn right and go 0.55 mile on CR-653; walk 0.55 mile west to the Anderson Prospect.

Amazonite. Amethyst. Apatite. Beryl. Mica. Tourmaline. (21)

29. Amelia Court House area: 0.65 mile south; at the Wingo Mine.

Amazonite. Amethyst. Beryl. Mica. (2)

30. Amelia Court House area: just west on SR-360 to RR cut.

Amazonite. Apatite. Beryl. Manganocalcite. Mica. Muscovite. (4-29)

31. Amelia Court House area: on CR-609; at the Champion
    Mine (aka Bland Mine and Jefferson No. 4 Mine).
    Amethyst. Apatite. Beryl. Tourmaline. (2)

32. Amelia Court House area: southeast on CR-638; on the
    Duncan property.
    Amethyst. (1-13-29)

33. Denaro area: 0.4 mile east; at the Munden Prospect.
    Amazonite. Amethyst. Apatite. Beryl. Garnet. Mica.
    Moonstone. Muscovite. Topaz. Tourmaline. (21)

34. Jetersville area: 1.2 miles south-southeast on CR-640; walk
    0.7 mile south of road to the Houston Prospect.
    Amethyst. Beryl. Mica. (21)

35. Jetersville area: 1 mile west-northwest on SR-307; on north
    side of road; at the Schlegal Mine (aka Norfleet Mine).
    Beryl: dark green to greenish blue; crystals to 10″
    diameter. Perthite: white to salmon pink. (2)

36. Lodore area: 4.7 miles east on CR-635 to 0.3 mile east of
    the CR-651 junction; walk 0.25 mile south of road to the
    James D. Laurence Prospect.
    Amethyst. Beryl. Mica. (21)

37. Lodore area: 3.9 miles east on CR-635; walk 0.7 mile north
    of road to the Ligon Mines.
    Beryl: green; crystals to 3″ diameter. Cassiterite. Gar-
    net. Manganocalcite. Mica. Microlite. Perthite: white
    to salmon pink. Plagioclase. Quartz, star. (2-5-massive
    milky quartz)

38. Lodore area: 0.6 mile east on CR-635 to CR-636; 1.3 miles
    north on CR-636; walk 0.4 mile east of road to the Tinsley
    Prospect.
    Amethyst. Mica. (21)

39. Morven area: 1.5 miles northeast on CR-149; walk 0.4 mile
    southeast of road to the old Lee Goodman Prospect.
    Amazonite. Amethyst. Beryl. Mica. Muscovite. (21)

40. Paineville area: 1.2 miles southeast on CR-644; on right side of road; at the C. G. Wingo Prospect.

    Amazonite. Garnet. Mica. Muscovite. Tantalite. (21)

41. Rodophil area: 1.2 miles northwest on CR-620; walk 0.6 mile northeast to the Lambert Prospect.

    Amazonite. Amethyst. Mica. Quartz, rock. (21)

42. Rodophil area: 1 mile northwest on CR-620; walk 0.6 mile east of the road to the Foster Prospect.

    Amethyst. Beryl. Mica. (21)

43. Rodophil area: 1 mile northwest on CR-620; walk 0.3 mile east of road to the Ford Prospect.

    Amethyst. Mica. Quartz, rock. (21)

44. Rodophil area: 0.6 mile west; at the James Anderson Prospect.

    Beryl: dark green to greenish blue; crystals to 10″ diameter. Perthite: salmon pink. (21)

45. Truxillo area: 2.1 miles south on CR-639; walk 0.8 mile north of road to the Keystone Mine.

    Amazonite. Apatite. Beryl. Cassiterite. Garnet. Mica. (2)

AMHERST COUNTY

1. Amherst area: 0.3 mile east on CR-604 to CR-659; then south on CR-659 to the Buffalo Ridge area; at crossroad, follow signs to the Schaar farm (CF).

    Amethyst: singly terminated crystals; occasional clusters. (1-13-16-29)

2. Henley's Store area: 2.2 miles north on Fancy Hill.

    Amethyst: gem quality; pale purple to deep purple. (1)

3. Riverville area: 1.75 miles northwest; at the Old Copper Mine.

    Ilmenite. (2)

4. Sweet Briar area: 3.4 miles east-southeast; at an abandoned copper prospect.

Penninite: plates with good cleavage; to 3″ wide. (21)

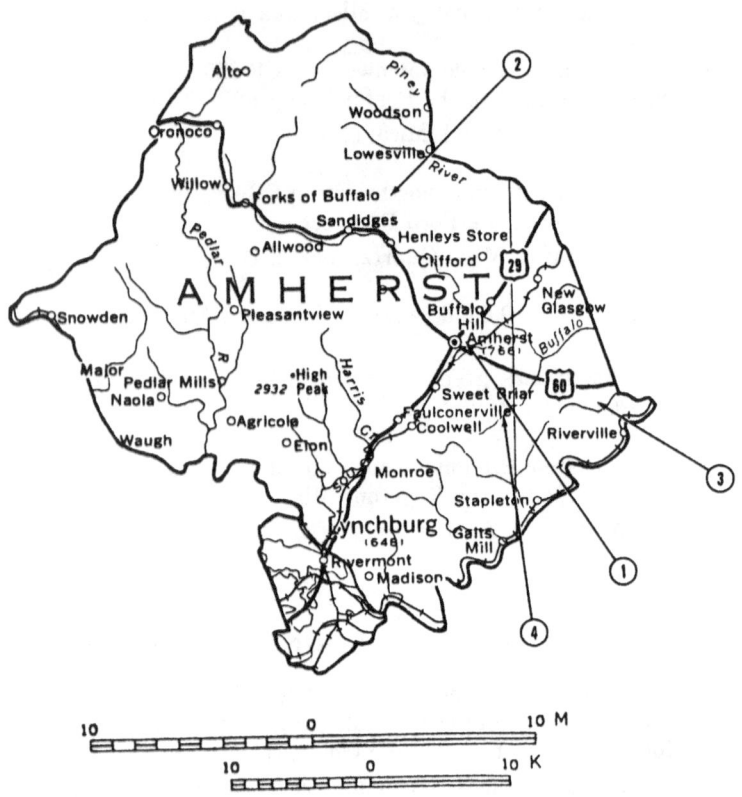

APPOMATTOX COUNTY

1. Bent Creek area: south; at the Enterprise Mine.
   Cryptomelane. (2)

2. Oakville area: 3.4 miles north-northwest; along Wreck Island Creek 2 miles upstream from its mouth at the James River; in the State Limestone Quarry.
   Bornite. Chalcopyrite. Sphalerite. (12-limestone -19-marble)

ARLINGTON COUNTY

1. Arlington area: in the Kirkwood Run valley; in the vicinity
   of I-66, SR-237 (Washington Boulevard) and SR-120 (Glebe
   Rd.).
   Jasper. (3-20)

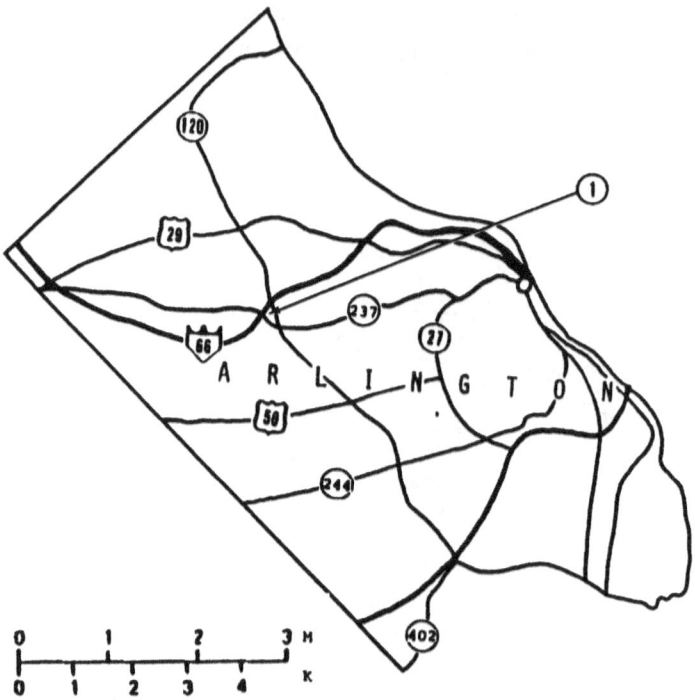

AUGUSTA COUNTY

1. Craigsville area.
   Calcite. (1)

2. Steeles Tavern area: 1.6 miles east; at the Vesuvius Mine.
   Manganite: good crystals. (2-5-massive cryptomelane
   -9)

BATH COUNTY

1. Yost area: on Chestnut Ridge; follow CR-640 to the third
   farm east of CR-629.

   Quartz, rock. (1-13-29)

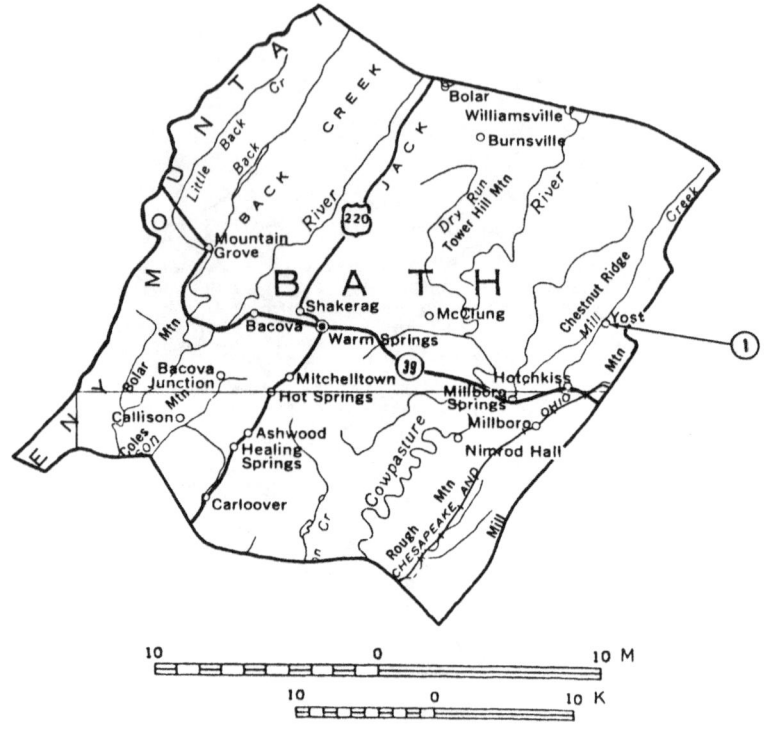

BEDFORD COUNTY

1. Bedford area: 2.9 miles north-northwest; 0.4 mile north-
   northwest of Peaksville.

   Fluorite: large masses. (2)

2. Irving area: 1.3 miles north; at the Saxon-McMillan Mines.

   Barite. (2)

3. Lowry area: 2.6 miles south-southwest; west bank of Little
   Otter River; in quarry.

   Garnet. (12)

4. Moneta area: 1 mile southeast; at the Young Mine.

   Amazonite. Garnet. (2)

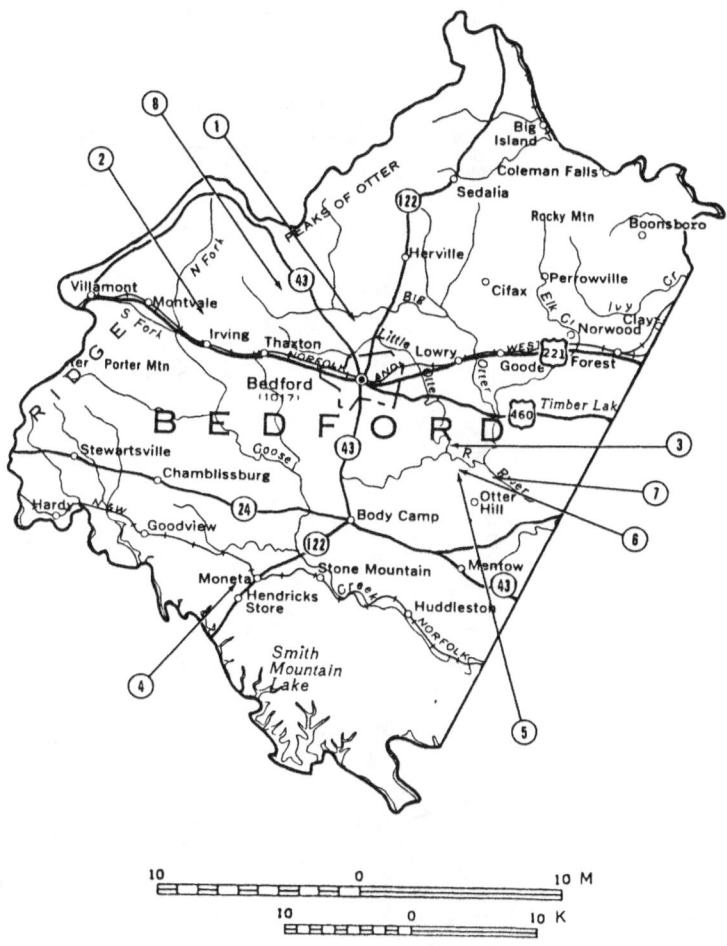

5. Otter Hill area: 1.6 miles north-northwest of Otter Hill; 350 yards southwest of the Little Otter River; on the John Mitchell property; at the Hottinger Mine.

   Biotite. Garnet. Muscovite. Perthite: white to pale green. Plagioclase. Pyrite. Quartz, milky. Quartz, smoky. (2)

6. Otter Hill area: 1.7 miles northwest; on the John Mitchell property; at the Mitchell Mine and adjacent areas.

Allanite: fine; tiny crystals. Apatite. Moonstone. Thulite. (1-2-13-24)

7. Otter Hill area: 1.3 miles northeast; 0.4 mile north of CR-714; on west side of the Otter River; in small quarry.

Garnet: crystals to 3″ diameter. (12)

8. Thaxton area: 3.3 miles north-northeast; at the Peaksville Mine.

Apatite. Quartz, rock. (2)

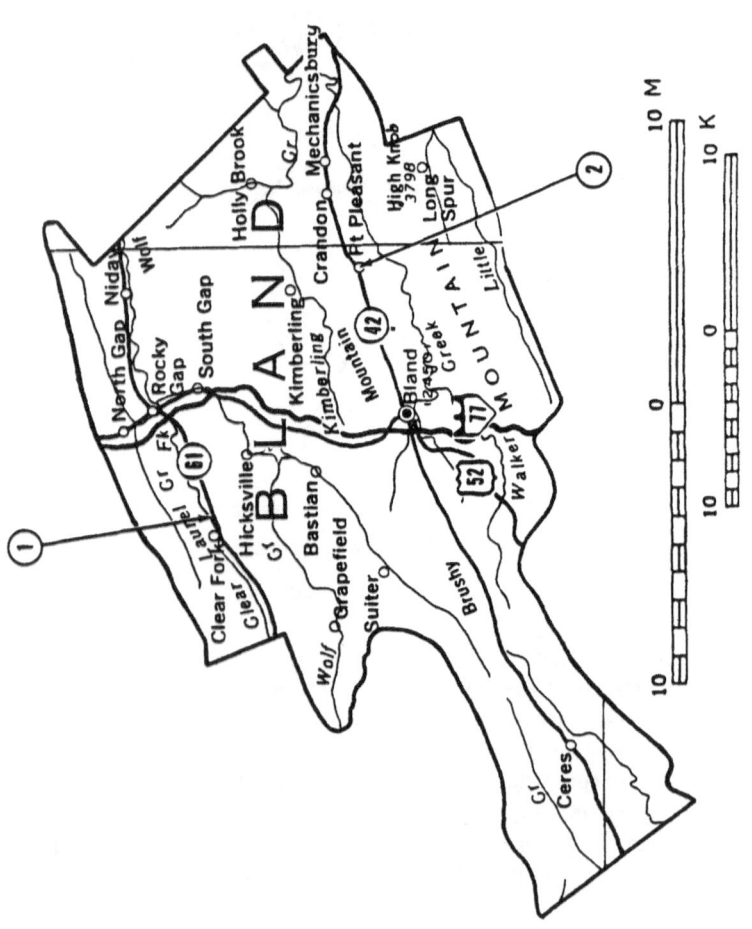

BLAND COUNTY

1. Clear Fork area: northeast; on Clear Fork.

    Manganite: fine crystals. (21)

2. Point Pleasant area: on the north side of Big Walker Mountain.

    Agate. (1-13)

BOTETOURT COUNTY

1. Buchanan area: 1 mile east; at the James River Hydrate & Supply Company Quarry.

    Calcite: good arborescent crystal clusters. (9-12)

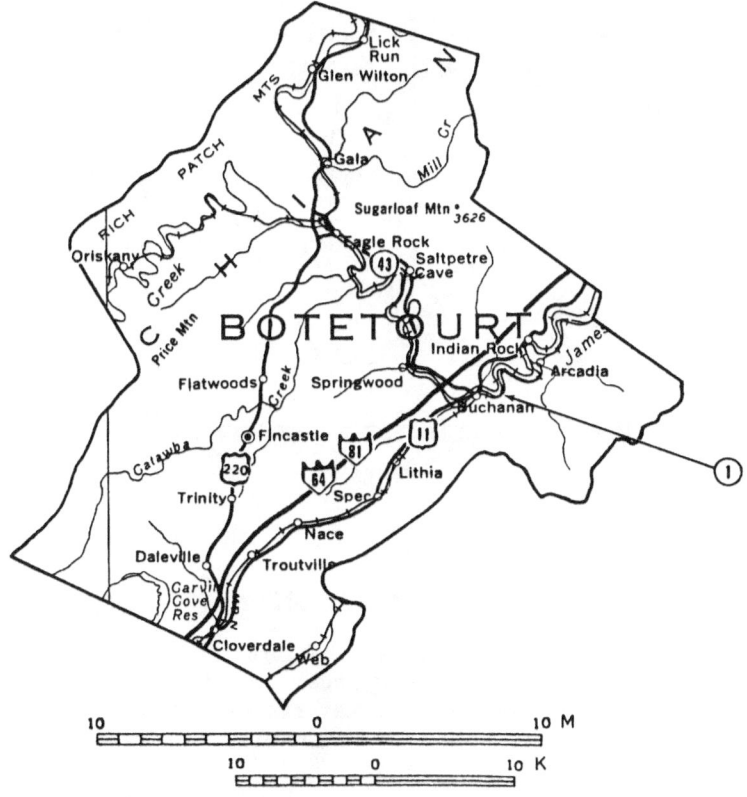

BUCKINGHAM COUNTY

1. Dillwyn area: 1.7 miles northeast; near Tower Hill.

   Actinolite. Anthophyllite. Garnet: red; abundant; crystals to 0.75" diameter. Kyanite. Sillimanite. Staurolite. (1-5-13-29)

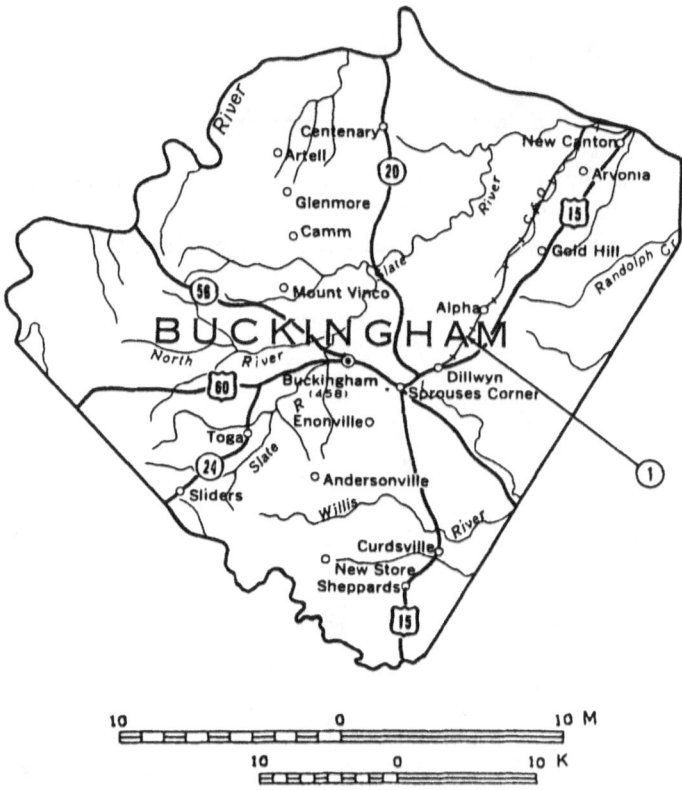

CAMPBELL COUNTY

1. Bocock area; 2.4 miles north to left turn on US-460; go just west on US-460 to where diggings are visible on side of hill between the road and Little Beaver Creek; at the Piedmont Mine. (aka Myers Mine and Lerner Mine).

   Cryptomelane. (2)

2. Brookneal area: 0.3 mile northeast; on the Lacey Rush property.

   Amethyst: fair; crystals to 4″ diameter and 10″ long. (1-13-29)

3. Evington area: 3 miles northeast; south of the west bend of Flat Creek; at the Mortimer Mine.

   Cryptomelane. (2)

4. Evington area: 2.5 miles northeast; at the Pribble Mine.

   Psilomelane. (2)

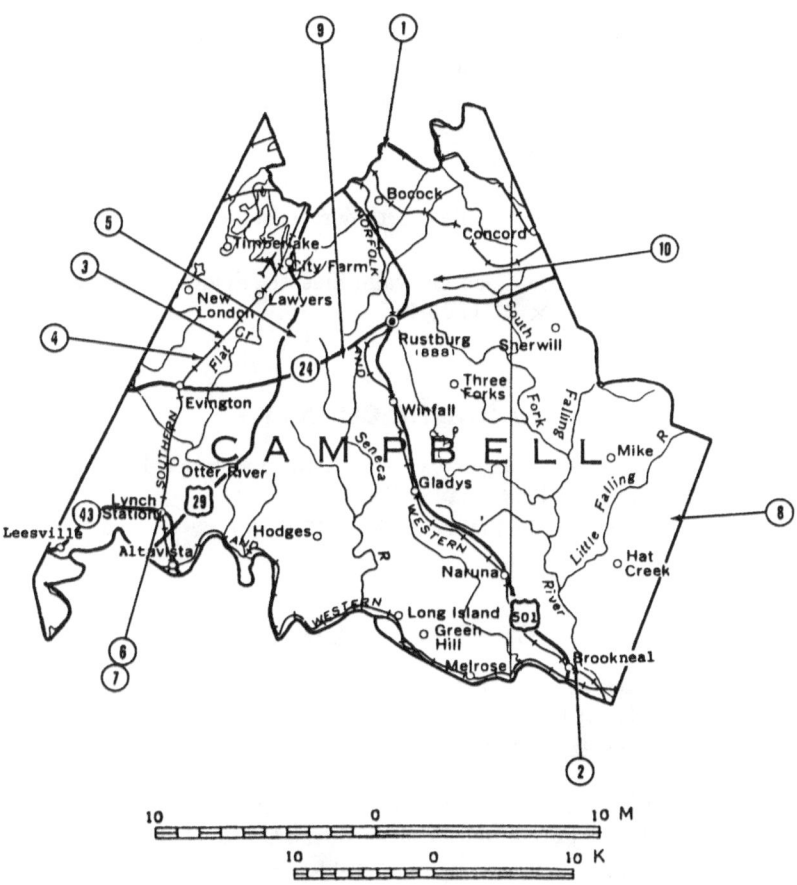

5. Lawyers area: 2.1 miles southeast; 1 mile west of US-29; at the Bell Mine.

    Cryptomelane. (2)

6. Lynch Station area: along small incline in copper prospect.

    Turquoise: crystals; only known site in the world for crystallized turquoise; occurs in drusy masses. (21-38)

7. Lynch Station area: just east of the old Bishop Mine.

    Cryptomelane. (21)

8. Mike area: 3.9 miles southeast; 3 miles west-southwest of Red Horse; on the Clay farm.

    Amethyst: gem quality; pale purple to deep purple; crystals to 4″ diameter and 10″ long. (1-13-29)

9. Rustburg area: 2.7 miles southwest.

    Kyanite. (1-13)

10. Rustburg area: 2.1 miles north-northeast; on the T. Graves property.

    Amazonite: good cleavage masses. Epidote: crystals to 0.75″ × 2″. (1-13-16)

CAROLINE COUNTY

1. Cedar Fork area: just south; on the Lawrence Beazley farm; at the Last Chance Mine.

    Beryl: individual crystals to 15 lb. (2)

CARROLL COUNTY

1. Dugspur area: 3 miles east; almost on the border of Floyd County; at the A. G. Vaughn farm.

    Quartz, rock. (1-13-29)

2. Dugspur area: 2.3 miles southeast; 1.9 mile west of the border of Floyd County; near Burks Creek.

    Copper, native. Epidote. Feldspar. (1-13-19-quartz)

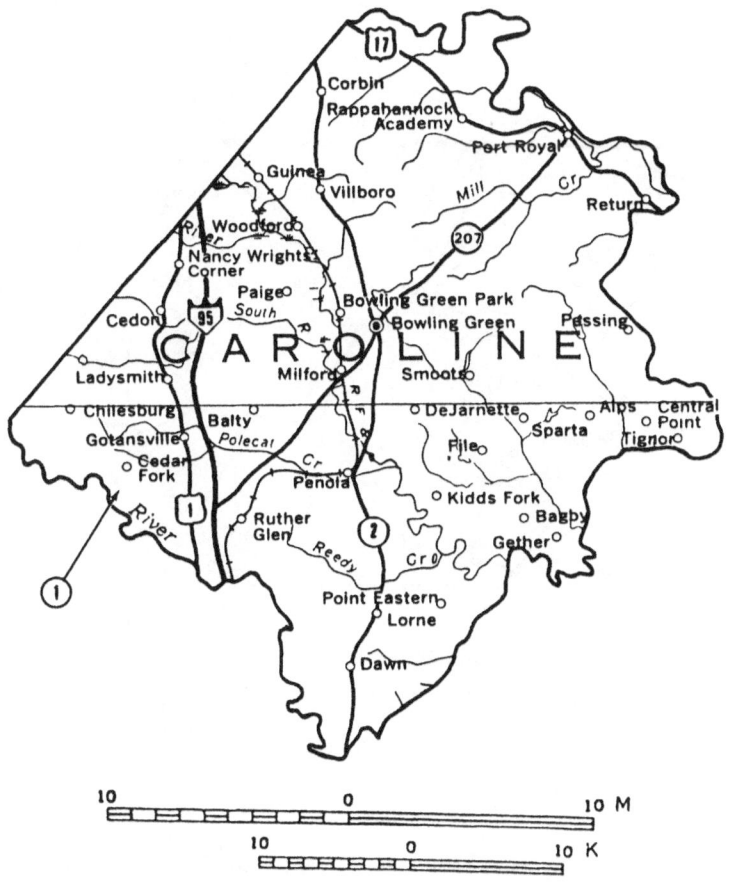

3. Hillsville area: 2.5 miles southeast; on the Early property.
   Copper, native. Epidote. Feldspar. (34)

4. Laurel Fork area: 3.75 miles east; 1 mile south; almost on
   the border of Floyd County; at the Guy Barnard farm.
   Quartz, rock. (1-13-29)

5. Laurel Fork area: 1 mile northwest; at the Marvin Marshall farm.
   Quartz, rock. (1-13-29)

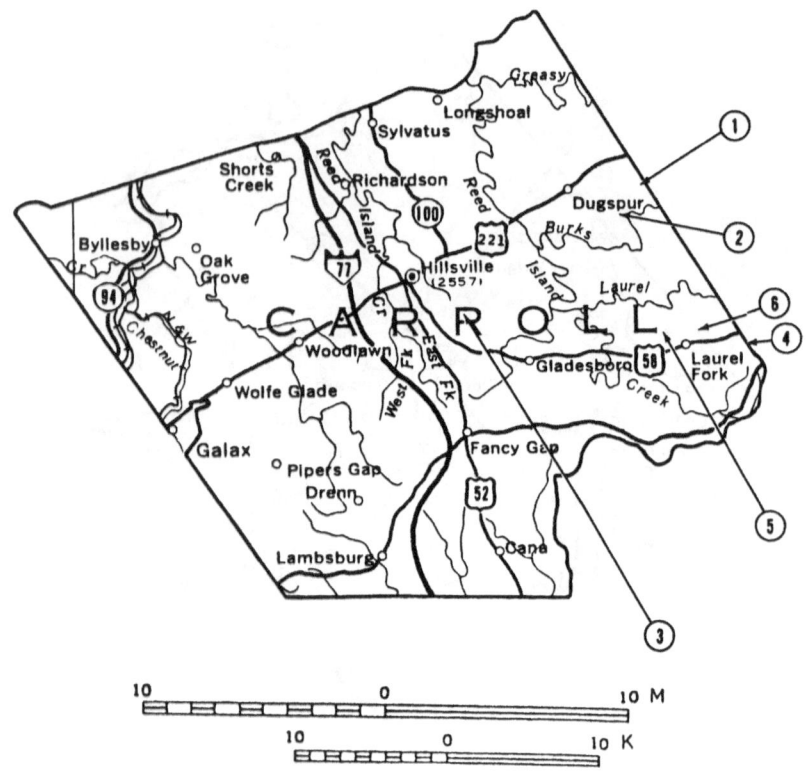

6. Laurel Fork area: 0.6 mile northeast; at the Marshall farm.
   Quartz, rock. (1-13-29)

CHARLOTTE COUNTY

1. Charlotte Court House area: 2.5 miles west on SR-40 to the
   Vassar property; then south on trail.
   Amethyst. (1-13-29)

2. Charlotte Court House area: broad area centered 2.25
   miles west-northwest; at the A. W. Donald plantation.
   Amethyst: fair to good. (1-13-29)

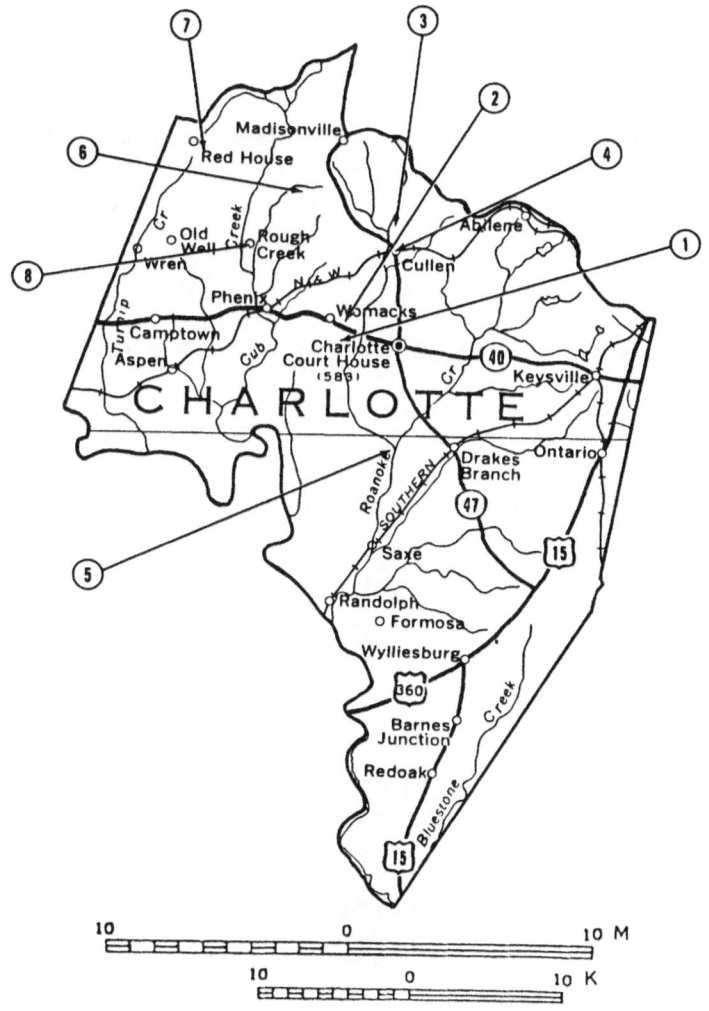

3. Cullen area: 1 mile north at the Crews No. 1 Mine.
   Beryl. Garnet. Quartz, rock. (2)

4. Cullen area: to the south; in the Mica Mines.
   Beryl. Garnet. Quartz, rock. (2)

5. Drakes Branch area: 2.3 miles west; broad area centered near fork of Roanoke Creek; on the Wingo property.
   Amethyst: gem quality. (1-39)

6. Madisonville area: 2.8 miles southwest; 2.5 miles west of the junction of SR-47 and SR-615.
   Monzanite. (1-5-granite -29)

7. Red House area: just southeast.
   Kyanite. (1-13-29)

8. Rough Creek area: 1.3 miles northwest; on the northeast side of SR-26.
   Monzanite. (1-5-granite -13-29)

CHESTERFIELD COUNTY

1. Itterdale area.
   Petrified wood: some opalized. (1-4)

2. Skinquarter area.
   Petrified wood: some opalized. (1-4)

CULPEPER COUNTY

1. Mitchells area mines.
   Azurite. Chalcocite. Chalcopyrite. Copper, native. Cuprite. Malachite. Pyrite. (2)

2. Rapidan area: at the Culpeper Mine.
   Gold. (2)

CUMBERLAND COUNTY

1. Guinea Mills area: 9.2 miles south on SR-45; on Prince Edward County border, just north of city limit of Farmville; on the east side of the road.
   Monzanite. (1-29-34)

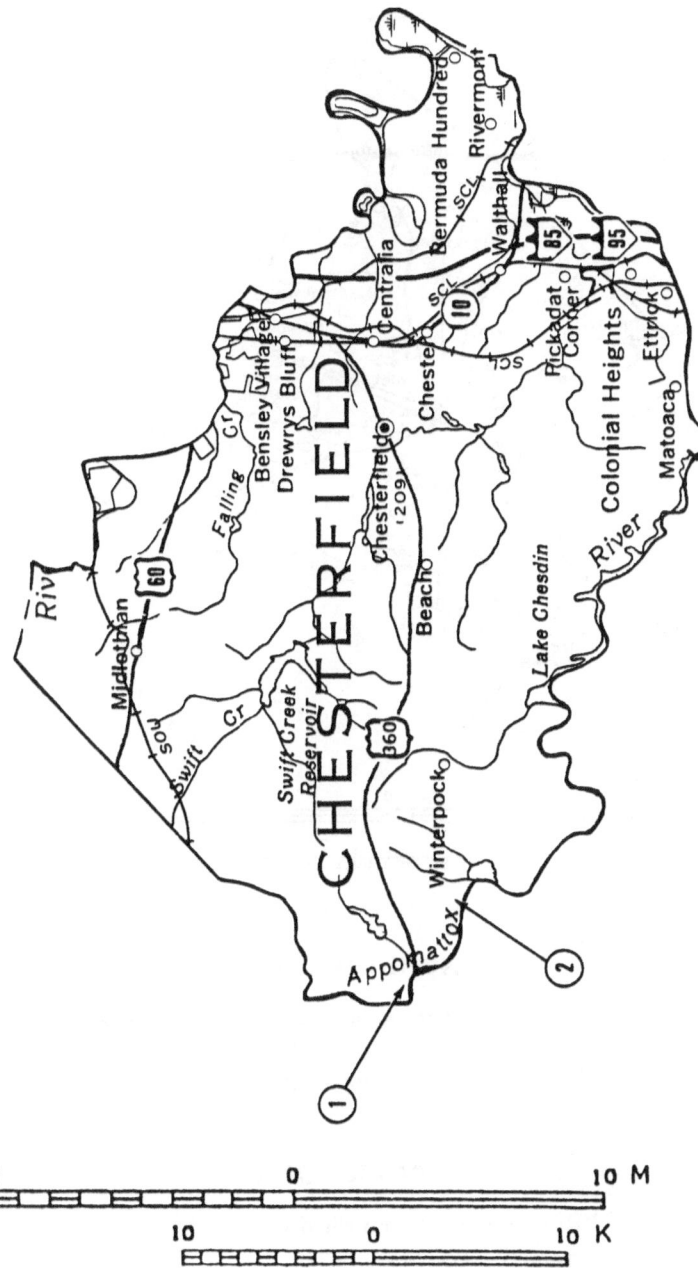

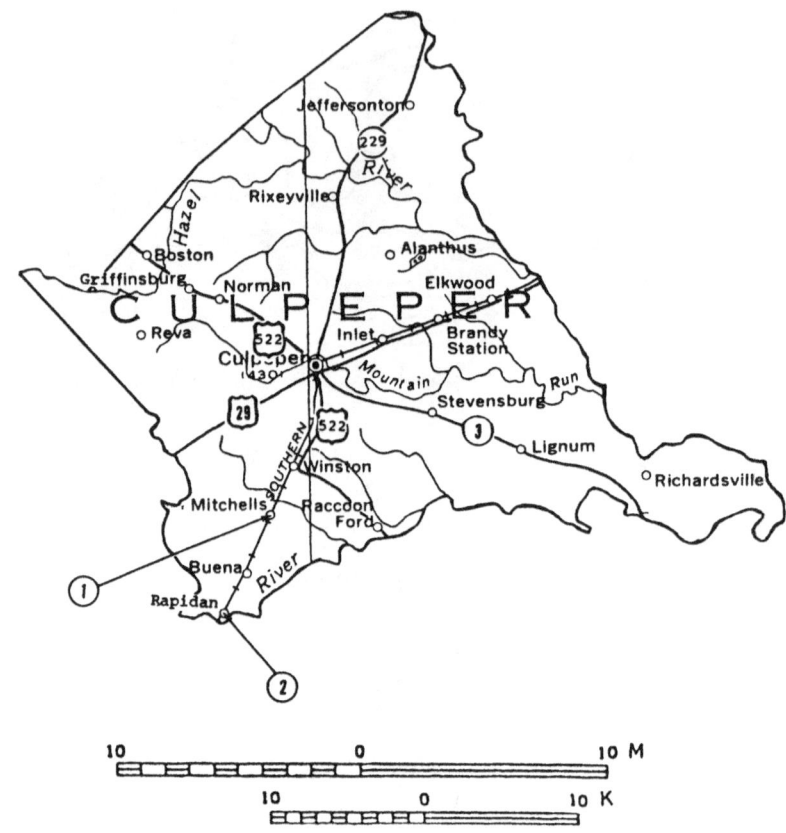

DINWIDDIE COUNTY

1. Wilsons area: 1.4 miles south, 64′ west; 0.8 mile east of border of Nottoway County; on south side of SR-460.
    Monzanite. (1-29-34-granite)

FAIRFAX COUNTY

1. Bull Run area: at the Fairfield Quarry.
    Prehnite. (12)

River
Lakeside Village
Cartersville
Tamworth
CUMBERLAND
Whiteville
Ashby
Trents Mill
Willis
Oak Forest
Cumberland
60
Hillcrest
Sunny Side
Guinea Mills
Stony Point Mills
Angola
Appomattox
45
1
10 0 10 Miles
10 0 10 Kilometres

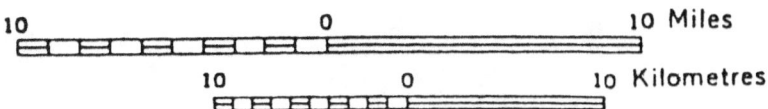

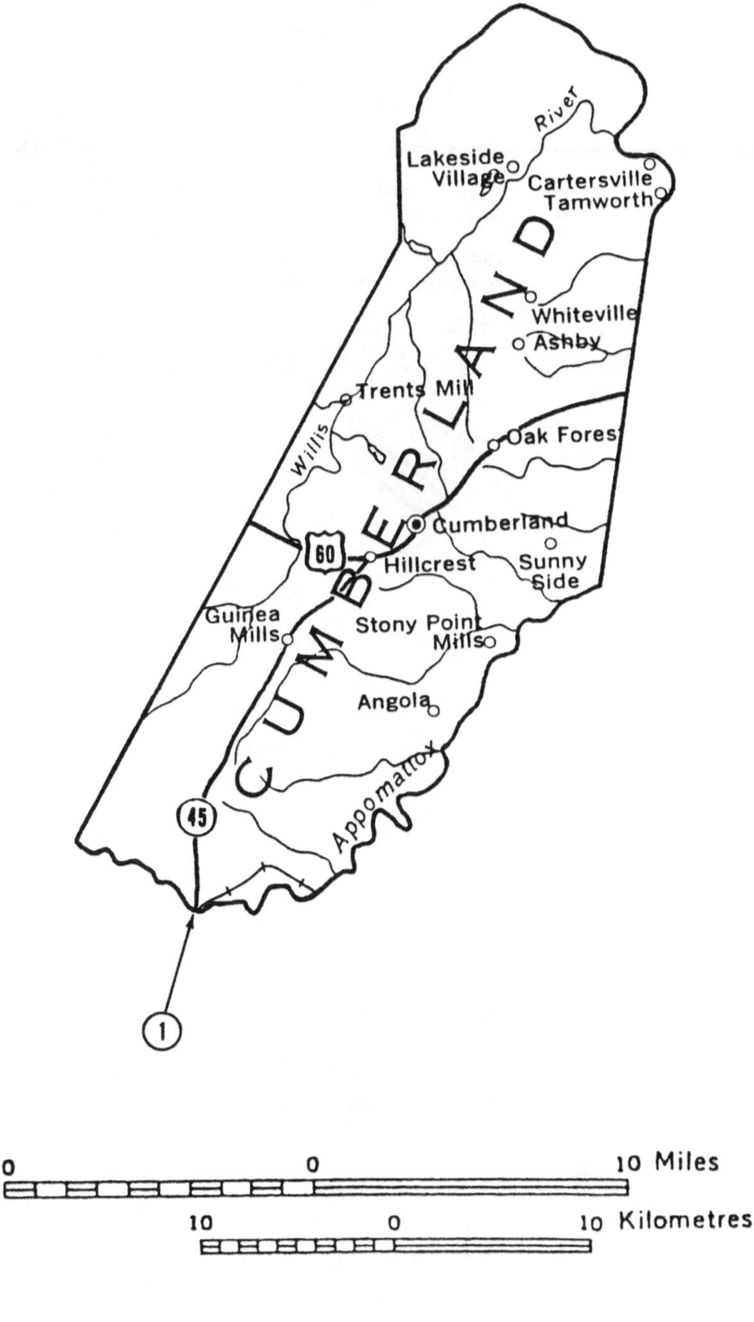

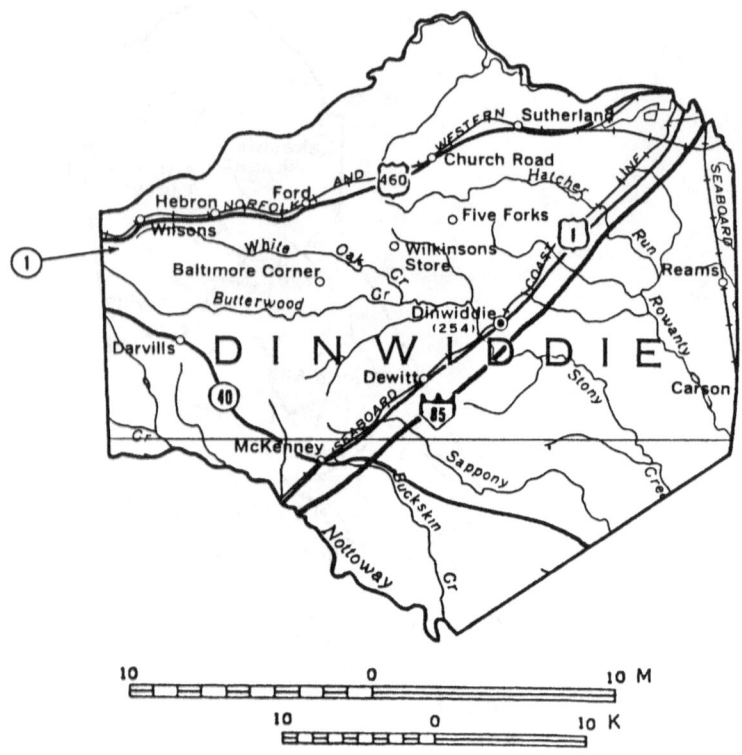

2. Centreville area: at the Bull Run Soapstone Quarry.

    Agate, moss: dark red. Apophyllite. Prehnite; fine gem quality. (12)

3. Centreville area: at the Centreville Quarry.

    Apophyllite: white; translucent. Prehnite: to 5″ × 18″ × 24″. (5-diorite -9-12-19)

4. Oakton area: 1.6 miles north-northwest; near Difficult Run.

    Andalusite. Kyanite. (5-quartz -29)

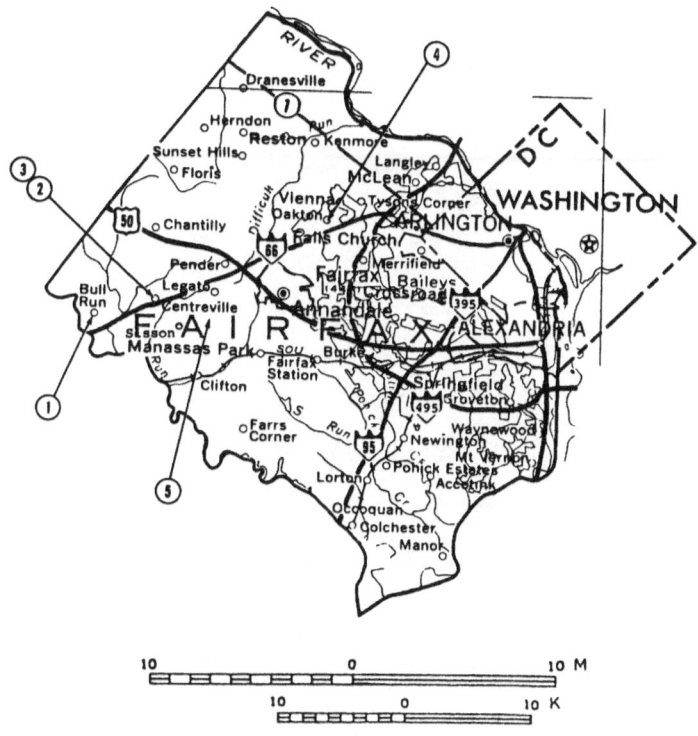

5. Sisson area: 1.25 miles east; in RR cut just south of SR-211.
    Andalusite. (4)

FAUQUIER COUNTY

1. Meetze area: 1.5 miles south; in diabase quarry.
    Calcite. Datolite: crystals. (12)

FLOYD COUNTY

1. Floyd area: 6 miles southwest; at the Toncrae Mine.
    Malachite. (2)

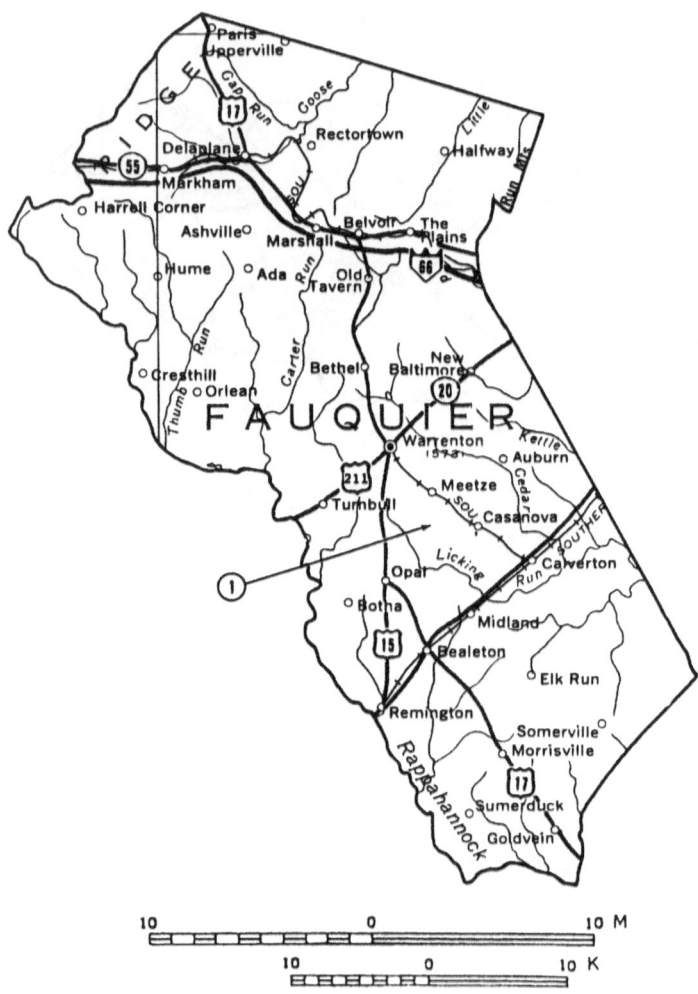

2. Terry's Fork area: 1 mile north; at the Brinton Arsenic Mine.

Arsenopyrite. Fluorite: lavender to deep purple. (2)

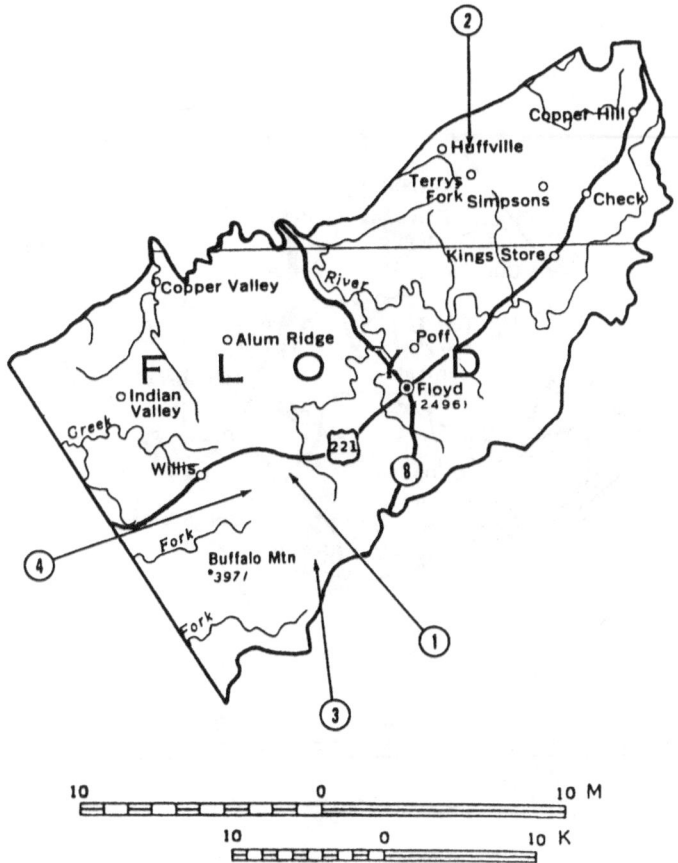

3. Willis area: 6 miles southeast; 1 mile east of Buffalo Mountain Church; at the A. T. Moles farm.

   Quartz, rock. (1-13)

4. Willis area: 2.2 miles east-southeast.

   Amethyst: gem-quality crystals to 5″ diameter and 10″ long. (1)

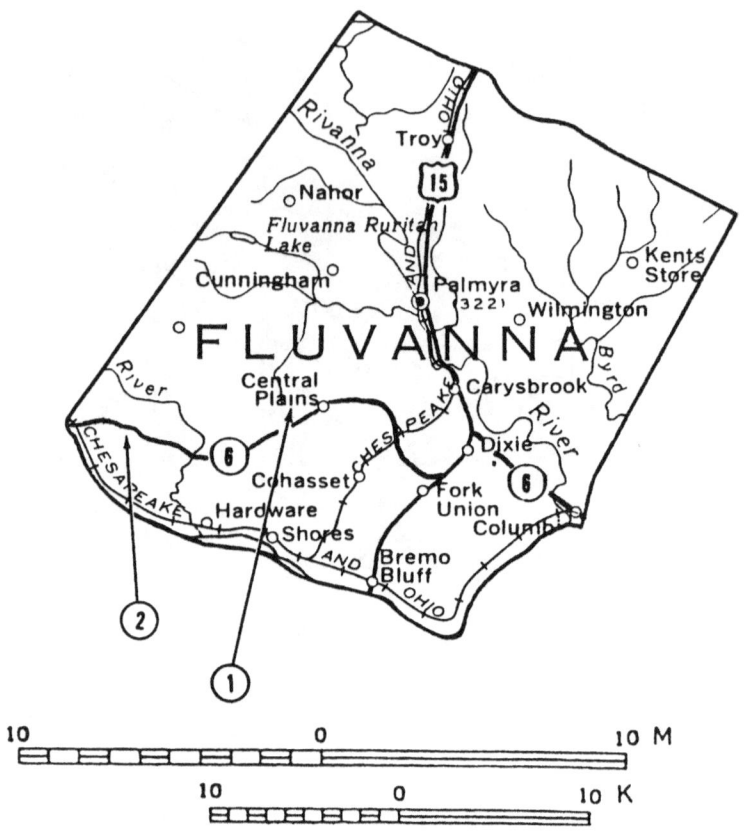

FLUVANNA COUNTY

1. Central Plains area: 1 mile west on SR-6 to fork at Kid's Store; bear right 0.5 mile to SR-620; take trail right (CF).
   Rhodonite. (1-13-29)

2. Hardware area: 4 miles northwest; 2 miles east of Albemarle County border.
   Chalcopyrite. Covellite. (19-quartz)

FRANKLIN COUNTY

1. Callaway area: on small ridge 1.8 miles west of the junction of CR-619 and CR-632; on the Lena W. Vines property; at the Klondike Mine.

    Beryl: pale blue; crystals and irregular masses. (2)

2. Callaway area: 1.9 miles west of the junction of CR-619 and CR-632; at the Simms Mine.

    Beryl: pale blue; crystals to 1″ diameter. (2)

3. Gladehill area: 1.3 miles south of Gladehill School.

    Apatite: yellow. (27)

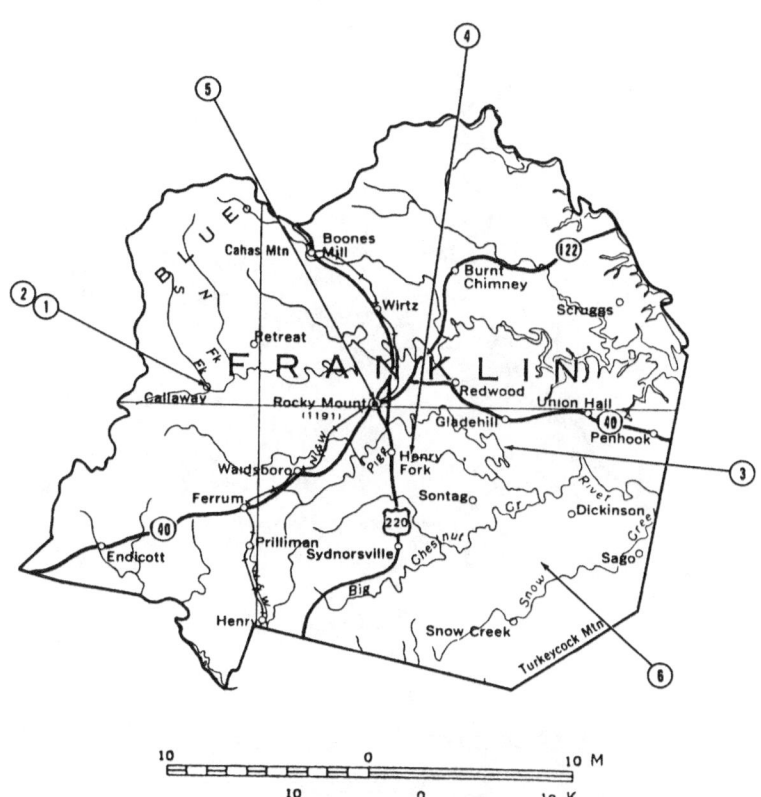

4. Henry Fork area: 0.7 mile east; at the King-Ramsey Quarry.

    Magnetite (Lodestone). (5-soapstone -12)

5. Rocky Mount area: just west; at the Clark Mine.

    Magnetite (Lodestone). (2)

6. Snow Creek area: 3 miles north-northeast; at the Chestnut Mountain Mine.

    Muscovite. (2)

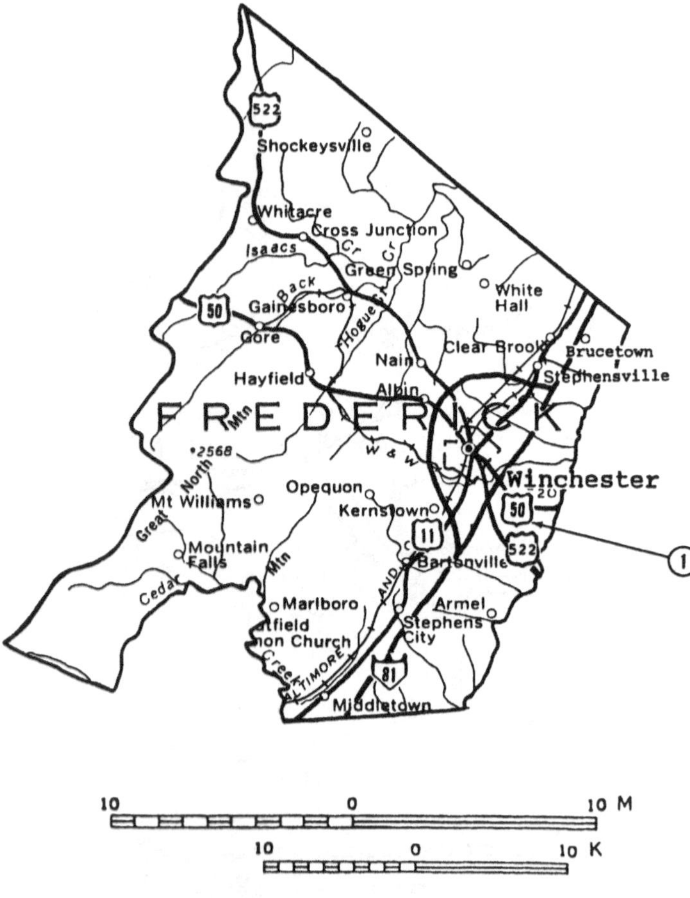

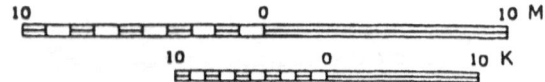

FREDERICK COUNTY

1. Winchester area: 4 miles southeast on US-50; at outcrop on west side of the road.

   Illite. (4-29)

GILES COUNTY

1. Goodwins Ferry area: 2.4 miles east; on Spruce Run.

   Quartz, smoky. (7-20)

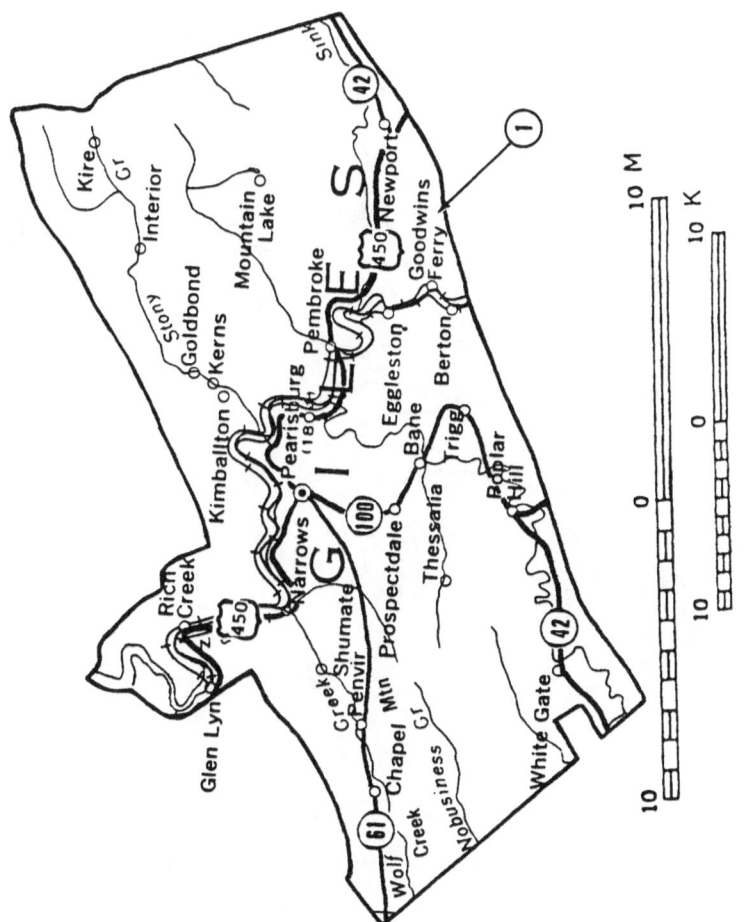

GOOCHLAND COUNTY

1. Manakin area: 4.3 miles east-southeast; material is along
   SR-738, in creek 0.75 mile from inoperative Saunders Mica
   Mine; at the O. W. Harris Mica Mine farm; and following
   creek downstream 1 mile to mouth at the Little River.

   Feldspar. Garnet. Kyanite. Mica. Moonstone; among
   the world's best. Quartz, rock. Quartz, smoky. Rutile:
   small masses. (7-25-26)

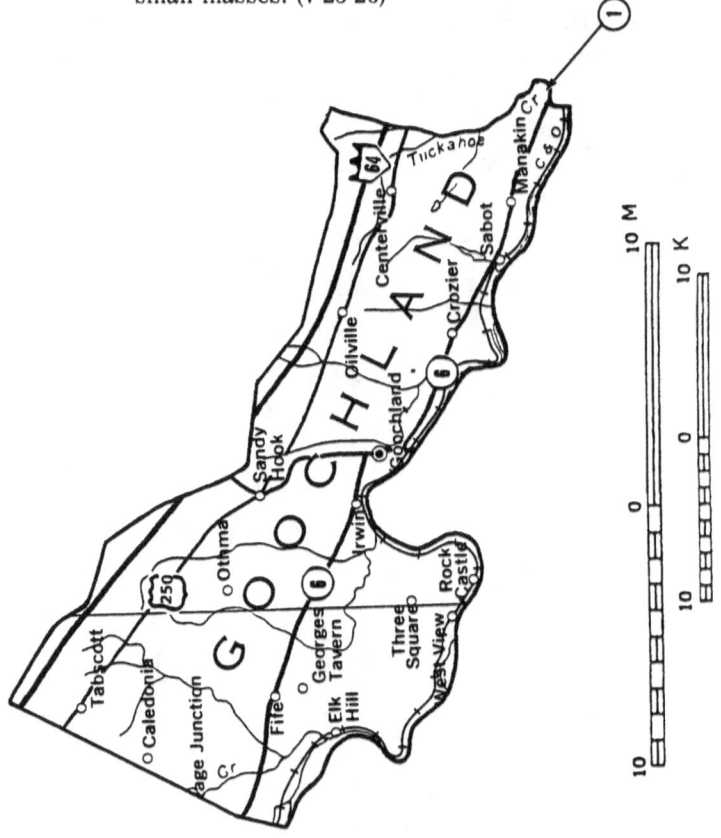

GRAYSON COUNTY

1. Baywood area: 3.1 miles southeast; northeast of CR-622;
   at the Higgins Prospect.

   Ilmenite. Quartz, rock. Spessartite. (21)

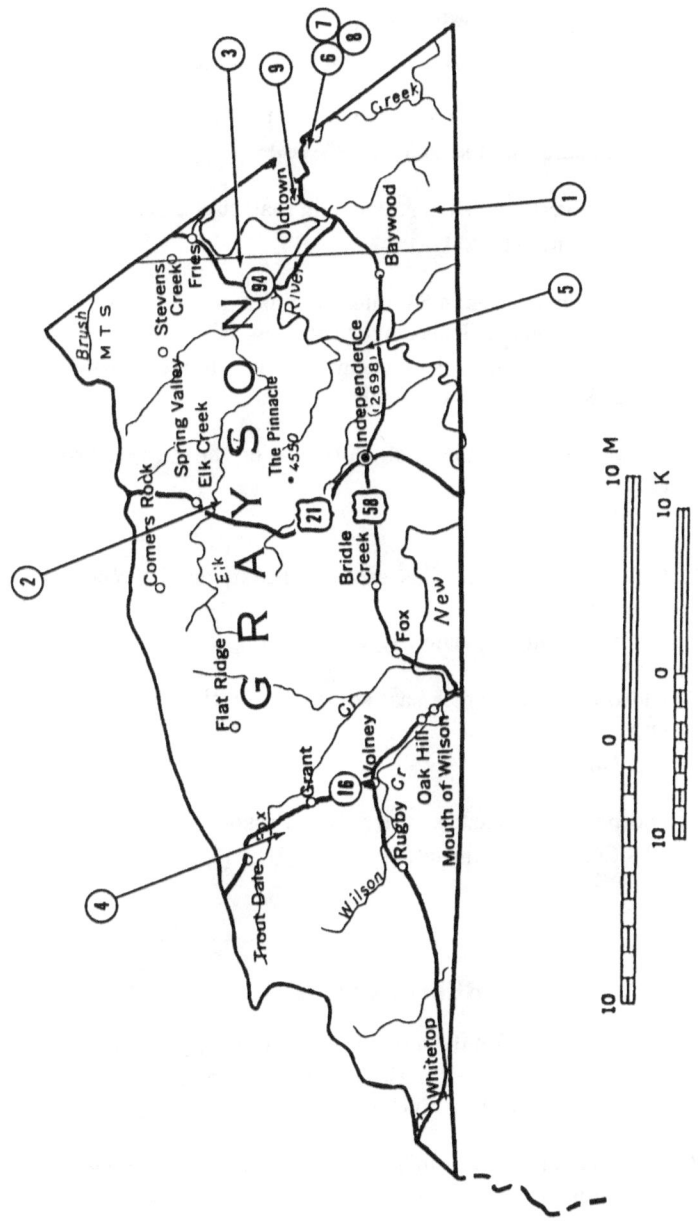

2. Elk Creek area: 1 mile south.
> Gold: flakes. Pyrite. (5-quartz -19)

3. Fries area: 2.1 miles southwest; near Adkins Church.
> Ilmenite: plates to 1" × 2". (16)

4. Grant area: 1.5 miles northwest; in road cuts and outcrops.
> Unakite. (1-4-29)

5. Independence area: 4.25 miles east; 0.2 mile northwest of Carico Memorial Bridge over the New River.
> Pickeringite: acicular crystals in stellate clusters; silky appearance. (4-29)

6. Oldtown area: 2 miles east-southeast; on the Higgins property.
> Garnet. (19)

7. Oldtown area: 2.1 miles east-southeast; on the Nuchols property.
> Corundum. Garnet. Kyanite. (1-13)

8. Oldtown area: 2 miles east-southeast; on the Phipps property.
> Garnet. (19)

9. Oldtown area: east; at the Pierce Prospect.
> Corundum. Garnet. Kyanite. (21)

GREENE COUNTY

1. Lydia area: 1.75 miles northeast; at the copper mines.
> Azurite. Chalcocite. Chalcopyrite. Copper, native. Cuprite. Malachite. Pyrite. (2)

2. Stanardsville area: 5 miles north; at the copper mines.
> Azurite. Chalcocite. Chalcopyrite. Copper, native. Cuprite. Malachite. Pyrite. (2)

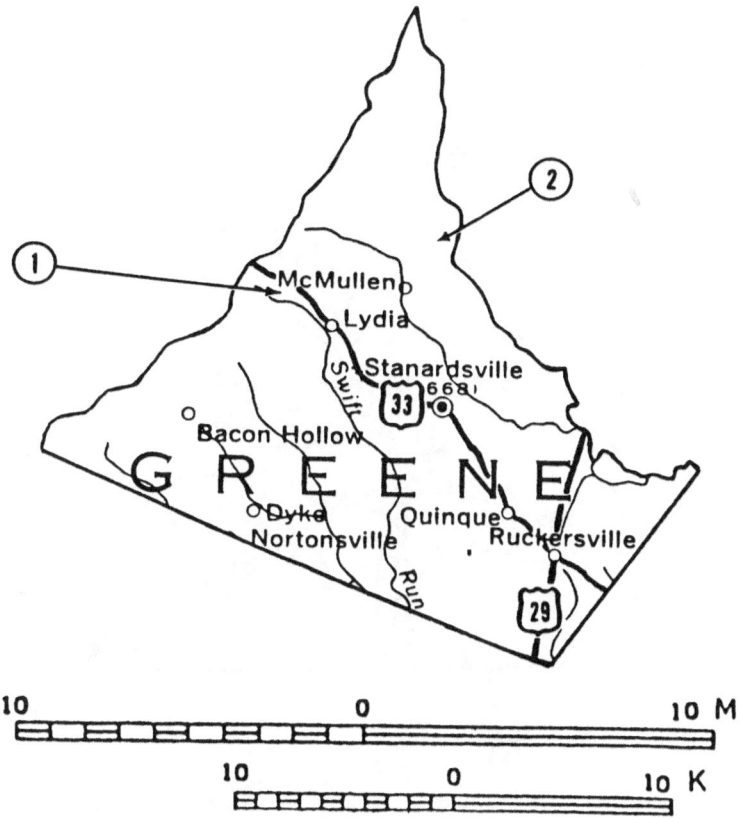

GREENSVILLE COUNTY

1. Brink area: 2.25 miles southeast; on the north side of CR-621; at the Trego Quarry.

    Laumonite. (12-19-quartz)

HALIFAX COUNTY

1. Meadville area: 1 mile southwest; in a ridge extending southwest from Sandy Creek.

    Kyanite: blue to pale green; crystals to 6" long. (16-24-29)

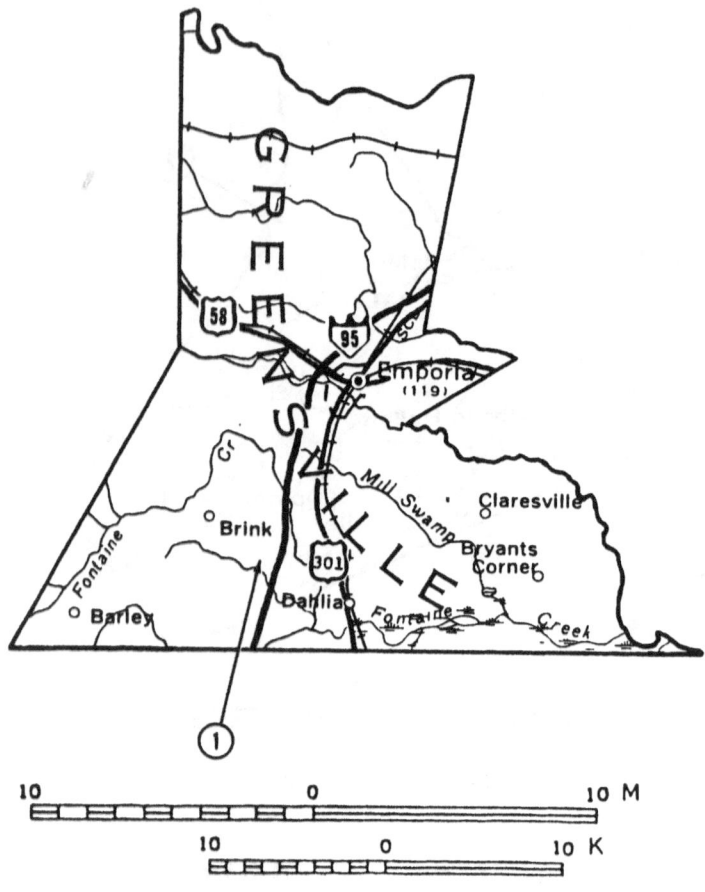

HANOVER COUNTY

1. Hewlett area: on south shore of the Little River; at the
   Saunders No. 2 Mine.

   > Garnet. Kyanite: pale blue to deep blue; fine translu-
   > cent crystals to 3″ long. (2)

2. Hewlett area: on CR-738; in stream north of house and in
   ground left of barn; at the Harris Mica Mine (CF).

   > Garnet. Moonstone. (1-7-20)

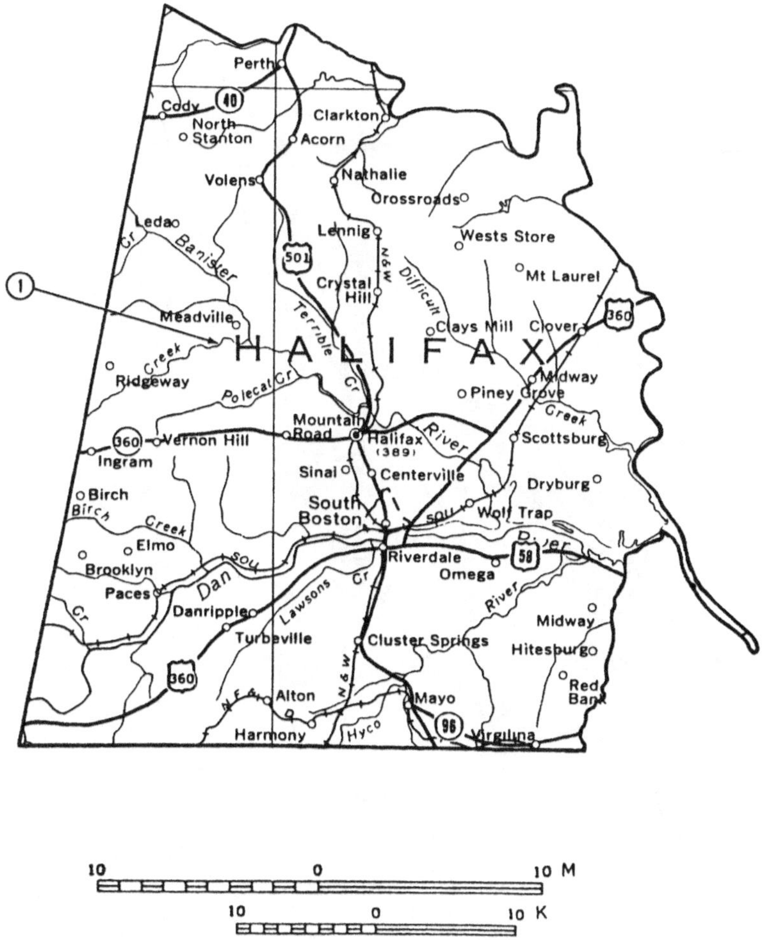

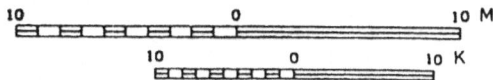

3. Hewlett area: on SR-51; along the east side of the road.
      Monzanite. (1-19-34)

4. Rockville area: 1.25 miles north; at the Metal and Thermite
   Company Mine.
      Ilmenite. (2)

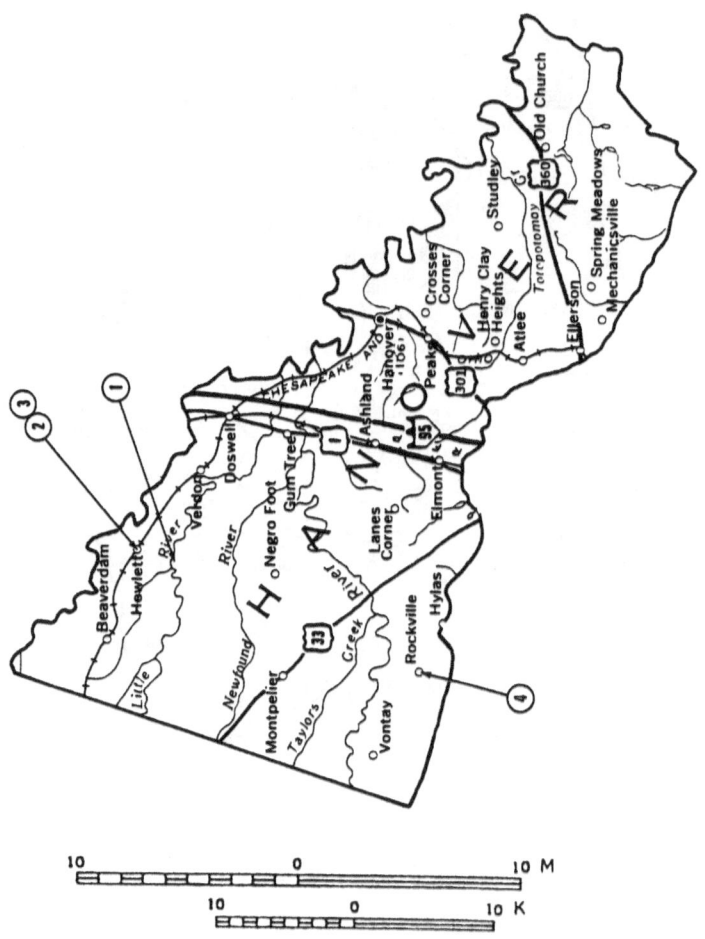

10   0   10 M

10   0   10 K

HENRICO COUNTY

1. Richmond area: in Manchester district within city limits.
   Diamond: 1 stone: 23.74 ct; named the "Dewey Diamond"; cut to a fine gemstone of 11.69 ct; found during construction excavating in 1885. (6)

HENRY COUNTY

1. Aiken Summit area: 3.1 miles southwest; at road cut near the south bank of the Smith River.

    Staurolite. (4)

2. Fieldale area; 3 miles west; in placer on southwest side of CR-687.

    Corundum. Ilmenite. Magnetite (Lodestone). Monzanite. Quartz, rock. Zircon. (21)

3. Fontaine area: along SR-229.

    Epidote. (1)

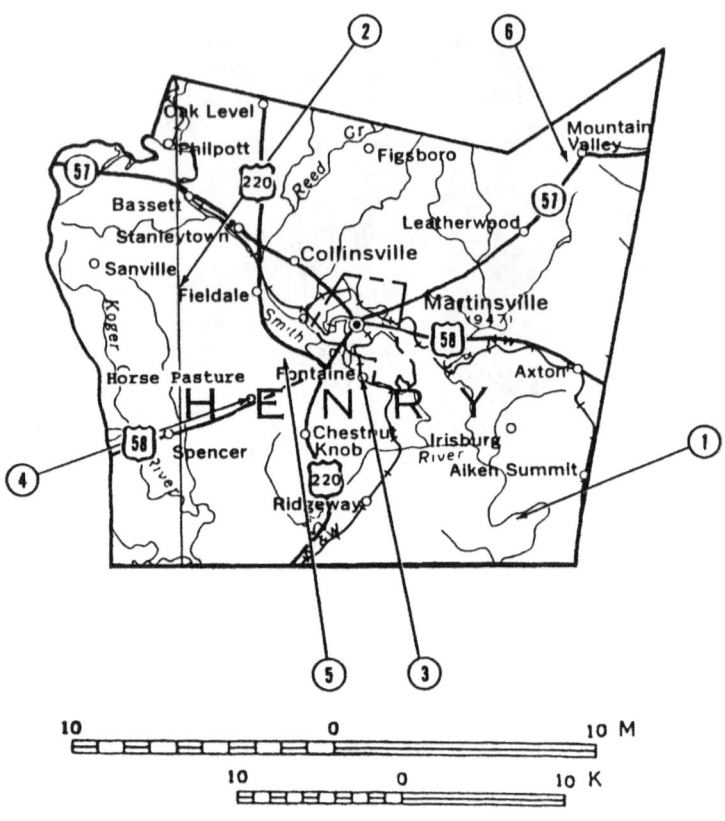

4. Horse Pasture area: north; at the Wilson Quarries.
     Garnet. (12)

5. Martinsville area: 3 miles southwest; 0.3 mile south of
     US-58; at the Williams Prospect.
          Beryl, golden: small, well-formed crystals; clear to
          golden; transparent. (1-21)

6. Mountain Valley area: 0.5 mile southwest on SR-57; just
     southwest of junctions of SR-57 and SR-647; on southeast
     side of road.
          Monzanite. (1-29-34)

LEE COUNTY

1. Pennington Gap area: in Gilley Cave.
   Hydromagnesite. (32)

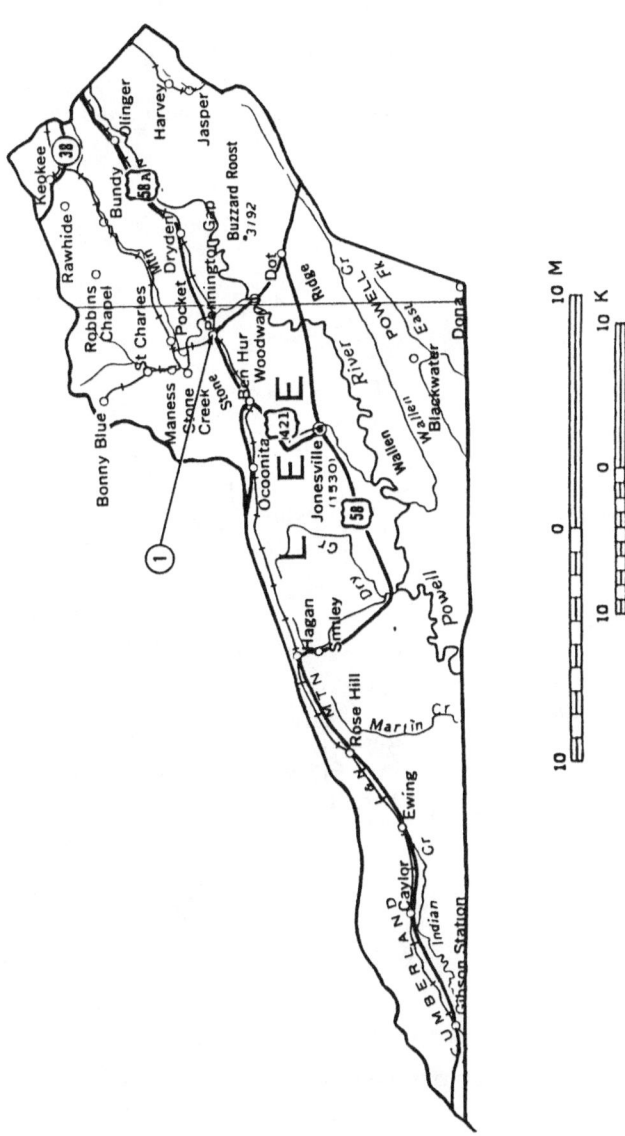

LOUDOUN COUNTY

1. Ashburn area; 2 miles northwest on Goose Creek; at the Belmont Traprock Quarry.

   Prehnite. (12)

2. Ashburn area: 2.4 miles northwest; on SR-7 near Goose Creek; at the Arlington Stone Quarry.

   Apophyllite: fine crystals; some tabular and some pseudocubic. Prehnite. (12)

3. Ashburn area: 2.4 miles northwest; on SR-7 near Goose
   Creek; at the State Quarry.

   > Chalcocite. Haydenite. Heulandite: small white crystals. Malachite. (12)

4. Leesburg area: at the Eagle Mine.

   > Hematite: crystals to 2" diameter. (2)

5. Mountville area: at the Virginia Lime and Stone Quarry.

   > Andradite: reddish brown crystals and masses. (12)

6. Mountville area: east to CR-7; then 1.5 miles to Highcamp
   Road; at marble quarry.

   > Marble. Serpentine. (12)

7. Oatlands area: along Goose Creek; in the Veasco Quarry.

   > Pyrite: crystals to 1.5" diameter. (12)

8. Oatlands area: along Goose Creek; in the Virginia Lime &
   Marble Quarry.

   > Serpentine. (12)

LOUISA COUNTY

1. Apple Grove area: 1.5 miles northwest; along the west side
   of CR-699.

   > Garnet: well-formed dodecahedral crystals to 0.6" diameter. (4-37)

2. Mineral area: at the old Armenium Mine.

   > Azurite: small amounts. Bornite. Calcite. Malachite. (2)

3. Mineral area: on CR-522 near Contrary Creek; at the Old
   Sulphur Mine.

   > Allanite: fine; tiny crystals. Actinolite: bladed crystals
   > and granular masses. Magnetite (Lodestone). (2)

4. Poindexter area: farm fields 1 mile northwest.

   > Amethyst: deep purple to colorless; to 1" diameter. (1)

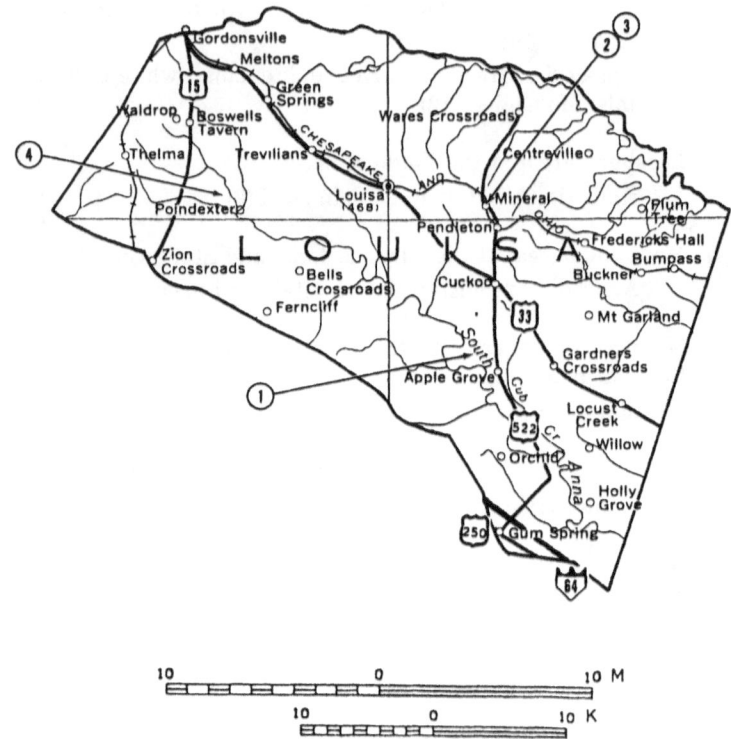

10   0   10 M

10   0   10 K

MADISON COUNTY

1. Syria area: in streams.
   Unakite. (7-20)

2. Syria area: on the Graves farm.
   Unakite. (1-13)

MECKLENBURG COUNTY

1. Forksville area: 1 mile northeast on US-1; 0.3 mile west-
   southwest of border of Brunswick County; on northwest
   side of road.
   Monzanite. (1-29-34)

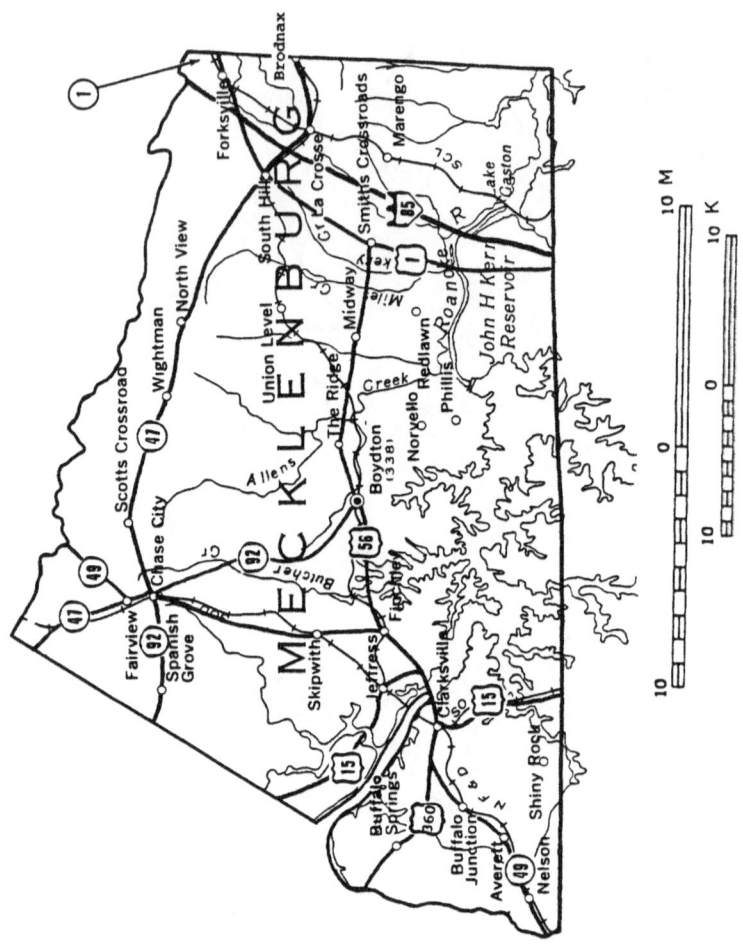

MONTGOMERY COUNTY

1. Blacksburg area: 1.25 miles northeast.
   Calcite: good crystals to 3.9″; well formed; white to nearly colorless; translucent to transparent. (12)

2. Ironto area: on the A. V. Smith farm.
   Barite: cauliflower nodules called "roses." (1-13-41)

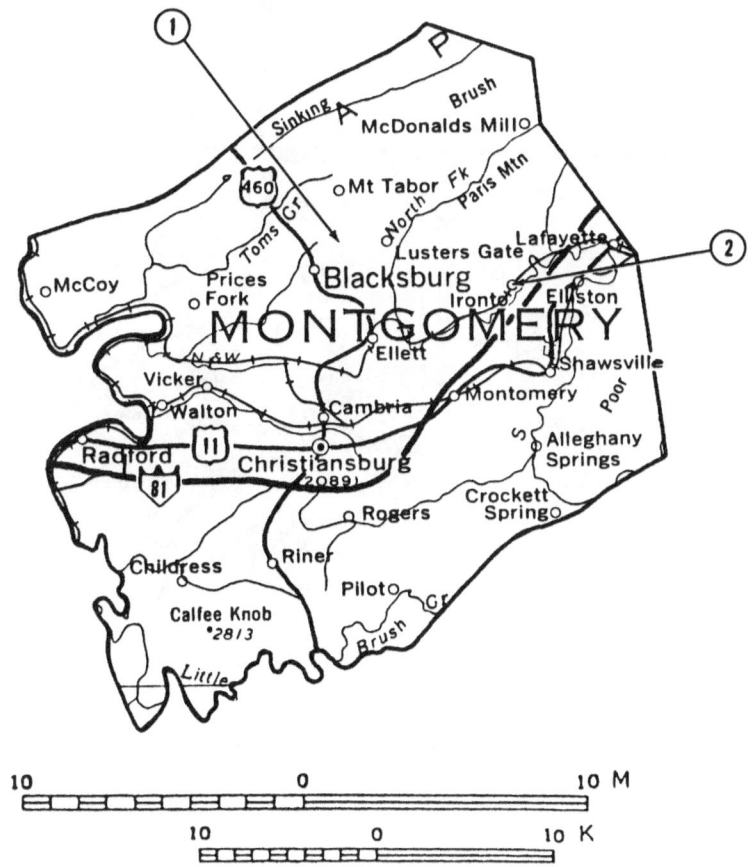

NELSON COUNTY

1. Alhambra area: just southeast; especially in old prospects
   and quarries.

   > Amethyst: extra-fine gem quality; very large; some
   > stones cut to gem weight of over 110 ct. (1-12-16-shat-
   > tered)

2. Faber area: at the old Faber Mine.

   > Cerrusite: small needles. (2)

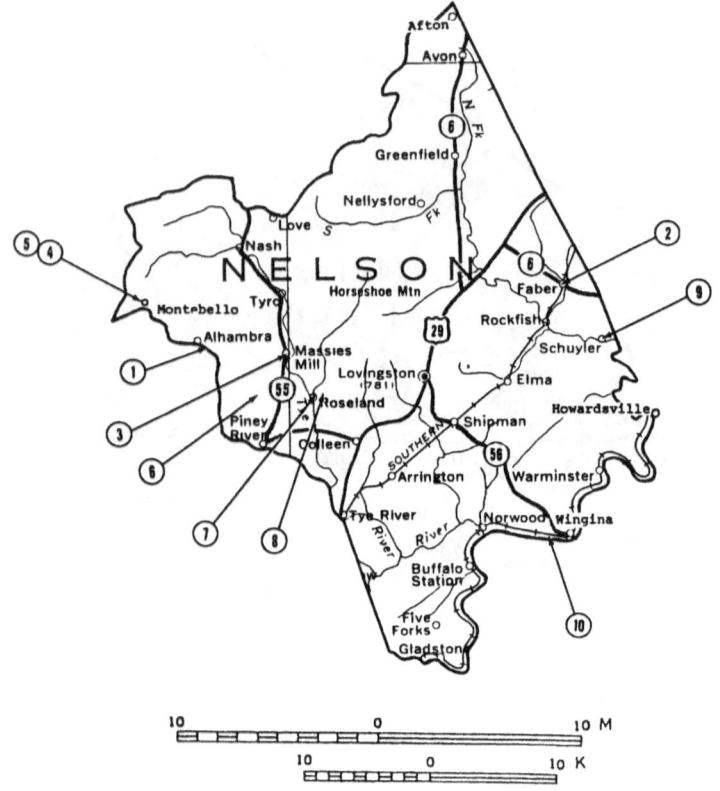

3. Massies Mill area: on the R. Saunders property; in a peach
   orchard.
   Amethyst. (1-13-29)

4. Montebello area: east of junction of Skyline Drive and
   SR-56.
   Unakite. (1-13)

5. Montebello area: southwest; at Irish Creek tin deposits.
   Moonstone: blue. (21)

6. Piney River area: 2.5 miles north.
   Amethyst. (1-13)

7. Roseland area.

   Apatite. Garnet. Ilmenite. Magnetite (Lodestone). (16)

8. Roseland area: 0.5 mile east; at the American Rutile Company Quarry.

   Apatite. Quartz, blue: asterism. (12)

9. Schuyler area quarries.

   Gypsum (Soapstone var.). (12)

10. Wingina area: 1.1 miles south on CR-647; along the road on the T. Viar property.

    Limonite. (1-13-29)

ORANGE COUNTY

1. Madison Run area: 225' east of RR station.

   Limonite: pyritohedrons. (1-13)

2. Unspecified locality: Vancluse Gold Mine.

   Diamond: a "first-water" stone; details not specified. (2)

PAGE COUNTY

1. Ida area: 0.9 mile to 1 mile northwest; along the road.

   Epidote: bright green. Hematite: black. Jasper. Quartzite. (1)

2. Ida area: north on SR-689; turn left just before stream and go west several miles; in streambed and on hills.

   Unakite. (1-7-20)

3. Ida area: northeast and northwest; along west side of Blue Ridge.

   Epidote: bright green. Hematite: black. Jasper: red. Quartzite: white. (1)

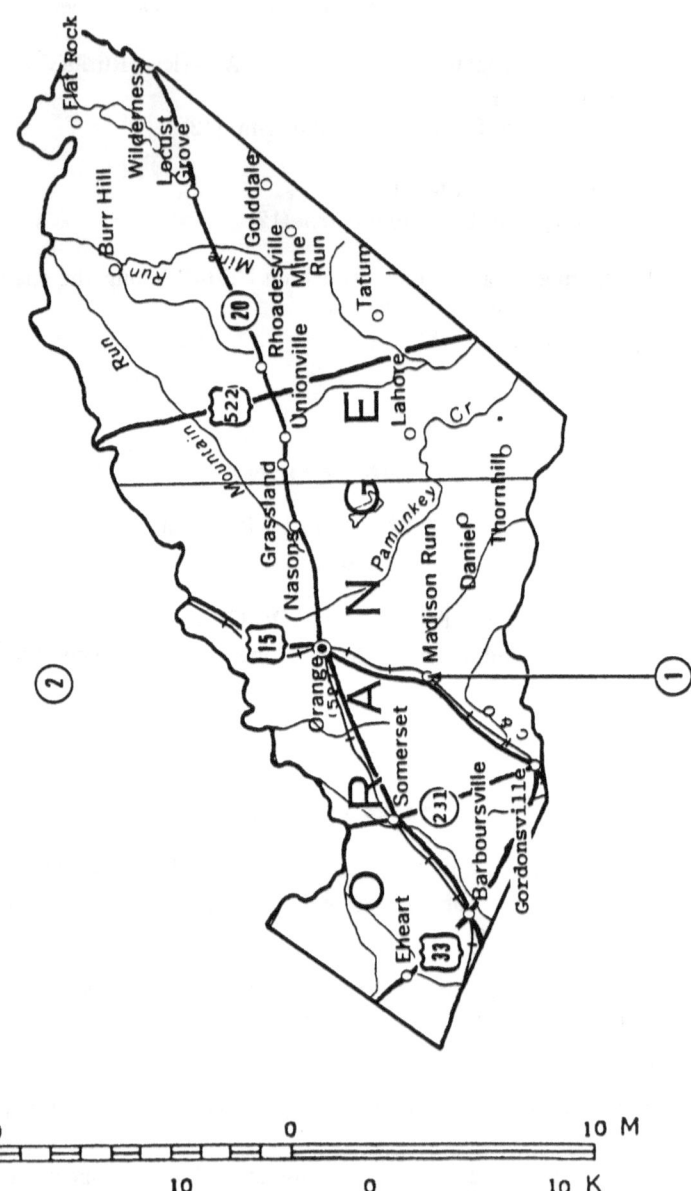

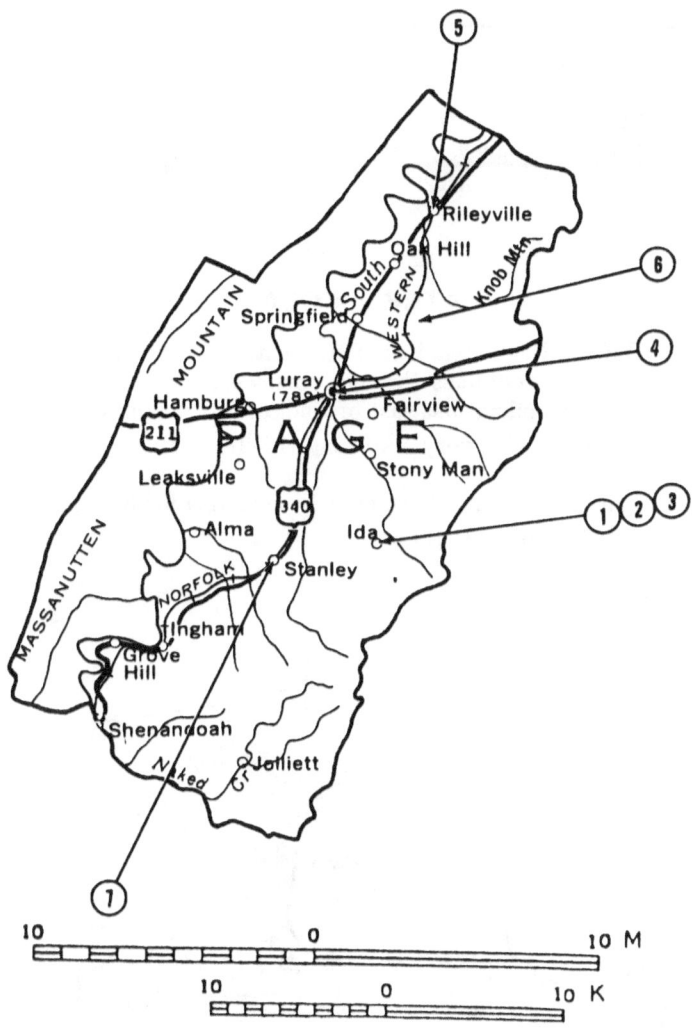

4. Luray area: in Bloodstone Creek.
   Bloodstone. Jasper. (7-20)

5. Rileyville area: 2.5 miles northwest on CR-662 to 1 mile
   east of CR-605; on west side of road; in copper prospect.
   Epidote. Serpentine. (21)

6. Springfield area: 2 miles east; in Jeremiah Creek.
   Jasper, orbicular. (7-20)

7. Stanley area: south of Marksville; on unmarked road from
   CR-680 to Bailey Mountain; at Fishers Gap.
   Unakite. (1-13-29)

PATRICK COUNTY

1. Buffalo Ridge area: belt of staurolite schist begins 4 miles
   east and continues through northwest corner of Henry
   County, ending just east of Endicot in Franklin County.
   Staurolite: very abundant. (1-3-27)

2. Fairy Stone State Park area: in area surrounding park.
   Staurolite. (1-13)

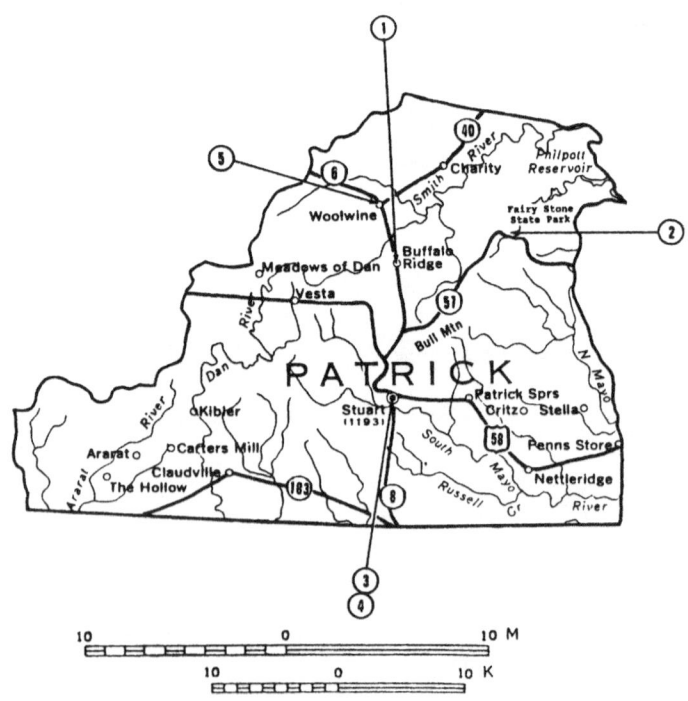

3. Stuart area: belt of staurolite-bearing schist begins 4 miles southwest and runs 20 miles northwest, ending 5 miles southwest of Floyd in Floyd County.

   Spodumene: very abundant. (1-3-13-17)

4. Stuart area: near Stuarts Knob; at the Hairston Mine.

   Magnetite (Lodestone). (2)

5. Woolwine area: east; 0.9 mile west of SR-663; on north side of SR-103.

   Monzanite. (1-29-34-granite)

PITTSYLVANIA COUNTY

1. Ajax area: 3.25 miles northeast; along shoreline of the Leesville Lake portion of the Roanoke River; 700' south of the mouth of Old Womans Creek.

   Chlorite. Cordierite: to 2" × 5". Muscovite. Paragonite. Staurolite. (1-3-20-27-29)

2. Museville area: east; 1.6 miles south of SR-40; near north shore of the Pigg River; at big bend.

   Limonite: pseudomorphs after pyrite. (1-13-29)

3. Sandy River area: at the Harston Mine.

   Muscovite. (2)

4. Sandy River area: at the Turner Mine.

   Muscovite. (2)

5. Sandy River area: at the Willis Mine.

   Muscovite. (2)

6. Sycamore area: just north along US-29.

   Beryl. (16)

7. Whittles area: 0.2 mile east of the junction of CR-777 and CR-778; just south of CR-778.

   Magnetite: fine octahedral crystals; small. (27-mica)

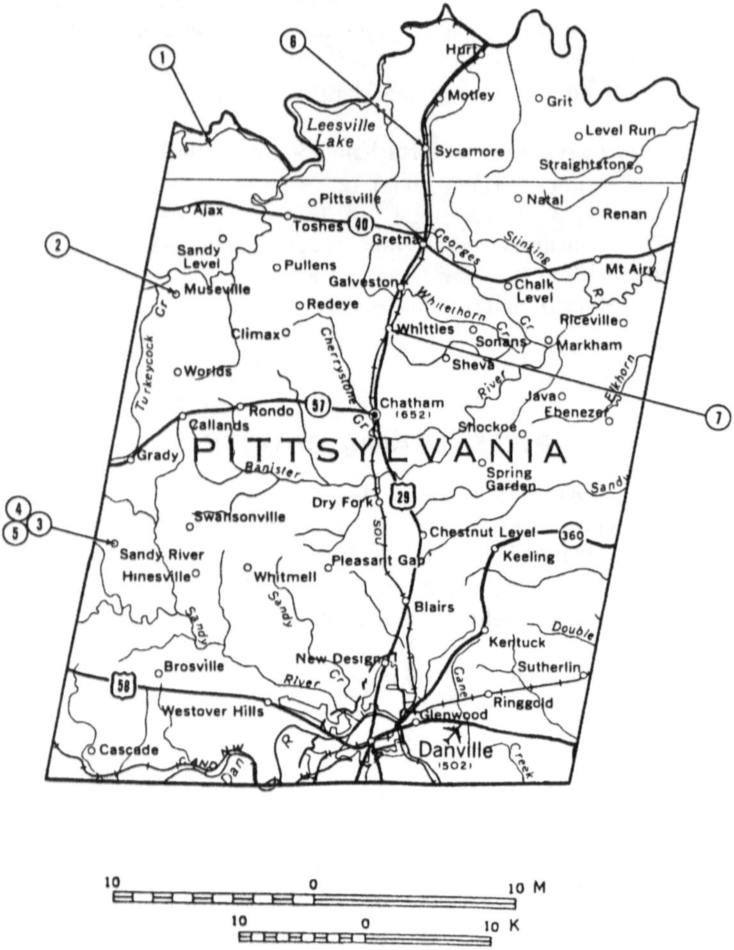

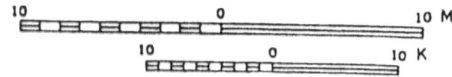

POWHATAN COUNTY

1. Belona area: on north side of SR-13.
   Monzanite. (1-29-34-granite)

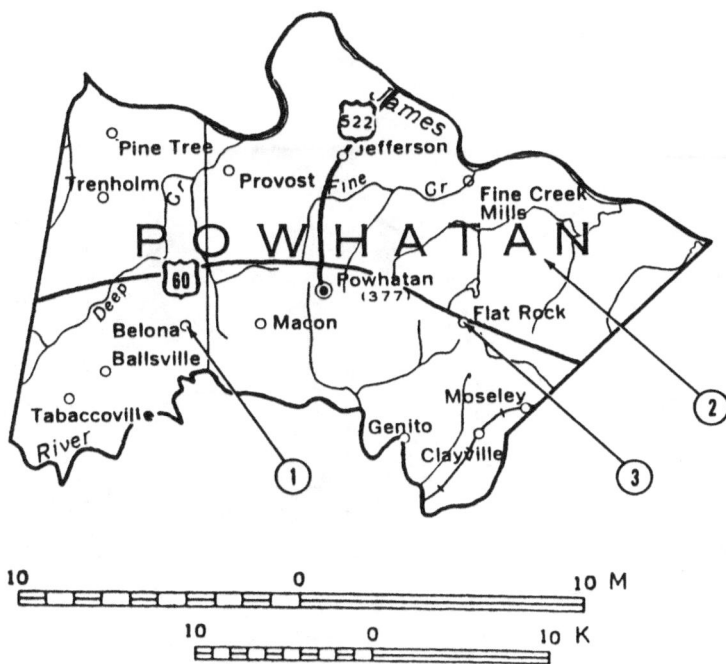

2. Flat Rock area: 3.6 miles northeast on CR-613; at the Herbb No. 2 Mine.

> Albite (Cleavelandite). Amazonite: blocky masses. Beryl. Quartz, rock. Topaz. (1-2-16)

3. Flat Rock area: northeast; at White Peaks Mine.

> Quartz, rock: asterism. (2)

PRINCE EDWARD COUNTY

1. Farmville area: at Willis Mountain.

> Kyanite. (1-7-13-29)

2. Rice area: 3 miles north on SR-619; in broad area of farm fields, centered at the George R. Smith farm.

> Amethyst: gem quality; crystals to 4″ × 6″. Quartz, milky. Quartz, rock. Quartz, smoky. (1-13-29)

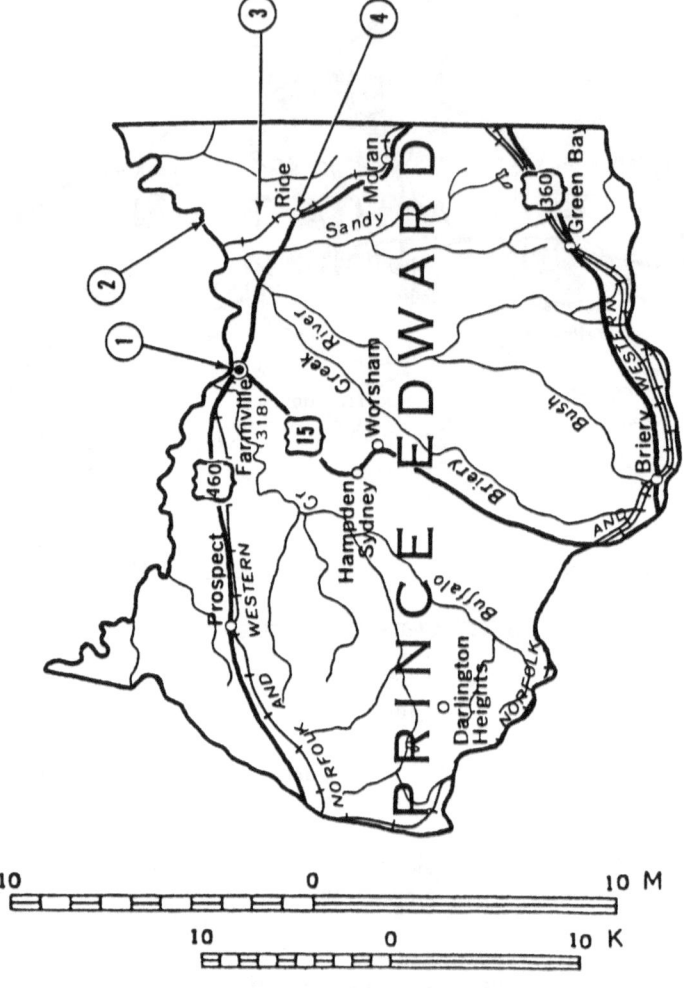

3. Rice area: 1 mile north; in old mica mines.
   Kyanite. (2)

4. Rice area: 0.2 mile on SR-307 to first road on left; follow
   it to the D. Hodges farm.
   Amethyst: loose crystals. (1-13)

PRINCE GEORGE COUNTY

1. Petersburg area: just east; in gravel pits.
   Chalcedony: silicified wood. (10)

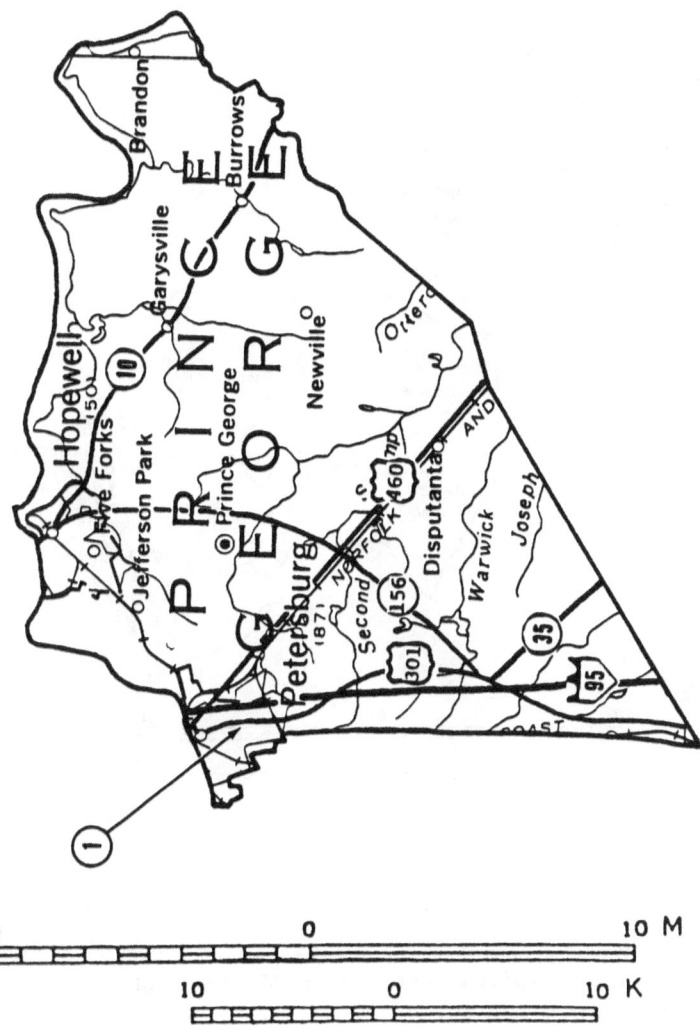

PRINCE WILLIAM COUNTY

1. Brentsville area: 1 mile northwest of Minnieville.

    Amethyst: gem quality; crystals to 5″ × 6″. (1-13-fields)

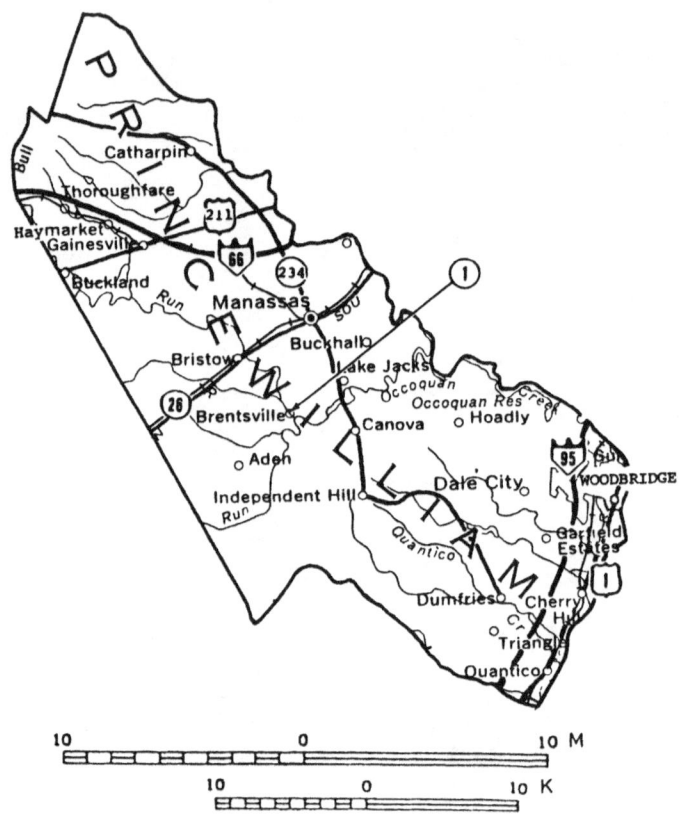

PULASKI COUNTY

1. Wurno area: 1.25 miles northwest on SR-100; on the northwest side of road; at the Salem Rock Corporation Quarry.

    Calciostrontianite: sharply terminated spearlike crystals. Calcite. Celestite: colorless; pale blue; pale blue-gray; tabular crystals to 2.75″. (9-12)

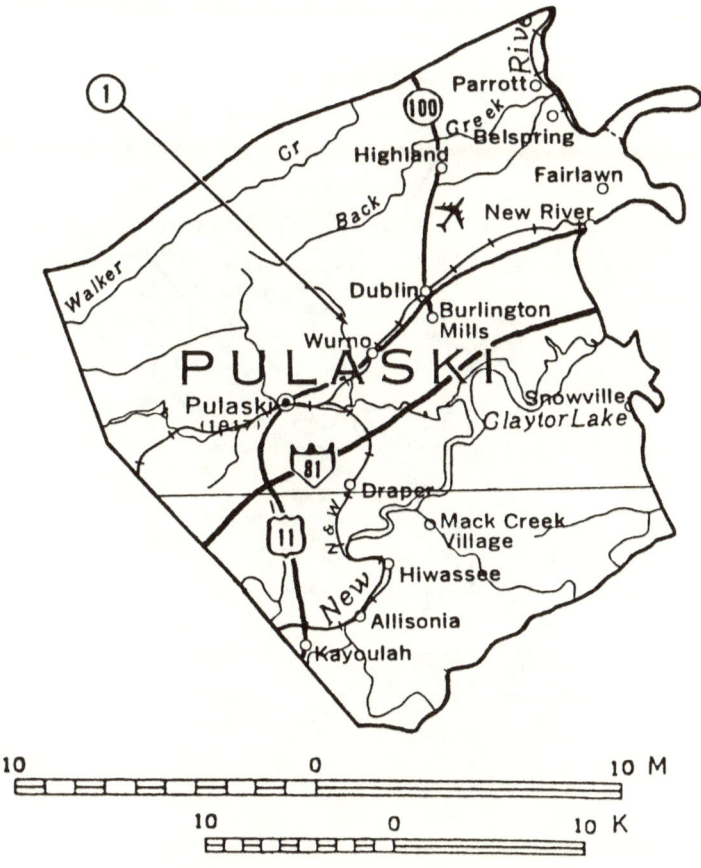

10   0   10 M

10   0   10 K

ROCKBRIDGE COUNTY

1. Buena Vista area: just south of Midvale Station along the Norfolk & Western RR; at the Midvale Mine.

    Manganite: excellent crystals. (2)

2. Glasgow: at the Lone Jack Quarry.

    Calcite: pink; coarse crystals. Chalcopyrite: sphenoidal. Chlorite. Dolomite: white, pink, gray, opaque. Fluorite: purple, yellow, colorless. Goethite: blood red

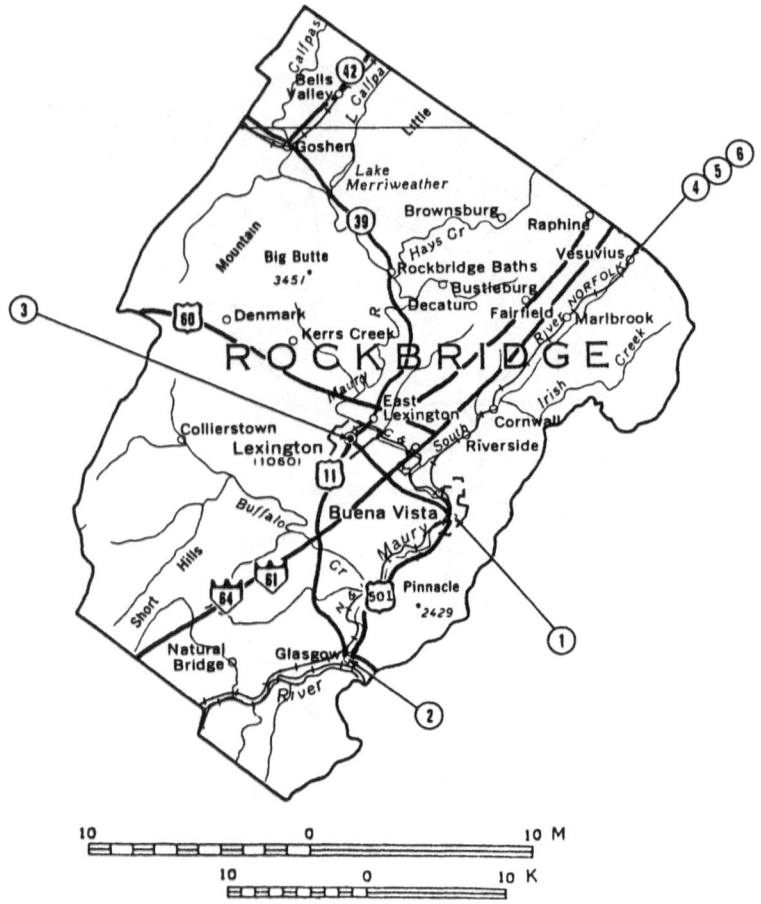

to coppery; small. Marcasite: striated rods. Melanterite: fluorescent coatings. Montmorillonite: white to ivory. Palygorskite: fibers to 0.25″. Sphalerite. (12)

3. Lexington area: just east on SR-60; at the Barger Quarry.
   Fluorite: purple masses. (12)

4. Vesuvius area: 2 miles west of Tye River Gap.
   Unakite. (12)

5. Vesuvius area: at the South River Mine.
   Chert. (2)

6. Vesuvius area: just below the Blue Ridge Parkway; in road
   cuts along SR-56.
   Unakite. (4)

ROCKINGHAM COUNTY

1. Bergton area: 2.75 miles north; at junction of CR-820 and
   CR-961.
   Illite. (28-29)

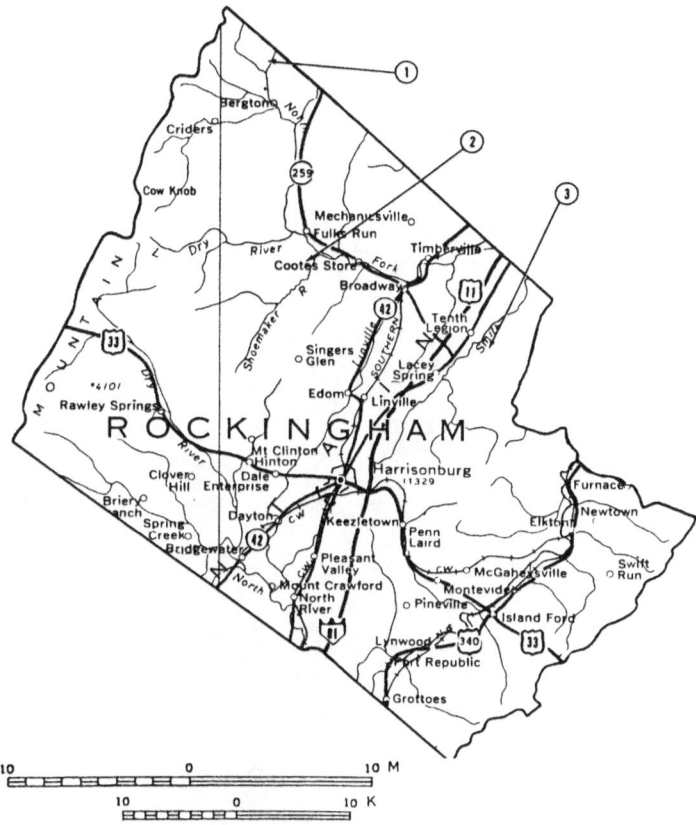

2. Cootes Store area: 3 miles west on SR-259; quarry on northeast side of road.

Illite. (12-28)

3. Tenth Legion area: 1 mile southeast on CR-798; on northeast side of the road.

Illite. (28)

RUSSELL COUNTY

1. Bolton area: north; northwest of the crest of Moccasin Ridge; in exposures of shale along SR-19.

Glauconite: blue-green. (4-28-29)

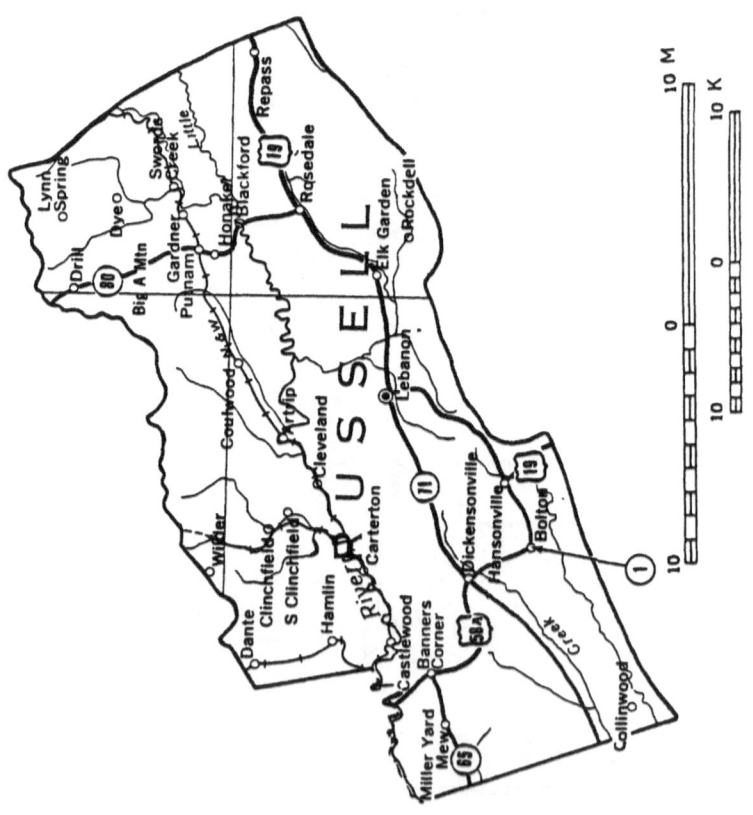

SHENANDOAH COUNTY

1. Woodstock area: at Powells Fort.
   Manganite: good crystals. (1-13-29)

2. Woodstock area: just east; at the Powells Fort Mine,
   Manganite: good crystals. (2)

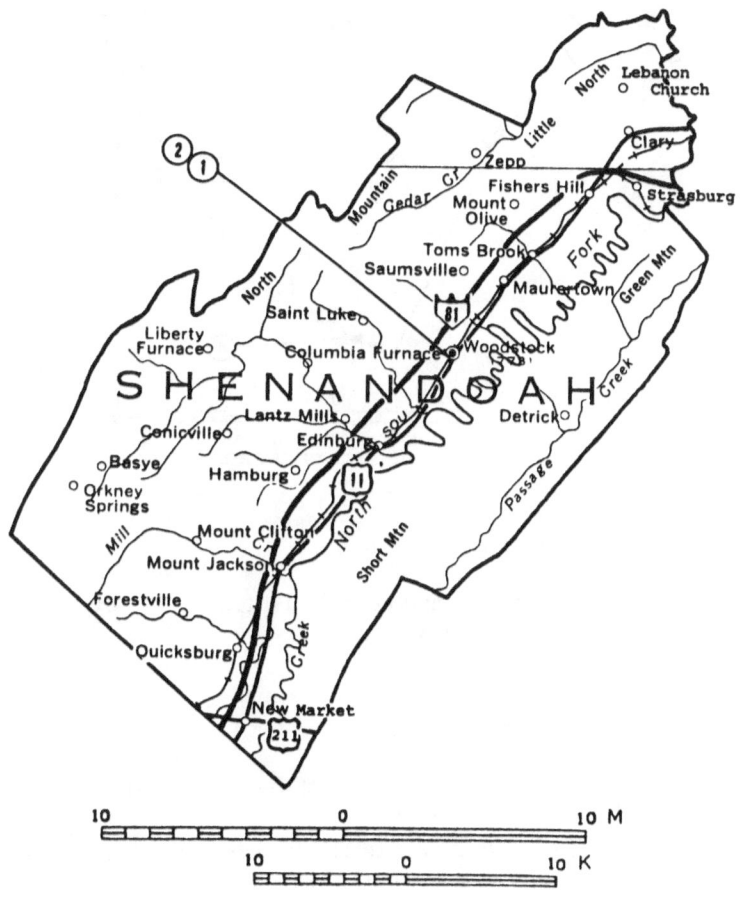

SMYTH COUNTY

1. Mount Rogers area: to north slope of Whitetop Mountain;
   in road cuts through the gap.

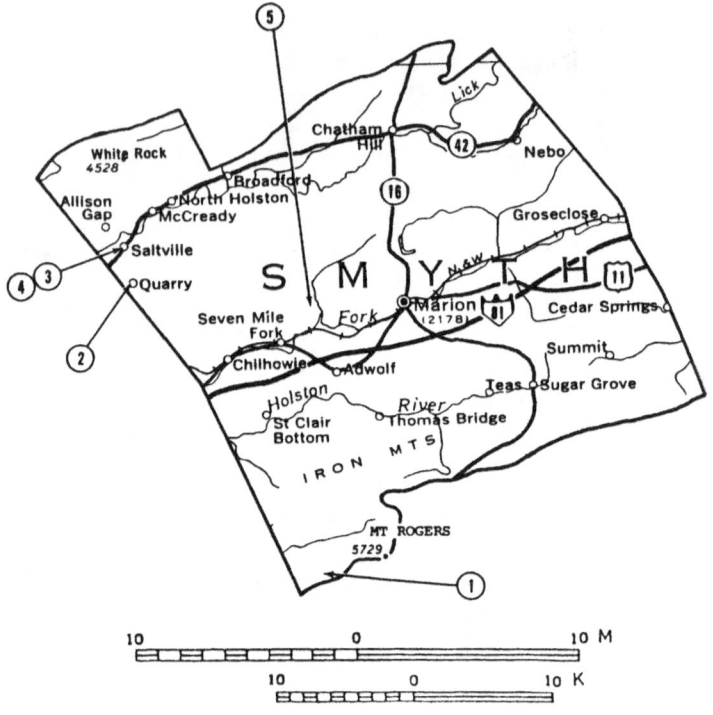

    Fluorite. (4-5-rhyolite)

2. Quarry area: at and near the Portersfield Quarry.
       Quartz, rock. (1-13)

3. Saltville area: at the Worthy Mine.
       Quartz, rock. Quartz, scepter. (2)

4. Saltville area: northeast; northwest side of road; along the
   North Holston River.
       Celestite: geodes to 11″ diameter; light blue crystals
   lining interior. (1-4-7-13-20)

5. Seven Mile Fork area: 2.1 miles northeast; at the Myers-Copenhaver Mine.

> Barite: prismatic crystals; radiating tabular crystal clusters. (2-5-limestone)

SPOTSYLVANIA COUNTY

1. Brokenburg area: 0.5 mile east of town on SR-208; at Edenton Mica Mine.

> Beryl. (2)

2. Fivemile Fork area: 1.3 miles west on SR-3; on south side of road.

> Monzanite. (1-29-34-granite)

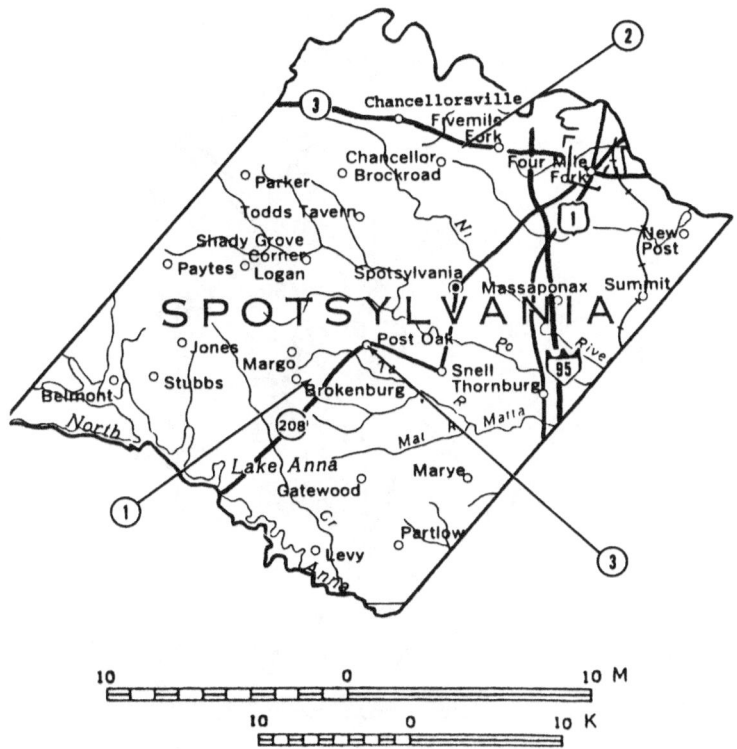

3. Post Oak area: 0.1 mile east-southeast on SR-606; on north
side of the road.
   Monzanite. (1-29-34-granite)

STAFFORD COUNTY

1. Hartwood area: at the Eagle Mine.
   Gold. (2)

2. Hartwood area: at the Rappahannock Mine.
   Gold. (2)

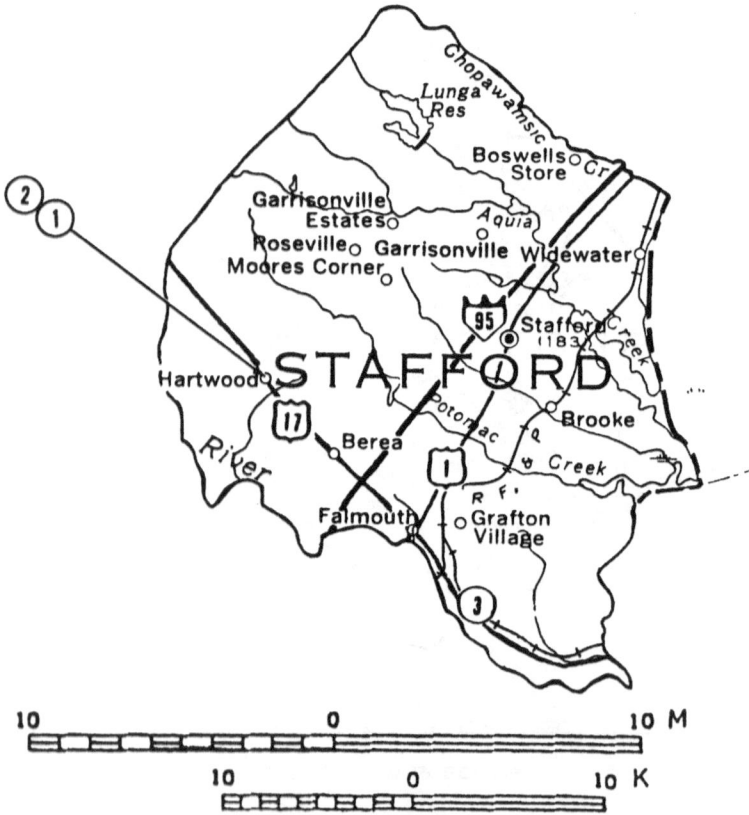

STATEWIDE

1. In many streams, including the James, Clinch, and Powell Rivers and their tributaries; also in many lakes.

 Pearl. (7)

TAZEWELL COUNTY

1. Pounding Mill area: at the Gillispie farm.

 Diamond: 1 stone; 0.83 ct; good color; cut into exquisite gemstone; found in cornfield in 1913. (1)

WARREN COUNTY

1. Browntown area: in surrounding fields and streams.
    Unakite. (1-7-13-20)

2. Front Royal area.
    Unakite. (7-19-granite -20-29)

3. Front Royal area: 3 miles west.
    Apatite. Chlorite. Dolomite. Epidote. Ilmenite. Leu-coxene. Magnetite. Perovskite. Phlogopite. Pyrite. (16)

4. Waterlick area: 2.25 miles northwest on SR-55; on south side of the road.
    Chlorite. (29)

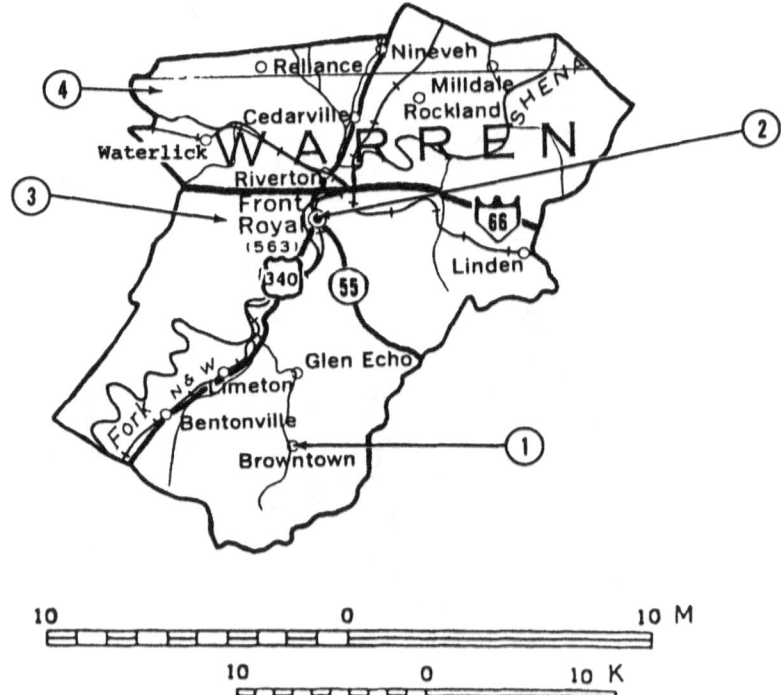

WASHINGTON COUNTY

1. Holston area: in vicinity of Alvarado; at the old Roverside Mine.

>   Hematite. Maghemite. Magnetite. (2)

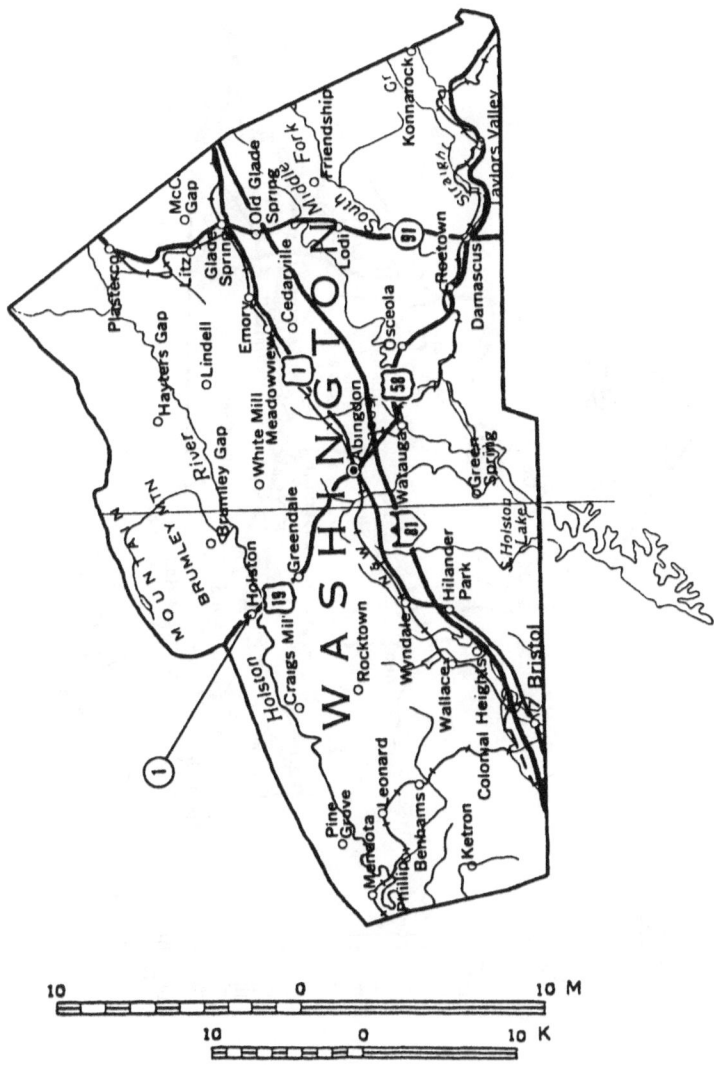

WISE COUNTY

1. East Stone Gap area: 0.5 mile east.
     Celestite. Hexahydrite. Leonhardtite. (12)

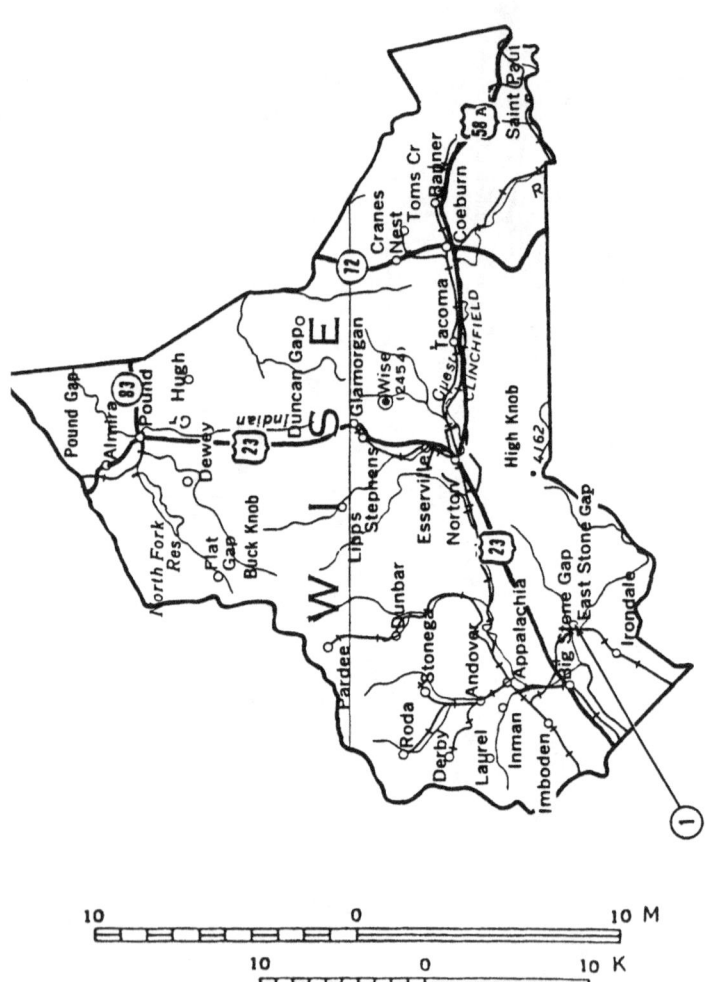

WYTHE COUNTY

1. Austinville area.
    Psilomelane. (1-13-29)

2. Wytheville area: in Speedwell Cave.
    Hydromagnesite. (32)

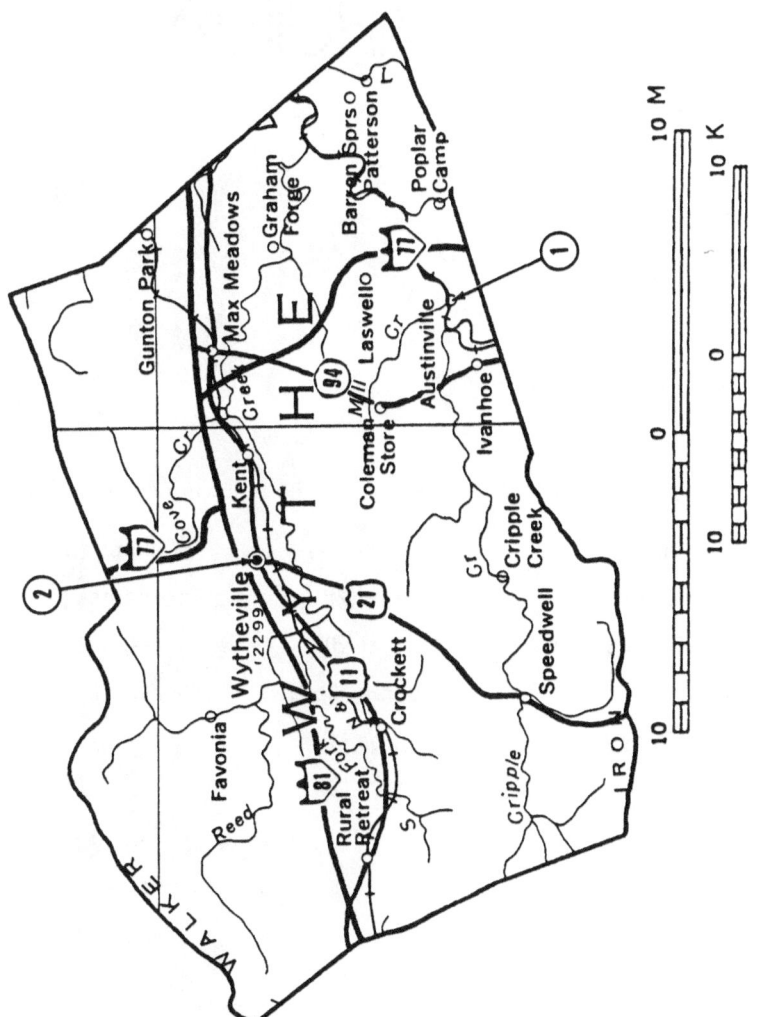

WEST VIRGINIA

Statewide total of 91 locations in the following counties:

Berkeley (3)	Kanawha (2)	Pendleton (8)
Grant (7)	Mercer (2)	Pocahontas (5)
Greenbrier (7)	Mineral (8)	Randolph (6)
Hampshire (4)	Monongalia (13)	Tucker (2)
Hardy (8)	Monroe (4)	Wetzel (2)
Jefferson (8)	Morgan (2)	

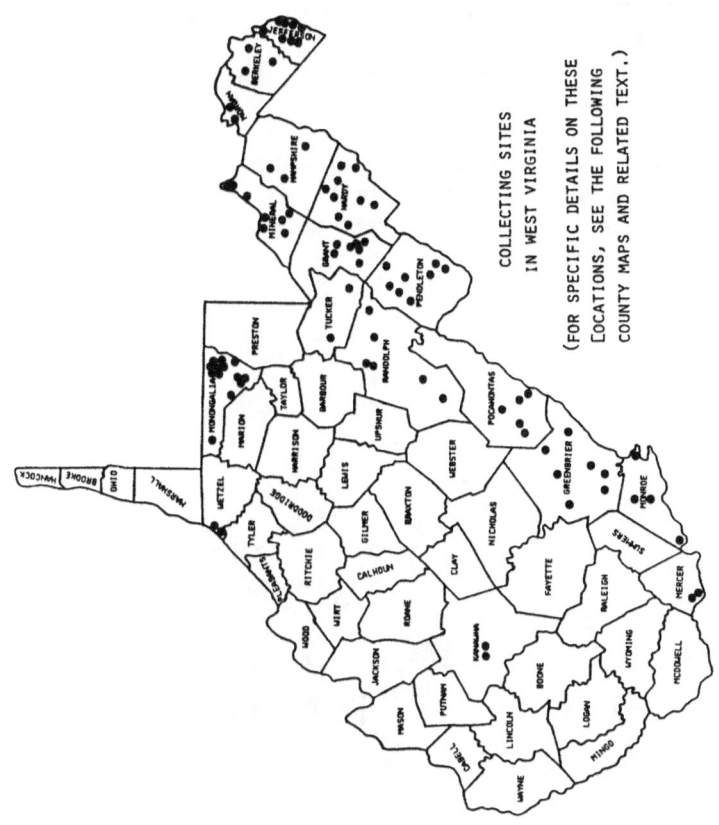

COLLECTING SITES
IN WEST VIRGINIA

(FOR SPECIFIC DETAILS ON THESE
LOCATIONS, SEE THE FOLLOWING
COUNTY MAPS AND RELATED TEXT.)

BERKELEY COUNTY

1. County-wide in limestone quarries.
   Calcite. Dolomite: crystals. Quartz, rock. (12)

2. Martinsburg area: 2 miles south; at the Martin Marietta
   Company Quarry.
   Calcite: transparent crystals. Dolomite: white; curved
   crystals. (12)

3. Martinsburg area: 3.5 miles northeast on US-11; on east
   side of road; at the Martin Marietta Company quarries.
   Calcite: transparent crystals. Dolomite: white crystals. Fluorite: purple crystals. Travertine: yellow;
   banded. (12)

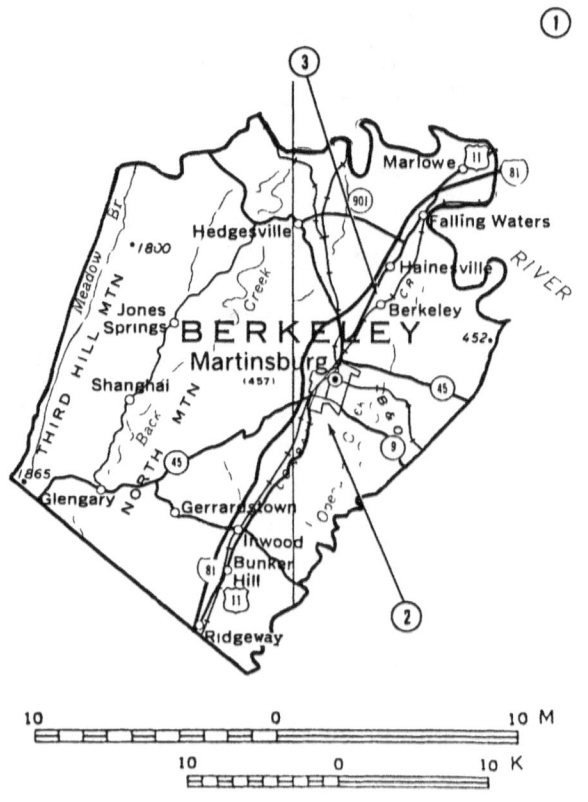

GRANT COUNTY

1. Maysville area.

    Hematite. (5-red quartzite-29)

2. Maysville area: on SR-42.

    Quartz, rock: glittering drusy on Tuscarora Sandstone. (12)

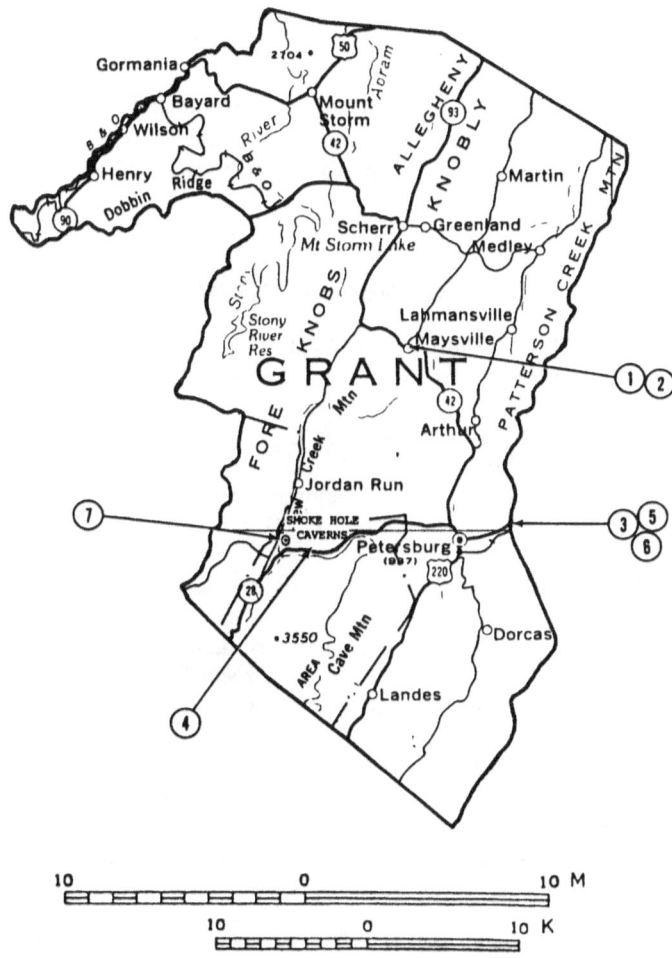

10   0   10 M

10   0   10 K

3. Petersburg area: 2 miles east; at abandoned quarry.
    Calcite. Chert. Dolomite. Quartz, rock. (12)

4. Petersburg area: 6 miles west on SR-28; 0.5 mile from road; in North Fork Gap.
    Calcite. Celestite. Dolomite. Travertine. (12-limestone)

5. Petersburg area: at Cosner Gap.
    Hematite: botryoidal clusters. (1-13)

6. Petersburg area: at the east end of Kline Gap.
    Hematite: botryoidal crystals. (1-13)

7. Smoke Hole Caverns area: in limestone cliffs.
    Chalcedony: geodes to 7" diameter. Fossils: flora and fauna; in limestone. Gypsum. Pyrite. (1-13-14-20)

GREENBRIER COUNTY

1. Alvon area manganese mines.
    Psilomelane: botryoidal clusters. (2-5-manganese ore)

2. Crawley area: just north on US-60 to RR tracks; in area of Eckle School.
    Calcite. Celestite. Dolomite: crystals. Oolite. Pyrite. Quartz, rock. Sphalerite. (12)

3. Fort Spring area: at the Acme Limestone Company Quarry.
    Calcite: white. Celestite: sky blue. Dolomite: brown. Limestone: oolitic. Quartz, rock: fine clear crystals; some doubly terminated. (12)

4. Lewisburg area: east on US-60; along little creek paralleling US-60; downstream to its mouth at the Greenbrier River.
    Quartz, smoky: transparent brown; many perfectly doubly terminated. (1-7-13-20)

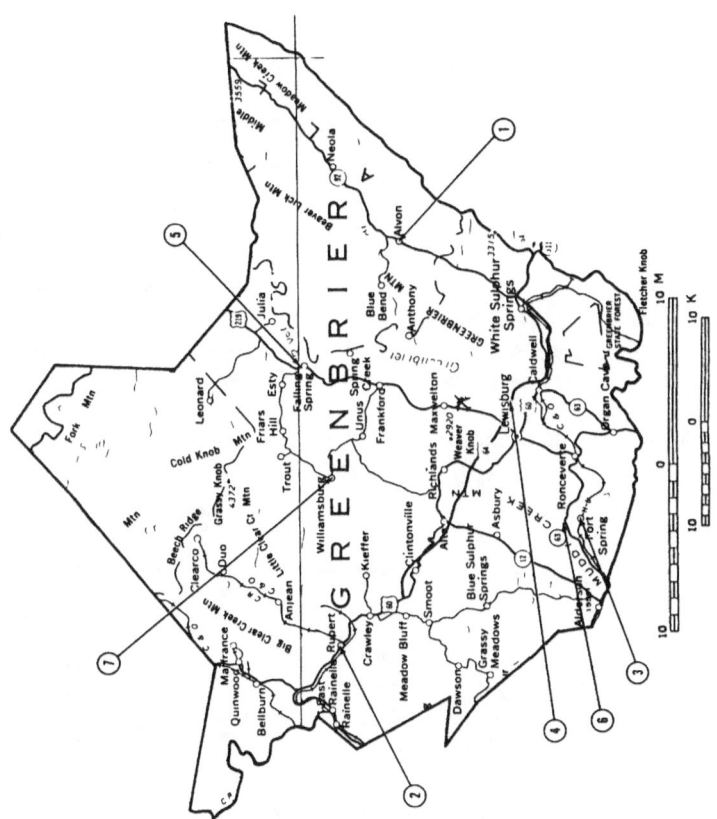

5. Renick area (aka Renick Station and Falling Spring): 1 mile east; just above RR tracks on steep hillside.
    Calcite: white. Fluorite: purple. (12)

6. Ronceverte area: 4 miles west.
    Quartz, rock. (1-12-13)

7. Williamsburg area.
    Fossils: marine; especially rose pink to deep red silicified corals. Hematite. Quartz, rock. (1-12-13)

HAMPSHIRE COUNTY

1. Hanging Rock area: 1 mile south.

    Quartz, rock: fine gem quality; transparent. (8-14-29-limestone)

2. Hanging Rock area: 2 miles northeast on US-50; in limestone quarry.

    Calcite: white. Fluorite: purple crystals. (12)

3. Romney area: 3 miles south; in gap of Mill Creek Mountain; at the Tonoloway Limestone Quarry.

    Calcite: white. Celestite: transparent; clear to pale blue. Chert. Quartz, rock: transparent. (12)

4. Wapocomo area: east on Poland Road to quarry.

    Barite. Calcite. Pyrite. (12-limestone)

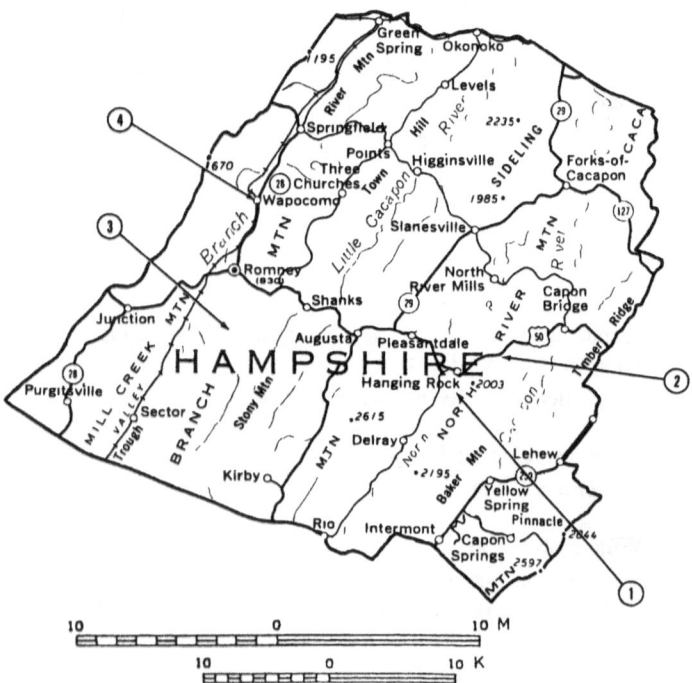

HARDY COUNTY

1. Baker area: 5 miles northwest; on the east slope of Branch Mountain.

   Hematite. Quartz, rock. Quartz, smoky: pale crystals. (1-13-14-29)

2. Baker area: 0.5 mile north of SR-55; near the Baker Lime Plant; on side of hill just north of Lost River.

   Calcite: massive; transparent; white opaque. Dolomite: cream-colored crystals. Travertine: yellow; banded. (12)

3. County-wide iron mines.

   Hematite. Jasper. (2)

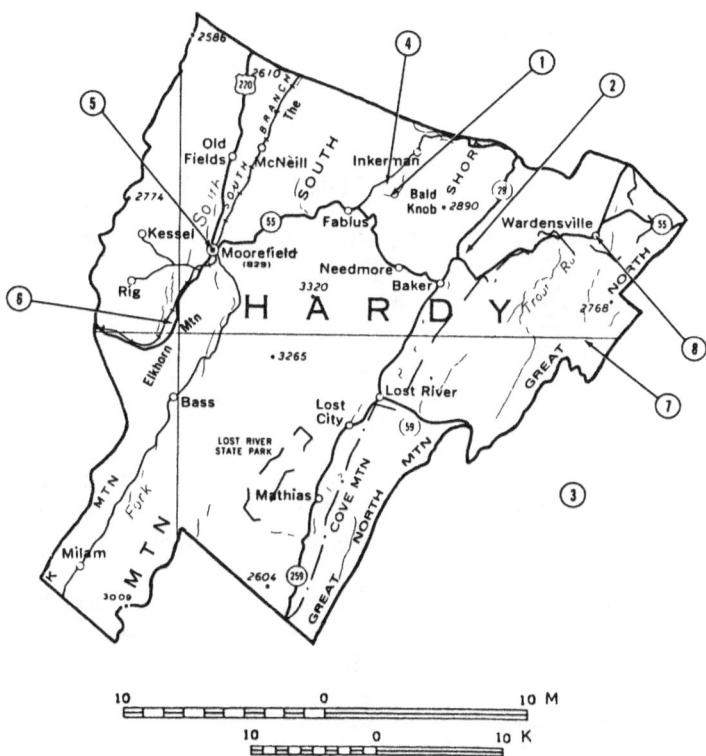

4. Inkerman area: 2.1 miles south-southwest; on east side of Branch Mountain.

>   Jasper: red. Quartz, rock. Quartz, smoky: very transparent. (1-13-29)

5. Moorefield area: 2 miles from Asbury Church; on sides of South Branch Mountain.

>   Quartz, milky; crystals. Quartz, rock. (5-massive quartz)

6. Moorefield area: 4 miles south on US-220.

>   Barite: white. Dolomite: yellow. Septarian: concretions to 16″ diameter. (4-14-41)

7. Wardensville area: 5 miles south on Waites Run Road.

>   Hematite: botryoidal crystal clusters. (1-13-20)

8. Wardensville area mines.

>   Hematite: botryoidal nodules. (2)

JEFFERSON COUNTY

1. Charles Town area: 7 miles southeast; on east side of Shenandoah River; at the Howell Zinc Prospect.

>   Dolomite: crystals. Galena. Sphalerite. (21)

2. Charles Town area: just east of city.

>   Dolomite: rhombohedral crystals. Pyrite: bright brassy cubic crystals. Quartz, rock. (9-12)

3. Halltown area: 0.5 mile southeast; at the Pennsylvania Glass Sand Corporation Quarry.

>   Calcite. Dolomite. Pyrite. Quartz, rock. (12)

4. Mannings area: 1.75 miles west on SR-9; on bank of Shenandoah River; at the Howell Zinc Prospect No. 2.

>   Dolomite. Galena. Sphalerite. Zinc. (2-21)

5. Mannings area: from SR-9 at Snickers Gap, turn north on poor, unmarked road and go 2 miles.

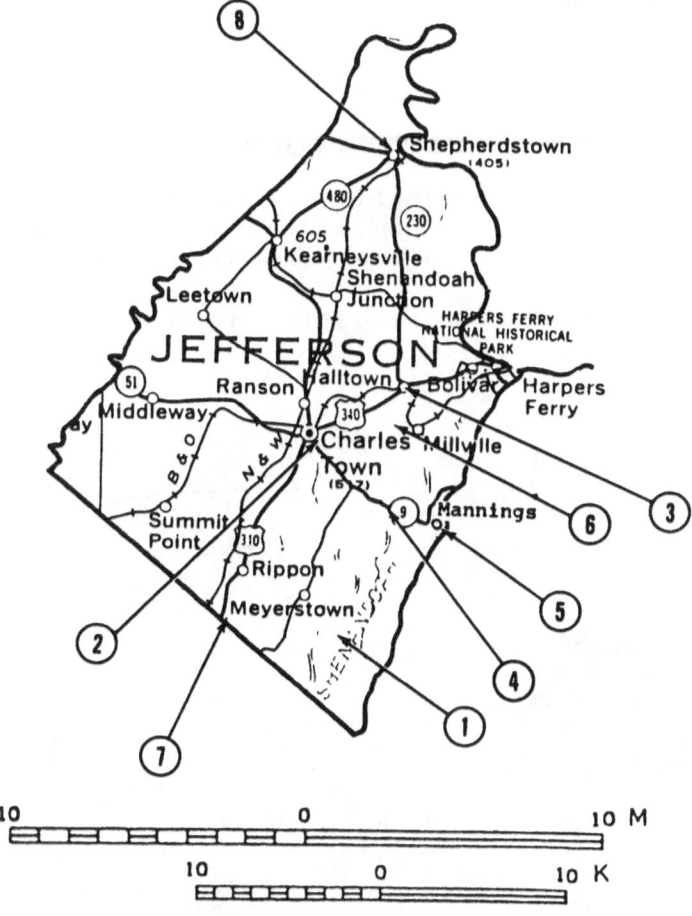

10   0   10 M

10   0   10 K

Epidote: green. Quartz, rock. (5-massive quartz -29-metamorphic rock)

6. Millville area: just west; at the Moler Quarry.
   Calcite. Dolomite. Pyrite. Quartz, rock. (12)

7. Rippon area: south on US-340 near the Virginia border; in road cut.
   Tourmaline. Zircon. (4-5-quartzite)

8. Shepherdstown area.

Calcite: white. Fluorite. (12)

KANAWHA COUNTY

1. South Charleston area: from the junction of Connell Road
   and Woodvale Drive, follow latter to sharp curve; search
   in woodlands beyond stone wall (wall built in part of petrified wood).

   Petrified wood: pieces to 18 lb. (1-4-13)

2. South Charleston area: on Berry Hills County Club Road; collect along north side of creek.

> Petrified wood: pieces to 6 lb. Quartz, rock: as drusy on petrified wood. (1-3-7-13-14-20-29-38)

MERCER COUNTY

1. Bluefield area: on US-21 Byp.
   > Chert: jet black. (12)

2. Willowtown area: in abandoned quarry.
   > Calcite (Onyx): banded; pastel green; translucent; high quality. (12)

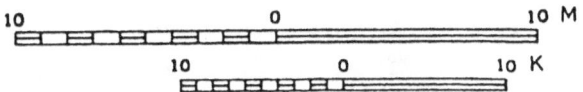

10   0   10 M

10   0   10 K

MINERAL COUNTY

1. Burlington area: 1.9 miles northwest; off Mill Creek; along Dry Run.
   Galena. (19-milky quartz -29)

2. Burlington area: 1 mile southeast; at mines.
   Hematite. (2)

3. Keyser area: east a short distance; at base of mountain; in two black limestone quarries.
   Calcite: white. Celestite: pale blue. Chert: white. Dolomite: brown. Fluorite: pink to purple. Galena. Sulphur: crystals. Travertine: banded. (12)

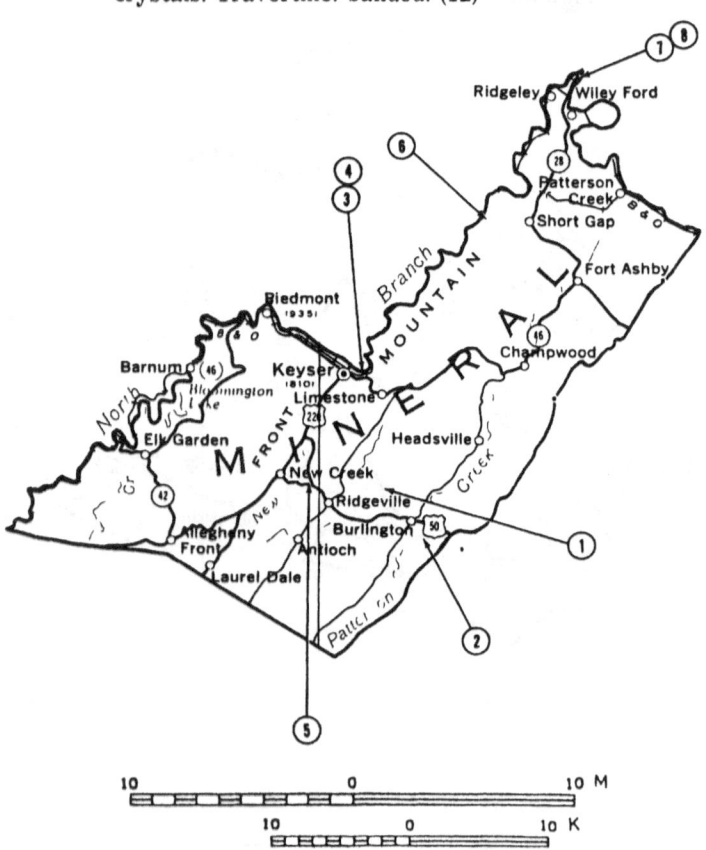

4. Keyser area: east on SR-46; at base and on face of Knobly Mountain.

   Chalcedony: geodes to 3″ diameter; bright quartz crystals linings. Chalcedony: massive; gray; with marine shell replacements. Sphalerite. (1-8-13-14-29)

5. New Creek area: 1 mile east on US-50 near its junction with US-220.

   Calcite: white crystals. Chert: snowy white. Hematite. (5-red quartzite -12)

6. Short Gap area: 1.6 miles west; just east of the crest of Knobly Mountain; at the Aurora Stone Company Quarry.

   Calcite: transparent crystals. Dolomite: opaque ivory; curved crystals. Fluorite: purple cubic clusters. Sulphur: crystals. Travertine: yellow. (12)

7. Wiley Ford area: 2.4 miles southeast; on west slope of Knobly Mountain; at Cedar Cliff.

   Celestite: blue: fine crystals. (5-9-limestone)

8. Wiley Ford area: 3 miles southeast; at west foot of Knobly Mountain; in RR cut.

   Celestite: blue; fine crystals. (4-5-9-limestone)

MONONGALIA COUNTY

1. Barker area: at the Connellsville Coal Company No. 1 Mine.

   Calcite. Marcasite. Petrified wood. Pyrite. (2)

2. Cassville area: just west; in headwaters of Scott Run.

   Opal: transparent to deep yellow-orange stalactites. (1-13-20)

3. Coopers Rock area: 2.5 miles north.

   Hematite: nodules to 3″ diameter. (1-13-29)

4. Coopers Rock area: underside of large slabs of sandstone.

   Opal: transparent to deep yellow-orange stalactites. (1-13-20)

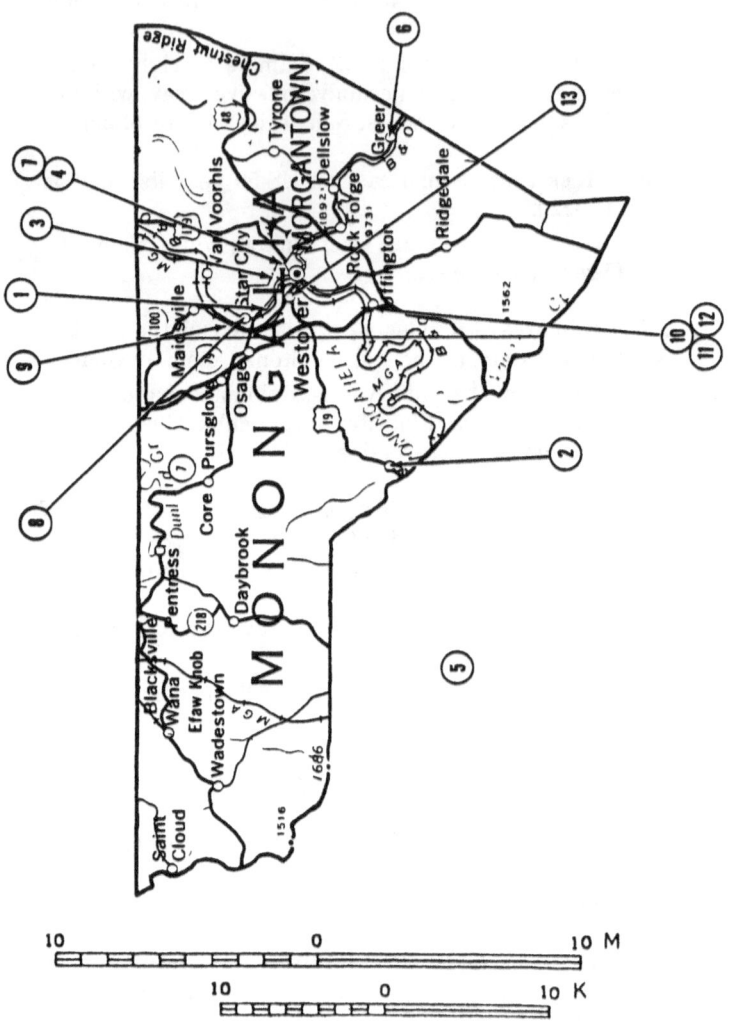

5. County-wide.
    Barite: crystals. Sphalerite; crystals. (1-13-14-20-28-
    29-41-siderite)

6. Greer area; along SR-7.
    Calcite: pink crystals. (1-13-29)

7. Morgantown area: just south of Booths Creek; in road cut through hillside.

    Melanterite. Pyrite. (4-19-coal)

8. Star City area: at the mouth of Scott Run.

    Pyrite concretions. (1-7-13-14-20-41)

9. Star City area: 0.8 mile north; upstream from the mouth of Scott Run.

    Pyrite: brassy crystals lining concretions. (1-13-14-20-29-41)

10. Uffington area: at exposures of Brush Creek Shale along SR-73.

    Siderite: nodules to 3" diameter. (4)

11. Uffington area: in road cuts south of Booths Creek.

    Melanterite: green. Pyrite. (4-5-coal)

12. Uffington area: upstream and down in Brush Creek.

    Siderite: nodules to 6" diameter. (28-29)

13. Westover area: follow Westover-Everettville Road to Grant Chapel; turn right and follow the road downhill to the river; park there and go 0.25 mile on foot to sandstone cliff above the RR track.

    Barite: crystals. Calcite: crystals. Melanterite. Selenite. Siderite: nodules to 4" diameter. (1-13-14-sandstone -20-29)

MONROE COUNTY

1. Peterstown area: vacant lot within city limits.

    Diamond: 1 stone; 34.46 ct; pale greenish gray; named the "Punch Jones Diamond"; kicked loose from soil by horseshoe during a horseshoe game in 1928. (6)

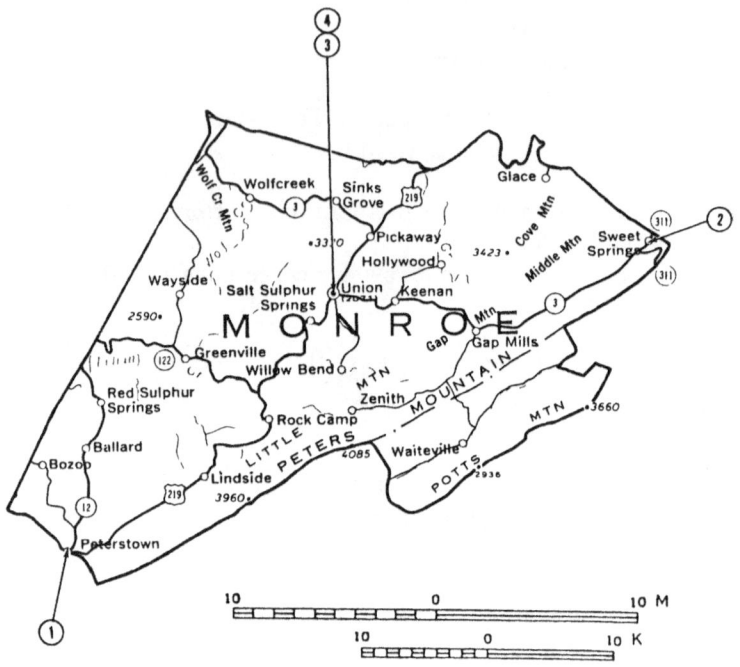

2. Sweet Springs area: on crest and southeast side of Moss
   Mountain.

   Chert: nodules to 3″ diameter. Psilomelane: lumpy
   nodules to 4″ diameter. Quartz, rock. Quartz, smoky.
   (1-8-13-14-29)

3. Union area: in both directions along Turkey Creek.
   Quartz, rock: some with clay inclusions. (1-7-13-20)

4. Union area: on the Fullen Brothers farm; in Turkey Creek.
   Quartz, rock. (1-7-13-20)

MORGAN COUNTY

1. Berkeley Springs area: for 4 miles northeast along Warm Springs Ridge; at a series of Pennsylvania Glass Sand Corporation quarries.

   > Calcite: white crystals. Japser: varied colors; pebbles. Pyrite: small crystals. Quartz, rock: excellent clear crystals, single and in clusters. Selenite. (12)

2. Rock Gap area: 2 miles north of Cacapon Mountain State Park entrance; on a side road turning off US-522.

   > Calcite: transparent crystals. Dolomite: yellow crystals. Quartz, rock. (12)

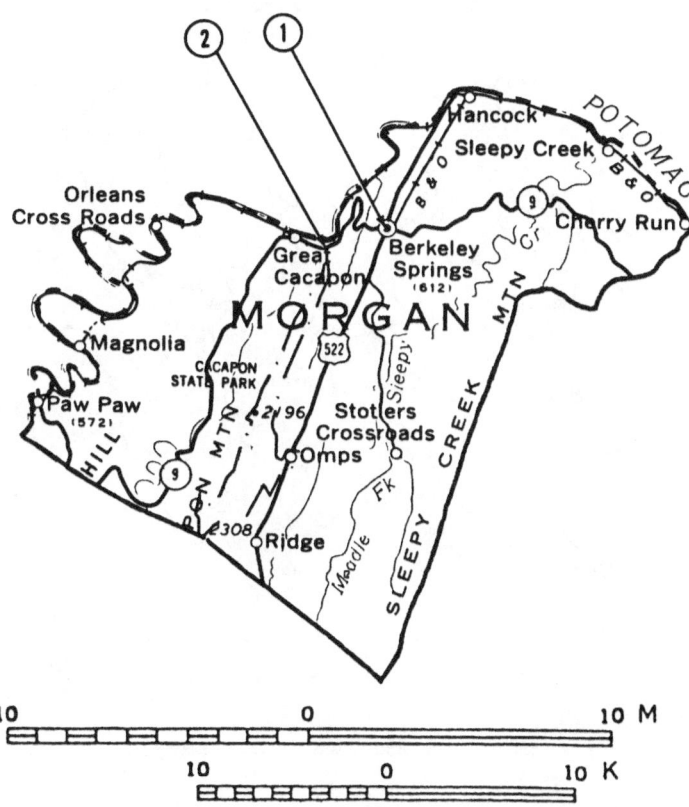

PENDLETON COUNTY

1. Franklin area: 4 miles west on US-33 near Friends Run.
   Chert. Quartz, rock: very clear. Quartz, smoky: faintly
   smoky enhancement in transparent crystals. (1-4-5-Oris-
   kany sandstone-29)

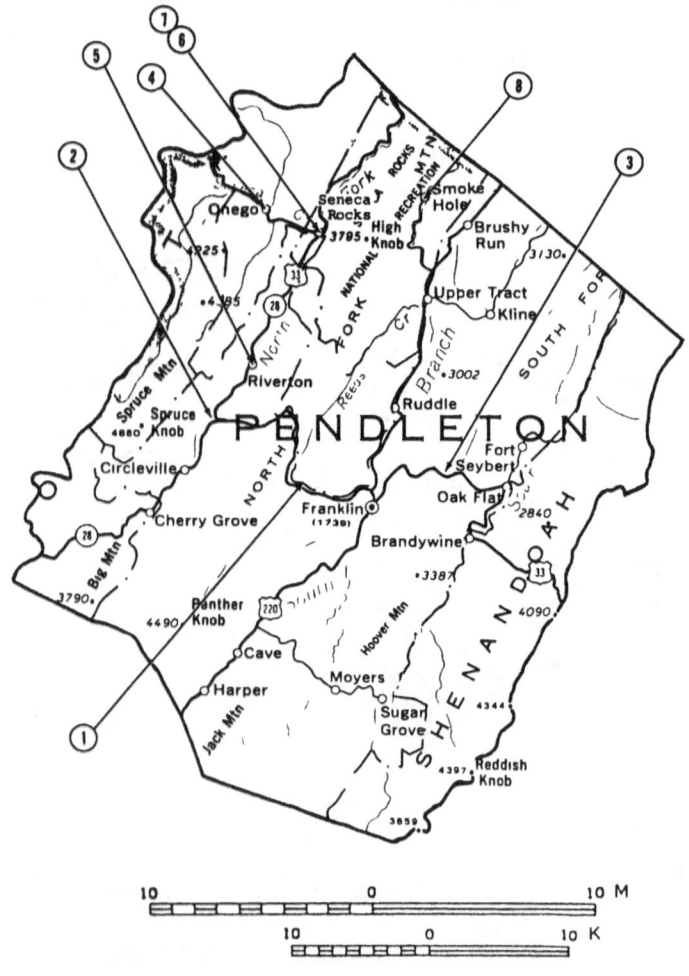

2. Judy Gap area: at limestone quarry.

> Calcite: white crystals. Dolomite: yellow crystals. Quartz, smoky: crystals with a silvery cast. Travertine: drusy on limestone. (12-38)

3. Oak Flat area: 1.2 miles west on US-33 in Hively Gap; on south side of highway; at downslope quarry.

> Calcite: crystals; white, clear. Celestite. Selenite: good bladed crystals. Sulphur, native: good crystals and crystal clusters; also nodules to 8″ diameter. (12)

4. Onego area: just east of town on US-33; at the Onego Quarry.

> Calcite: white. Fluorite: lilac to purple; cubic crystal clusters. Travertine: banded; yellow and white. (12)

5. Riverton area.

> Barite: crystal linings and fillings in concretions. (1-7-13-14-20-29-41)

6. Seneca Rocks area: where SR-28 splits off from US-33; at the German Valley Limestone Company Quarry No. 1.

> Travertine: banded; yellow, white. (12)

7. Seneca Rocks area: take SR-28 to Mill Creek Road; follow latter for 1.6 miles to the German Valley Limestone Company Quarry No. 2.

> Calcite: white crystals. Fluorite: lavender cubic crystals. (12)

8. Smoke Hole area: in limestone cliffs.

> Fossils, marine: invertebrates. Gypsum. Pyrite. Quartz: geodes to 8″ diameter. (1-13-14-29-limestone)

POCAHONTAS COUNTY

1. Edray area: north on US-219 to north edge of town; turn right on Clover Lick Road; east to West Virginia Road Commission Stone Quarry.

> Anhydrite: gray. Calcite: white, yellow. Celestite:

blue. Dolomite: brown. Fluorite: violet. Gypsum: tan; chatoyant. (12)

2. Hillsboro area: 2 miles southwest to left turn onto Locust Creek Road; 1.5 mile to stone bridge over Locust Creek; in creek.

Fossils: marine; especially silicified corals; pastel blue to deep rich blue. (1-7-13-20)

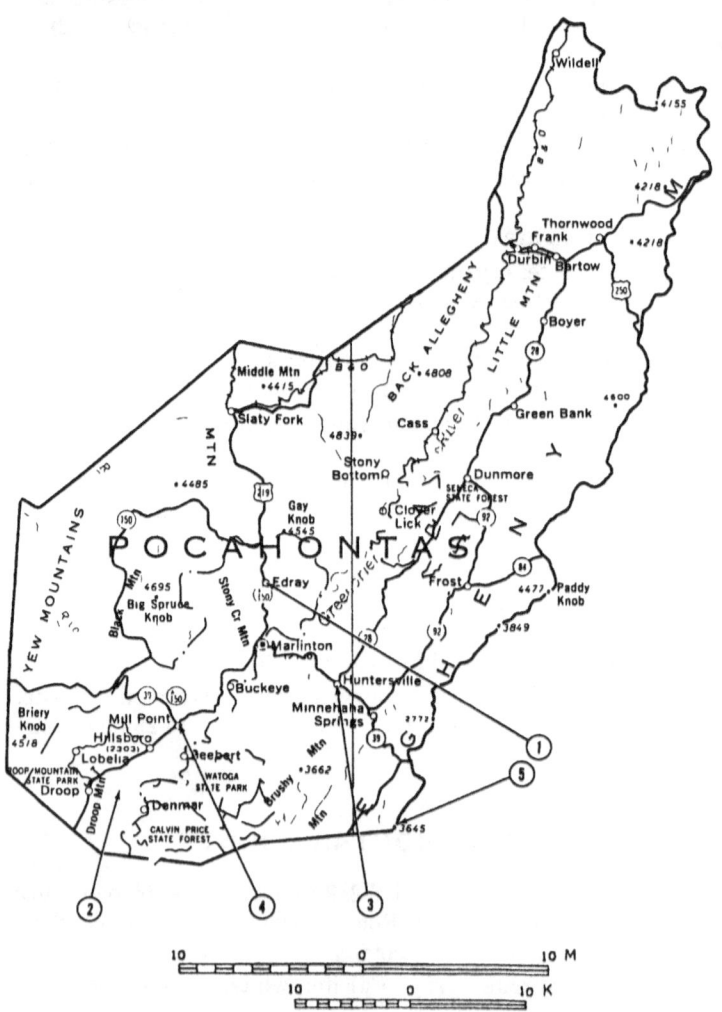

3. Huntersville area: 1.5 miles southeast of Browns Creek; 1 mile off SR-39; at Possum Hollow strip mines.

> Chert: patterned nodules to 2″ diameter. Hematite: oolitic. (1-2-13)

4. Mill Point area: just north of the junction of SR-39 and US-219; at the Mill Point Quarry.

> Fossils: marine; especially silicified corals; pastel pink to deep red. (1-7-12-13-20)

5. Minnehaha Springs area: 9 miles south; on the southeast slope of Beaver Lick Mountain; at iron mines.

> Psilomelane. (2)

RANDOLPH COUNTY

1. Bowden area: just east; near fish hatchery.
> Dolomite: brown, pink. (12)

2. Cheat Bridge area: 3.2 miles west on US-250.
> Calcite: crystals of various pastel colors. Pyrite. Quartz, rock. (12)

3. Elkins area: on US-33 at the Paulina Limestone Quarry.
> Calcite: pink, white; crystals. (12)

4. Elkins area: on the Simmons farm; just off US-219.
> Pyrite: bright brassy cubic crystals. (1-13-20)

5. Harman area: along SR-32 and US-33.
> Calcite. Dolomite. Quartz, rock. (19-limestone-29)

6. Valley Head area: 2.5 miles south on US-219.
> Fossils: flora; some replaced by pyrite. Pyrite. (29-gray sandstone)

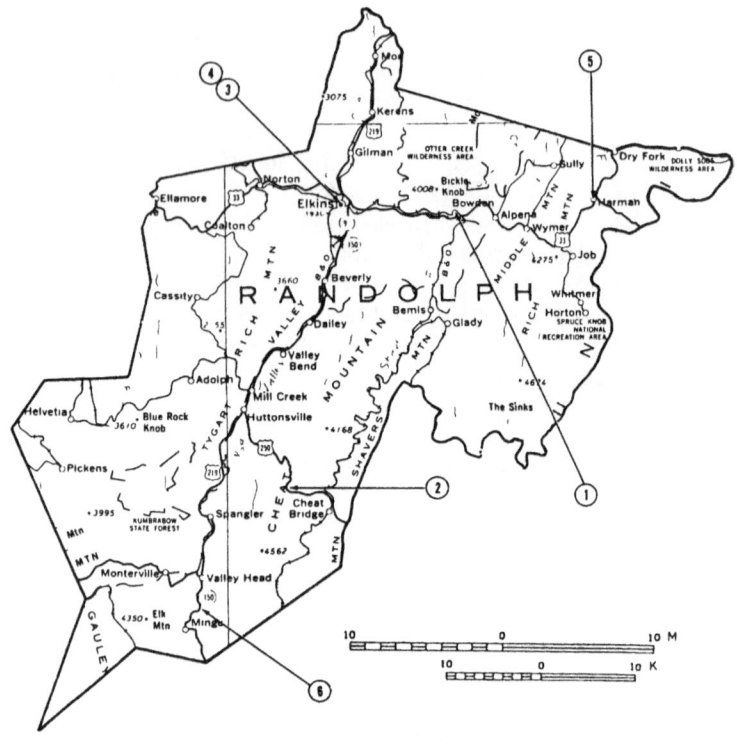

TUCKER COUNTY

1. Laneville area: 3 miles northeast; in Seneca Creek.
   Chalcedony: geodes to 7″ diameter. Fossils: flora and
   fauna; in limestone. Gypsum. Pyrite. (1-7-13-14-20)

2. Parsons area: in Sissaboo Hollow.
   Pyrite. (29-quartz)

WETZEL COUNTY

1. New Martinsville area quarries.
   Calcite: crystals; pink. (9-12)

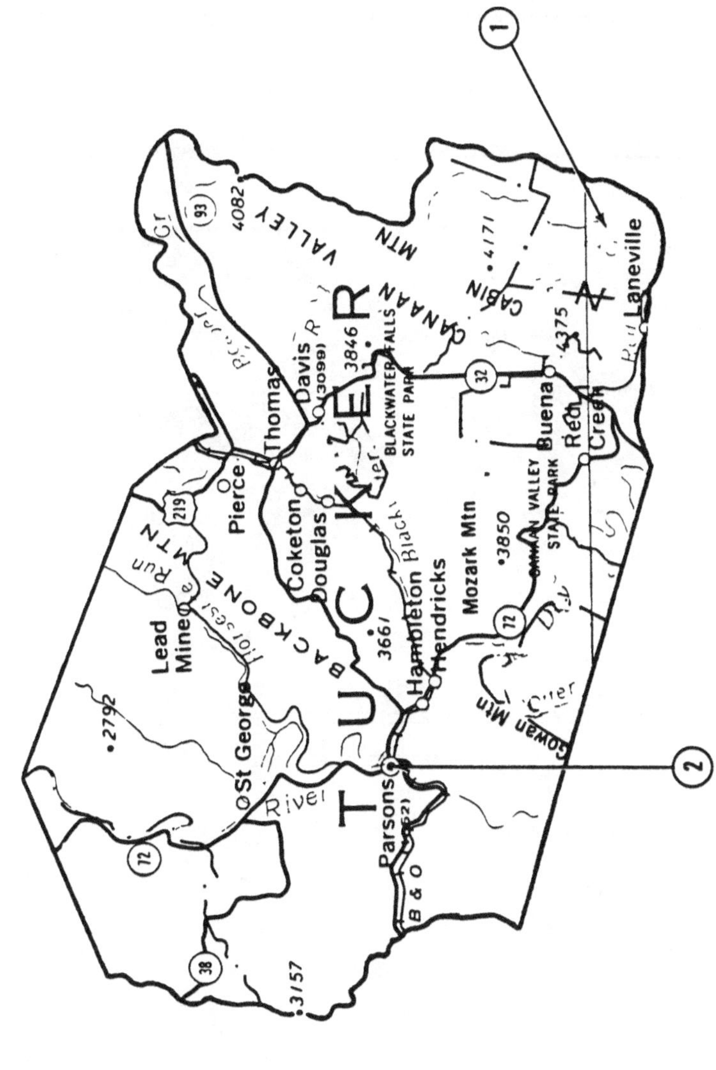

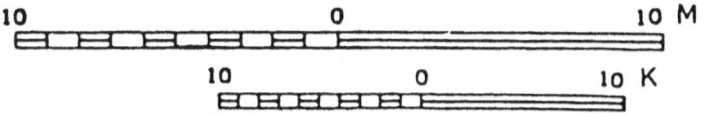

TUCKER

93 4082

Cr.

VALLEY MTN

CABIN 4171

Laneville

Run

Davis R.

3846

CANAAN

Thomas

3099

Blackwater Falls STATE PARK

32

4375

Buena
Rich
Creek

Run Creek

N

Pierce

219

Coketon

Douglas

Black

3661

Mozark Mtn

3850

CANAAN VALLEY
STATE PARK

72

Dry R.

Lead
Mineo

Run

BACKBONE MTN

Hambleton

Hendricks

Gowan Mtn

Cowan Creek

2792

St George

River

Parsons

2621

72

B & O

38

3157

Scale:

10   0   10 M

10   0   10 K

2. New Martinsville area: 3 miles southwest; along the Ohio River.

   Muscovite. Petrified wood. Pyrite. (1-13-20)

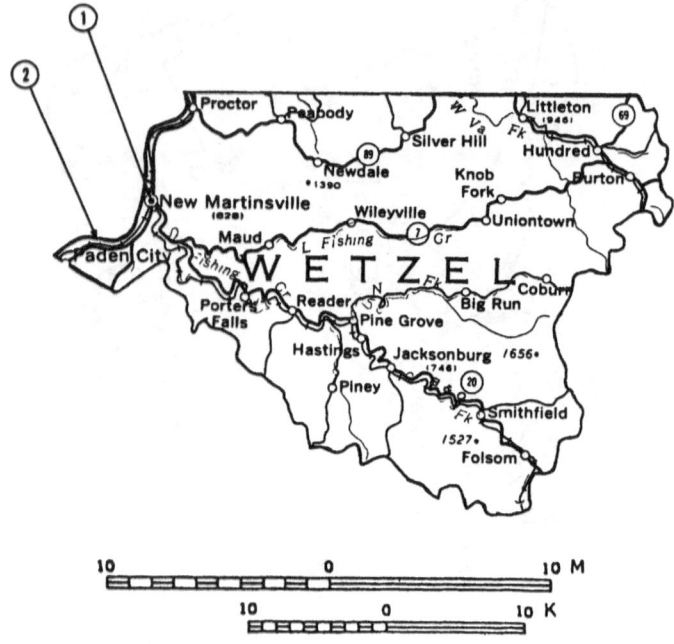

GLOSSARY

Accessory minerals Minerals usually present in igneous rocks, but only in minor abundance, their presence not being essential to the size, shape, or definition of the rock.

Accidental minerals Minerals found (sometimes in large quantities) in a particular rock type where they are not normally associated.

Acicular Describing a crystal that is considerably elongated, often to the point of being needle-shaped.

Adamantine Exhibiting a bright luster, as the diamond does.

Adit A mine entryway from outside; usually horizontal or only slightly inclined.

Aggregate A mass of rocks or minerals naturally cemented together.

Alluvial Pertaining to sand, silt, clay, and gravel deposits laid down by running water, most often in areas where the water flow no longer exists or is periodic.

Ameboid Featuring a coating of minute lobate crystals imparting a distinct texture.

Amorphous Describing a mineral or rock mass lacking crystal structure; literally, a formless mineral mass (e.g., opal).

Amygdule A usually round cavity in volcanic rock formed by cooling of the fluid rock around a gas bubble; the cavity is very often lined or filled with secondary minerals, often crystallized.

Anhedral crystals Crystals that, due to varying causes, do not exhibit their usual crystal form.

Argillaceous Largely comprised of (or containing) clay (e.g., some shales and slates).

Arroyo The channel, usually with steep banks, cut by a creek; its bed is ordinarily dry or has only rare and intermittent flowage.

Ash Very tiny volcanic cinders that have not become solidly cemented.

Asparagus stone   A chartreuse (yellow-green) apatite.

Asterism   The property of reflecting light in starlike patterns of 4, 6, or 12 rays (and, rarely, 24 rays). Sapphire and ruby are examples. (See also *stellate.*)

Badlands   A region, usually of desert characteristic, essentially barren of vegetation, in which erosion (by wind and, less often, water) has carved the soft sedimentary deposits into rather incredible pinnacles, formations, and gullies (e.g., the South Dakota Badlands).

Banded   Describing a form of agate (and often other minerals) in which a pattern of arced (rather than angular) stripes occur, usually in concentric manner and either of different colors or gradations of the same color. (See also *fortification.*)

Bar   A deposit of gravel, sand, or mud in streams or rivers where currents are less swift, as on the inside of bends and at the mouths of streams.

Basalt   A very dense dark rock formed by solidified lava and mainly comprised of plagioclase feldspar and pyroxene; often contains crystal-producing gas cavities; frequently found in flowage columns as the result of contraction during cooling, creating a tree-trunk appearance (e.g., Wyoming's Devils Tower).

Batholith   A vast mass of igneous rock (cooled molten rock) that formed deep underground.

Bed   Rocks (usually of a single type) laid down in strata forming roughly parallel layers.

Bladed   A crystal structure that is thin, flat, and elongated.

Bolson   A relatively level, often roughly circular valley draining to a central basin or pan, surrounded by high terrain. (See also *playa.*)

Botryoidal   Formed into masses of small symmetrically spherical bulges; often in grapelike clusters.

Breccia   A mass of jagged or angular mineral fragments naturally cemented together.

Brittle   Easy to break.

Butte   A steep-sided and usually isolated large hill or small mountain.

Calcareous   Composed of (or containing) calcite or calcium carbonate.

Capillary   Extremely elongated hairlike crystal form.

Carat   A gemstone weight unit equivalent to 200 milligrams (0.20 gram); there are 141.75 carats to the ounce (avoirdupois).

Cast   A mineral that takes the shape of a cavity in rock left by the disintegration of plant or animal matter.

Cat's-eye   A translucent stone whose inner fibers reflect light in a sharp streak across the surface.

Cattle-guard   A barrier for cattle, constructed across an opening in a fence or at a gate through the parallel placement of slightly separated pipes (usually over a ditch or trench) upon which cattle cannot secure safe footing.

Cephalopod   An ancient fossil invertebrate, cylindrical in shape and segmented (often internally silicified) and distantly related to the modern squid.

Cerro   Hill or (sometimes) mountain.

Chatoyant   Having the property of reflecting light in an undulating or moving streak across the surface.

Cinder cone   The naturally expelled debris, usually in a conical pile, surrounding the vent of a volcano.

Claim   A piece of land legally filed upon and held for the purpose of mineral prospecting and exploitation; the maximum size a lode claim can be in the United States is 600′ × 1500′.

Clast   A rock fragment or mineral fragment, the fragmentation having occurred naturally, through volcanic explosion, earthquake, or meteoritic impact or unnaturally during transportation.

Clay   Exceptionally fine-grained earthy substance derived from the weathering of granite and other aluminous rock.

Cleavage   The smooth plane surface along which a mineral splits or parts.

Colloform   Having a smoothly rounded kidney shape. (see *reniform.*)

Color zone   The unblended layering, segregation, or striping of different colors within a crystal (e.g., tourmaline).

Columnar   Describing a crystal that has the shape of a column.

Conchoidal   Describing a concave fracture in a glassy mineral, the sort of fracture made by a pellet or BB striking plate glass; obsidian is markedly conchoidal in its fracturing.

Concretion   A generally spherical formation (often so precise as to appear artificial) caused by the natural cementing together of particles of gypsum, silica, and other minerals in sedimentary rock; easily freed from matrix material; regularly weathers free.

Conglomerate   Also called *puddingstone;* rounded pebbles of essentially the same size that have been naturally cemented together.

Contact   The zone where differing rock types meet.

Coprolite   The fossilized excrement of reptiles, birds, and mammals.

Core, pegmatite   The central mass of rock (often a bean-shaped or disk-shaped mass of quartz) found within a pegmatite body.

Coulee   In the northwestern states, a very steeply walled river valley or dry riverbed.

Country rock   The rock mass that surrounds ore bodies, veins, dikes, and the like.

Cove   A relatively small, level area intruding upon a mountainous area.

Crater The natural depression marking the vent of a volcano or impact site of a meteorite.

Crystal A mineral mass, solid and having a regular geometric shape, with smooth flat planes (crystal faces or, erroneously, facets).

Craze The network pattern of fine cracks that develop on the outer surface of some gemstones.

Crinoid An ancient invertebrate fossil animal having the appearance of long-stemmed, blossoming plant material; the stem is segmented and frequently separates in disklike pieces.

Crust The thin, outermost rocky layer of the earth.

Cryptocrystalline Describing a rock surface or solid rock mass comprised of extremely minute or microscopic crystals.

Cubic Describing an isometric system crystal in the form of a cube.

Cuesta The high exposed end of a sloping plain.

Cueva A cave or grotto.

Delta A deposit (often very large) of sand, gravel, and silt at the mouth of a river.

Dendritic Describing the branching or treelike pattern formed by a foreign mineral within another mineral; also descriptive of some minerals that generate themselves into such branching forms without the involvement of other minerals.

Deposit Mineral or ore concentrations laid down naturally in a specific area.

Diatom Fossilized (often silicified) microscopic marine plant frequently found in sedimentary rocks.

Diggings The excavations made by a prospector or miner; also, the worked mineral site generally.

Dike A flattened vertical body of igneous rock that has intruded through a major mass of country rock without being influenced by country rock layering.

Dip The downward angle of inclination of bedrock from the horizontal.

Divide The crest of land that separates the direction of stream flow in a given area.

Dome A rounded knob of rock, usually large and relatively isolated.

Drift The collection of rock and mineral debris deposited by a glacier at the terminus of its advance or by the runoff waters of that glacier; also, in mining, a horizontal tunnel following a vein or main ore body.

Druse A pocket, vug, or cavity in rock that is lined with crystals of the same minerals of which the rock is comprised (as opposed to amygdule or geode, in which the filling is not related to the country rock).

Drusy A crust of uniformly tiny crystals filling a cavity or coating a matrix.

Dump The piles of waste rock that are the residue of mines and prospects.

Eluvial A deposit formed by the *in situ* disintegration of rocks.

Enhydro A crystal that contains a fluid-filled cavity, sometimes with a bubble visible through the crystal walls.

Escarpment An extensive bank of sheer cliffs.

Estuary A bay that has formed at the mouth of a river where tide flows meet river current.

Evaporite A chemically originated sedimentary rock or mineral that was formed by the process of evaporation (e.g., halite: rock salt).

Exfoliation The process of weathered rock sheets splitting away in layers from larger underlying rock masses, leaving the remaining masses essentially rounded.

Extrusives Igneous rocks that reach the earth's surface while still in a molten state.

Fault A major fracture in rock along which ordinarily the opposing sides have slid in different directions on the same plane.

Ferruginous Containing iron.

Fibrous Containing or consisting of needlelike or threadlike crystal fibers, usually laid in parallel planes (e.g., asbestos).

Fire The intrinsic brilliance of a gemstone resulting from its characteristic of splitting light rays.

Fissure An extensive crack or break in a matrix material.

Fissure vein A band of mineral materials that intruded a fissure while in a molten state.

Flat A markedly level, usually treeless area.

Float A general term applied to pieces of rock some distance detached from their outcrop source.

Flow The hardened residue of a single expulsion of lava.

Flow banding A structure similar to folding where, in some volcanic rocks, alternating layers of unlike mineral material form and lie adjacent to one another as the result of lava flowing at intervals.

Flower A form of agate showing a bouquet pattern of variegated colors.

Fluorescence The emission of brightly colored light by a substance when exposed to ultraviolet rays.

Foliated Basically describing schist minerals in thin, parallel, sheetlike layers like the pages of a book (e.g., mica).

Fool's gold Amorphous pyrite.

Formation A related deposit of rocks formed during the same time period and under generally the same circumstances.

Fortification A form of agate with an angled striped pattern of generally concentric configuration. (See also *banded.*)

Fossils   Plant and animal remains preserved in rock.

Fracture   The type of break that occurs in any rock or mineral.

Friable   Crumbly material that is broken, pulverized, or separated with facility.

Fumarole   A hole or vent through which a volcano issues hot gases.

Gabbro   A composition of pyroxene and plagioclase feldspar in a dark granitelike igneous rock.

Gangue   The waste material, or material of no commercial value, in a mineral deposit being worked; quartz is often a gangue material.

Geode   A hollow mineral shell formed within a cavity of sedimentary rock, often lined with mineral substances including fine crystals of varying types and colors and seldom related to the shell material; easily separable by weathering or artificial methods from the enclosing rock.

Glaciated   Subjected to glacial action.

Gneiss   A crystalline metamorphic rock that, due to the segregation of its component minerals, has a banded appearance; most frequently comprised of mica, quartz, and a dark mineral such as hornblende, these minerals to some extent laid out in streaks. Gneiss and schist are similar, but gneiss, having less mica in its composition, is stronger.

Gouge   A soft material occurring as a layer between country rock and the walls of a vein.

Grain   The unit of weight for pearls; equivalent to 0.25 carat.

Gulch   A mountain erosion channel or small canyon normally formed by intermittent torrential flows.

Habit   The shape in which a crystal characteristic occurs.

Hackle   A jagged, noncleavage fracture in a rock or mineral.

Hanging wall   The upper portion of country rock bordering a vein or dike; opposite to the foot wall.

Hardness   Degree of resistance of a rock or mineral surface to being scratched; the relative 1–10 *Mohs scale* is a measure of hardness, with talc and diamond representing opposite properties of resistance.

Hemimorphic   A distinct difference in opposing ends of a crystal due to a lack of transverse planes of symmetry (e.g., zincite and tourmaline).

Hexagonal   The crystal system wherein the vertical axis is intersected by three axes at right angles and by a fourth at 60 degrees.

High-grade   Used as a verb to connote deliberate selection of the high-value mineral, choicest ore, or gemstone material.

Horizon   The particular stratum in which minerals or fossils of similar characteristics appear throughout a sedimentary formation.

Horse A separated mass of country rock enclosed in a vein.

Hyaline A mineral substance that has a glassy texture.

Hybrid rock A rock variety formed when one magma combines with another.

Igneous rocks Rocks formed through crystallization of lava or magma.

Incline A narrow, oblique shaftlike mine excavation.

Inclusion The presence of a foreign body (including gas or liquid) enclosed within a crystal. (See also *enhydro.*)

Indurated Hardened by pressure and/or heat (e.g., indurated clay).

Inorganic Derived from or pertaining to material of inanimate origin.

Interpenetrant twins Two or more twin crystals that pass through one another.

Intrusive Molten material squeezed into cracks or crevices or between already existing rock layers to form an igneous rock that has hardened before reaching the earth's surface.

Iridescence A rainbow display of colors in a mineral.

Iris A form of agate that exhibits a distinct spectrum when light passes through it.

Isometric Crystal system wherein all three axes are of equal length and intersect at right angles.

Joint A divisive of a plane within a rock mass, occurring in series and intersecting at right angles, with the result of separating the rock into blocks.

Kidney An oblong, rounded mass of ore or massive gemstone material that in outline form resembles a kidney.

Knob Any relatively isolated rock hill or mountain.

Lamellar Comprised of thin, sheetlike layers, scales, or plates.

Laminated rocks Numerous overlapping layers of sedimentary rock.

Lava The superheated liquid rock discharge of a volcano.

Ledge The term used to designate an outcrop of mineral-rich material or to designate a pegmatite outcropping in districts where pegmatites occur.

Lens A lens-shaped (roughly circular, thick in the middle, and tapering to thin edges) mineral deposit within a body of ore.

Lenticular Having a flattened, oblong, rounded shape like a lima bean or a lentil.

Level A horizontal passage (or tunnel or drift) within a mine; such levels are numbered in descending order from the adit.

Lithoclase A crystal-filled fracture in a rock.

Lithophysae Distinctive hollow spherulites occurring in rhyolite, obsidian, and similar rocks of glassy character.

Locate   To mark out the boundaries of a claim, establishing possession of same.

Lode   Essentially the same as a vein.

Loess   Wind-deposited drifts of silt.

Luster   The distinctive characteristic of light reflection from a particular mineral.

Magma   The subcrust molten material (called lava when it reaches the surface) from which igneous rocks derive.

Malleable   Possessing the nonbrittle characteristic that permits a mineral material to be shaped by hammering.

Mammillary   A rounded aggregate of minerals having a breastlike form.

Marl   Calcareous loam or clay that may contain glauconite.

Massive   Having no definite crystal form or structure.

Matrix   The base material (clay, rock, etc.) upon which or within which minerals and/or fossils are perched or embedded.

Mesa   A generally isolated tableland or flat-topped elevation.

Metamorphic rocks   Rocks that have, through action of heat, solutions, and/or gases, become altered in a significant manner.

Meteorite   A celestial body that has fallen to earth.

Mica schist   A basic matrix material whose most conspicuous ingredient is mica.

Mineral deposit   Economically valuable minerals in heavy concentration.

Monoclinic   Crystal system wherein the axes are unequal in length, and two meet at right angles while the third intersects at an oblique angle.

Moraine   An accumulation of debris deposited by the outermost edge of a glacier.

Mother lode   The principal vein or lode; the richest and largest source of a desired mineral.

Nacreous   Exhibiting an opaque pearly luster.

Native   Occurring in a pure or uncombined state, as native copper, native gold, native silver, etc.

Nodule   A roundish lump (generally fist-sized or smaller) of mineral, often formed within a cavity in rock.

Octahedron   Crystal system composed of eight triangular faces.

Oolite   A limestone comprised of numerous small spherules that may be silaceous, calcareous, or ferruginous; these spherules are spherical grains that are less than 2 mm (.08") in diameter, and the organic or mineral nucleus is successively encased in concentric layerings on various minerals; the term can also be applied to a rock comprised of oolites.

Opalescent   Exhibiting a lustrous milky appearance with a faint but distinct bluish milky cast.

Opaque   Describing a rock or mineral through which light cannot pass.

Opencut   The workings into a mineral deposit via a trench excavation.

Orbicular   Containing a number of solidly encased orbs or spherules with eyelike patterns of alternating mineral material and colors, occurring in metamorphic and igneous rocks. (See also *pisolite*.)

Ore   A deposit of mineral material that is sufficiently rich and of a quantity to be mined with profit.

Ore body   The general continuous mass of material (as distinguished from country rock) in which minerals of value occur.

Organic   Having to do with compounds produced in animals and plants rather than of chemical or otherwise inorganic origin.

Orthorhombic   Describing a crystal system wherein the axes are all of unequal length but wherein all meet at right angles.

Outcrop   The portion of a body of ore, a stratum, or another massive rock form that becomes exposed above the earth's surface.

Overburden   The essentially valueless surface material over a mineral deposit.

Oxidized zone   The portion of an ore body that has been altered through the action of surface waters that carry oxygen, carbon dioxide, etc.

Pan   The process of washing earth, gravel, and sand in a graduated circular pan in search of gold and other heavy materials; also, the center of a basin declivity in the land.

Parting   The occasional tendency of a crystal to separate in an area that is not a true cleavage plane.

Patented claim   A claim legally filed upon and to which permanent rights have been obtained through a conveyance or *patent.*

Pearly   Exhibiting a rich luster similar to that of a pearl. (See also *nacreous.*)

Pegmatite   A body of coarse-grained granite ordinarily occurring in sheetlike masses that have intruded into country rock or formed within the granite itself; most pegmatites exhibit clearly oriented color zones or *units* within which there are characteristic mineral growths.

Perfect   Term for cleavage that is easily instigated along a characteristic plane and that after cleaving leaves a smooth surface.

Perlite   An obsidianlike volcanic rock that has been rendered light and friable due to a multitude of spheroidal and ellipsoid fissures; "Apache tears" are small obsidian nodules that have weathered from perlite masses.

Petrifaction (permineralization)   The process of fossilization; the transitional metamorphic state during which the original cell structure remains but the cells become filled with mineral that itself is often silicified.

Phantos A visible "ghost" crystal that is completely enclosed within another perfectly formed transparent crystal.

Phenocyst A prominent crystal surrounded by considerably smaller grains of similar mineral in prophyritic rock.

Phosphorescence A distinctive form of luminescence that persists after the light stimulus is removed.

Phyllite A lustrous and exceedingly compact schistose rock derived through metamorphic process from clay sediments.

Piedmont The generally shallow-sloping area at the base of mountains.

Pillow basalt Dark basaltic rock flows in which large rounded masses of "pillows" of firm rock were created when the molten material met water; the spaces between these masses filled with friable rock in which occur "pillow" cavities that often contain crystals.

Pica The summit of a mountain.

Pipe A vertical volcanic tube of considerable diameter containing a friable, easily decomposable, cylindrical mass of kimberlite or other igneous rock extending downward to great depth.

Pisolitic Describing a mineral formation of rounded beanlike or pealike masses. (See also *orbicular.*)

Pitch The dip or inclination of a vein or mineral bed at right angles to its strike.

Placer An alluvial deposit often containing gold.

Plateau An extensive flat-topped region of considerable elevation.

Playa A central flat area or depression in a bolson; the dry lakes of California are playas. (See also *pan.*)

Plug A solid lava core located in the neck of an extinct volcano.

Plutonic rocks Rocks formed from molten material that cooled some distance below the earth's surface and generally show distinct grain textures like granite.

Pocket Generally, a small body of ore; in pegmatites pockets are central openings lined with crystals including gem species; term can also be applied to a natural cavity in any rock.

Point A minute diamond weight equivalent to 1/100 carat.

Porphyry An igneous rock comprised of large grains or crystals occurring in fine-grained groundmass; most porphyries are rhyolitic.

Precious Of considerable value. (Accurately, the only gemstones that can be termed precious are diamond, emerald, ruby, sapphire, and opal.)

Prismatic Describing a narrow crystal shape with sides covered with rectangular planes joining in parallel edges having the property of breaking light into its spectrum.

Prospect A "dig," superficial pit, or more extensive working for the purpose of locating the presence of valuable minerals; many mines originate as prospects.

Pseudomorph   A mineral crystal having the faces and angles of a different mineral species; the term is also broadly applied to other misleading substitutions such as petrified wood, wherein quartz or opal have become pseudomorphs of wood.

Pumice   A bubbly lava that has solidified in so frothy and light a state that it often floats on water.

Punta   A headland or point.

Quarry   An extensive, deep, steep-sided open excavation worked with machinery and explosives to remove rock, usually for commercial purposes.

Quartzite   A conglomerate or sandstone that is composed primarily of coarse quartz grains or pebbles converted to solid rock through metamorphic action.

Radiating   Describing a crystal structure in which individual crystals (often acicular) fan outward in a starburst pattern from the center. (See also *stellate.*)

Ravine   A natural erosion trench or valley with steep sides; generally speaking, smaller than a canyon but larger than a gully.

Reniform   A kidney-shaped mineral form.

Replacement   The process through which one mineral assumes the place and (often) the shape of another. (See also *pseudomorph.*)

Reticulated   Having a lattice or network formation.

Rhyolite   Volcanic rock species that is the equivalent of granite and contains tiny orthoclase and quartz crystals, along with biotite and hornblende, or a pyroxene; most rhyolites are light in color.

Rough   Uncut, unpolished gemstone material.

Rutilated   Crystal with inclusions of rutile needles (titanium oxide) laid out most often in a helter-skelter jackstraw pattern rather than lying parallel.

Sagenitic   Describing a crystal that contains smaller needlelike crystals of a foreign mineral.

Schist   Foliated (book-sheet) metamorphic rocks characterized by long, drawn-out streaks and containing considerable mica, causing it to be brittle as well as readily separable into larger sheets along the layers.

Seam   A thin layer, stratum, or vein of mineral.

Sectile   Describing a mineral soft enough and malleable enough that it can readily be cut into shavings.

Sedimentary rocks   Rocks formed in layers through the accumulation and solidification of animal, mineral, or vegetable matter; water, wind, and glaciers are the principal means by which the loose sediments are transported and deposited.

Semiprecious   Nonprecious gemstone minerals that can be cut, polished, and sometimes worn as gems (including jasper, agate, chalcedony, etc.).

Shaft A deep and narrow mining excavation sunk vertically.

Shale A sedimentary rock formed by mud or silt under great pressure.

Shear zone A rock mass in which a zone of shearing has occurred and this shear zone has become filled with crushed and brecciated rock.

Sheet A usually extensive bed of eruptive rock that has intruded between layers of preexisting rock; also, a similar body of volcanic rock that has outpoured upon the earth's surface in a layer of relatively uniform thickness.

Shingle Generally uniformly sized pebbles of waterworn beach gravel.

Siliceous (silicified) Describing a mineral in which silica or quartz is present; may also be applied to any animal, vegetable or mineral substance that has been replaced by chalcedony or opal.

Silky Imbued with the luster and tactile sensation of silk.

Sill A broad, flat ribbon or thin sheet of igneous rock that has intruded parallel to the bedding of other rocks and lies in a more or less horizontal plane.

Sinkhole A very steep-walled craterlike depression; usually occurs in limestone regions due to the roof collapse of a subterranean cavern.

Slate A rock that is similar in character to shale but forms in a more highly compressed state and is more easily split along the bedding planes; small scales of mica are often present, giving the slate a distinctive silvery luster.

Sluice (sluice box) A narrow wooden inclined trough with a bottom ribbed with properly spaced bars called *riffles* over which gravel is poured and washed down in order to extract the heavier minerals at the riffles.

Spall To break off slabs or pieces of rock from a much larger rock mass.

Specific gravity The comparison of the weight of a mineral substance with that of the water it displaces upon being immersed.

Spherulites Distinctly rounded aggregates or rosettes of very fine, delicate, needlelike crystals radiating from common centers; orbicular jaspers tend to exhibit some spherulitic structure.

Splendant Brilliantly reflective.

Spoil Debris or waste materials removed from a mine, prospect, or other excavation.

Stalactite A tapered column of mineral matter depending from the ceiling of a cavern; its growth is determined by mineral particles contained in dripping water.

Stalagmite The opposite of a stalactite; a tapered pillar rising from the floor instead of depending from the ceiling; its growth is often determined by the dripping of water from a stalactite directly above.

Star An astirated gem, or one that has the quality of reflecting a four- or six-rayed star from embedded fibers (e.g., star ruby, star sapphire).

Stellate Having a radiate crystalline structure forming starlike patterns.

Stock (stockwork) A mass of country rock in which many small veins have developed into a complicated network.

Stope The series of "steps" above or below a mine level where ore has been extracted by excavating an inclined ore body.

Stratum A layer of rock, not necessarily horizontal but of similar constituent character throughout.

Streak The colored mark that is left behind when a mineral is rubbed on an unglazed white porcelain tile; a method of mineral identification.

Striation A minute groove or channel on a crystal face.

Strike The horizontal direction or trace of an outcrop.

Stringer An especially narrow vein (or veinlet) that is generally irregular in its thickness and follows an erratic course.

Structure The general visible character of a rock specimen or outcrop.

Symmetry The regular, balanced arrangement of the features, forms, and properties of any given crystal relative to its axes.

Tabular Describing broad, flattened crystals of tablet shape (e.g., Wulfenite).

Talus A sloping heap of (usually) naturally fallen debris at the base of a mountain, cliff, or promontory.

Tenacity The ability of a mineral to resist breaking or separation.

Termination of a crystal The group of planes completely enclosing the crystal's end; a doubly terminated crystal has both ends covered with faces and thus ordinarily has its entire surface comprised of crystal faces.

Tetragonal Describing a crystal system wherein only two axes are equal in length and all three axes intersect at right angles.

Thunderegg A nodule (usually fist-sized or smaller) having a colored agate core and a surface structure of rhyolite or jasper.

Translucent Describing a crystal with the ability to pass light but not the image of an object.

Transparent Describing a crystal with the ability to pass light so clearly that objects can be observed through it.

Traprock Basalt and numerous other dark, fine-grained igneous rocks.

Triclinic Describing a crystal system wherein all three axes are unequal in length and in which they meet at oblique angles.

Trilobite Ancient Paleozoic fossil crustacean that bears a resemblance to the modern sow bug or wood louse but was much larger.

Trisoctahedron Crystal system formed of 24 faces so arranged that three take the place of each face of a regular octahedron.

Tuff A well-cemented but easily breakable volcanic ash.

Twin A compound crystal that is comprised of two or more crystals (or parts of crystals) in reversed positions with respect to each other; also, a specimen that consists of two or more single crystals of the same mineral intergrown in a definite systematic arrangement.

Vein An irregular twisting mineral deposit of considerable length that is very thin in proportion to its length and breadth; veins are normally formed as a result of hydrothermal activity.

Vesicle A small cavity found in glassy and volcanic igneous rocks; such vesicular cavities are formed through the expansion of gases.

Vitreous Describing a rock or mineral that is like glass in physical properties, in appearance, or both.

Vug A cavity, usually small, in rock; it is normally lined with crystals firmly fastened to the walls or scattered on the bottom that are related to the minerals found in the country rock; vugs are believed to be caused by extensive shrinkage of the enclosing rock.

Wash The loose surface debris that is found in the bottoms of canyons and intermittent streams; a wash is also the generally broad and shallow depression in which such debris occurs.

Xenolith Country rock enclosed in magma.

INDEX